FUNDAMENTALS OF
ANIMAL
SCIENCE

FUNDAMENTALS OF
ANIMAL
SCIENCE

COLIN SCANES

DELMAR
CENGAGE Learning™

Australia • Brazil • Japan • Korea • Mexico • Singapore • Spain • United Kingdom • United States

Fundamentals of Animal Science,
by Colin Scanes

Vice President, Career and Professional Editorial:
Dave Garza

Director of Learning Solutions: Matthew Kane

Acquisitions Editor: Benjamin Penner

Managing Editor: Marah Bellegarde

Product Manager: Christina Gifford

Editorial Assistant: Scott Royael

Vice President, Career and Professional Marketing: Jennifer Baker

Marketing Director: Debbie Yarnell

Marketing Manager: Erin Brennan

Marketing Coordinator: Jonathan Sheehan

Production Director: Carolyn Miller

Production Manager: Andrew Crouth

Content Project Manager: Anne Sherman

Art Director: David Arsenault

Technology Project Manager: Tom Smith

Production Technology Analyst: Thomas Stover

For product information and technology assistance, contact us at
Cengage Learning Customer & Sales Support, 1-800-354-9706

For permission to use material from this text or product,
submit all requests online at **www.cengage.com/permissions.**
Further permissions questions can be e-mailed to
permissionrequest@cengage.com

Library of Congress Control Number: 2009931912

ISBN-13: 978-1-4283-6127-0
ISBN-10: 1-4283-6127-8

Delmar
5 Maxwell Drive
Clifton Park, NY 12065-2919
USA

Cengage Learning is a leading provider of customized learning solutions with office locations around the globe, including Singapore, the United Kingdom, Australia, Mexico, Brazil, and Japan. Locate your local office at:
international.cengage.com/region

Cengage Learning products are represented in Canada by Nelson Education, Ltd.

To learn more about Delmar, visit **www.cengage.com/delmar**

Purchase any of our products at your local college store or at our preferred online store **www.ichapters.com**

NOTICE TO THE READER

Publisher does not warrant or guarantee any of the products described herein or perform any independent analysis in connection with any of the product information contained herein. Publisher does not assume, and expressly disclaims, any obligation to obtain and include information other than that provided to it by the manufacturer. The reader is expressly warned to consider and adopt all safety precautions that might be indicated by the activities described herein and to avoid all potential hazards. By following the instructions contained herein, the reader willingly assumes all risks in connection with such instructions. The publisher makes no representations or warranties of any kind, including but not limited to, the warranties of fitness for particular purpose or merchantability, nor are any such representations implied with respect to the material set forth herein, and the publisher takes no responsibility with respect to such material. The publisher shall not be liable for any special, consequential, or exemplary damages resulting, in whole or part, from the readers' use of, or reliance upon, this material.

Printed in Canada
1 2 3 4 5 6 7 12 11 10 09

Dedication

To the people who have been my
sources of inspiration, in particular my
wife Cate, my parents, and my brother.

TABLE OF CONTENTS

This book is intended as a textbook for introductory animal science courses regardless of whether the course is called Animal Science 101 (or 101 and 102), Introductory Animal Science, Animal Production, Introduction to Animal Production, Principles of Animal Science, or Domestic Animal Biology. The goal is that the book will be appropriate for either majors or nonmajors in animal science or preveterinary science. Moreover, the objective is that students will find the book interesting and useful regardless of whether they come from a farming background, or from cities, suburbs, or rural locations. A strong focus is placed on underlying principles, production systems, the basic biology, and societal issues.

There are three major thrusts. There is a focus on livestock, that is, cattle, pigs, sheep, and goats. In addition, specific chapters cover poultry, aquaculture, and minor agricultural animals such as bison, deer, and rabbits. There is a parallel emphasis on horses and companion animals. The latter focuses on dogs and cats but with consideration given to other companion animals. The book addresses the fundamental science of domestic animals, including animal reproduction; growth; genetics, including genomics; behavior; health; and environmental interactions. In addition to the applied science of production, the importance and safety of animal products are included.

The book is divided into five sections. The first section, "Animals and Society," encompasses chapters on the importance of agriculture to the growing world population and reducing hunger, the origin and development of agriculture, an overview of the livestock industry around the world, careers in animal science, horses and the equine industry, and companion animals. The second section is on livestock production, including cattle, with sections on both dairy and beef cattle (the dairy area largely reflects a contribution from Howard Tyler), and pigs (with major contributions from Tom Baas), production and utilization of sheep and goats, alternative mammalian livestock (buffalo, llamas and alpacas, rabbits, and deer), poultry, and aquaculture. The third section covers the biology of domestic animals with chapters on cells, organs, and systems, (including digestion and the gastrointestinal tract, cardiovascular systems, and the endocrine system), animal nutrition (feeds and foods), animal genetics and breeding, animal reproduction, animal growth and development, and lactation. The fourth section covers animal health, welfare, and the environment, with chapters on animal behavior and welfare; animal diseases, including metabolic, environmental, and infectious ones; animal health, including therapeutics (drugs), immunology, and vaccines; and the impact of animal agriculture and the environment. The final section discusses animal products, with chapters on food safety, meats, milk and dairy products, and alternative animal products.

ACKNOWLEDGMENTS

I greatly appreciate the contributions from my former faculty colleagues from Iowa State University, Tom Baas and Howard Tyler, who are class instructors for senior classes in, respectively, swine and dairy. I thank my editors for the confidence they have placed in me and for their numerous areas of help. The ongoing tremendous support from my wife is gratefully acknowledged.

The author and Delmar Cengage Learning would like to thank the following individuals for their review of the manuscript throughout the development process:

Frank White – Cameron University
John Marchello – University of Arizona
John McNamara – Washington State University
Brian Rude – Mississippi State University
Edward Bonnette – University of Findlay
Cindy Wood – Virginia Tech

ABOUT THE AUTHOR

Colin Scanes is dean of the Graduate School at the University of Wisconsin, Milwaukee. For 30 years, he has been active in teaching and research in animal sciences, holding faculty positions at a number of universities, including Rutgers–The State University of New Jersey (Professor II and chair), Iowa State University, and Mississippi State University. He is a longtime member of the American Association for the Advancement of Science (fellow), the American Society of Animal Science, and the Poultry Science Association (fellow). He has authored/edited 12 books and 340 research papers and reviews.

Also available for the instructor:

Instructor's Manual
ISBN: 1 4283 6128 6
ISBN 13: 978 1 4283 6128 7

The *Instructor's Manual* is made up of the chapter objectives, answer key to the Issues for Discussion and Review Questions, and References and Further Reading.

Instructor Resource to Accompany Fundamentals of Animal Science
ISBN: 1 4283 6129 4
ISBN 13: 978 1 4283 6129 4

This powerful electronic resource is an essential tool. In it an instructor will find everything needed to prepare for, teach, present, and test the concepts of animal science.

This resource includes a PDF of the full *Instructor's Manual* with an answer key to the Issues for Discussion and Review Questions; a computer test bank containing more than 900 questions with answers and Internet testing capability; PowerPoint presentations containing over 400 slides that focus on each chapter's key points to facilitate classroom presentations; and an image library, including all figures appearing within the text.

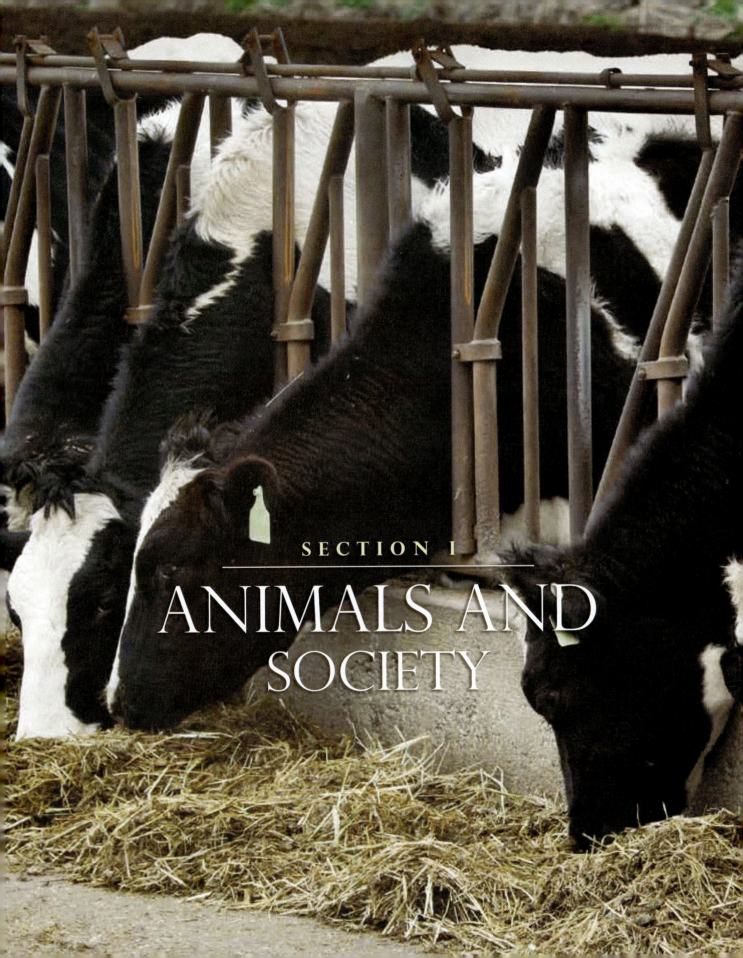

SECTION I
ANIMALS AND SOCIETY

CHAPTER 1

Introduction: World Population, Hunger, and the Importance of Agriculture

OBJECTIVES

This chapter will consider the following:

- The historical trends and projections for the world population.
- The population trends that have occurred in the United States, and the trends that are likely to occur in the future.
- The extent of hunger and inadequate nutrition in the world.
- The critical importance of agriculture in feeding the growing world population.

WORLD POPULATION

The population of the world is over 6.6 billion people (July 2007) and is growing at 1.2% per year. Over human history, the world population grew much more slowly until about 1800, when population growth started accelerating. Today, the rate of growth of the human population is slowing, but the total population may still reach 10 billion people in your lifetime by the year 2050. The changes in world population are shown in Table 1-1 and Figure 1-1.

TABLE 1-1 Changes in world population during human history

YEAR	POPULATION
8000 BC (−10,000 y)	5 million
1000 BC (−3000 y)	50 million
500 BC (−2500 y)	100 million
AD 1 (Christian Era)	300 million
1000	310 million
1500	500 million
1800	1 billion
1850	1.2 billion
1900	1.65 billion
1927	2 billion
1960	3 billion
1974	4 billion
1987	5 billion
2000	6 billion
2007	6.6 billion

Source: http://www.census.gov/ipc/www/idb

Between 50,000 and 100,000, and 10,000 years ago, the human population increased from as little as 10,000 to 5 million, spreading out from Africa to Asia, and then to Europe, Australia, and the Americas. The population of the world increased 10-fold as a result of the Neolithic revolution and the beginning of agriculture between 8000 BC and the year 1 (AD or Christian Era).

Between the year 1 (AD or Christian/Common Era) and 1500, the global population increased from 300 to 500 million. The population increased from 500 million to 1 billion between 1500 and 1800. Between 1800 and 1927, the population increased from 1 to 2 billion, and between 1927 and 1960, it increased from 2 to 3 billion. The population increased from 3.0 to 6.8 billion between 1960 and 2009. Put another way, between 1960 and 2009, world agriculture had an additional 3.8 billion mouths to feed.

Put another way, between 1960 and 2007, world agriculture has had an additional 3.6 billion mouths to feed. Not only are there more people, but they are also eating more. In 1960 (3 billion people), the average caloric intake was 2,000 kilocalories per day in the developing world; in 1998 (6 billion people), it had increased to 2,700 kilocalories per day in the developing world.

> **INTERESTING FACTOID ON COMPOUND INTEREST OR POPULATION GROWTH**
>
> Rule of 72 for compound interest: doubling time = 72 divided by the percent increase (or 72/%). This applies to population growth, growth of a company, or interest on a savings account or other investments.

FIGURE 1-1 Changes in the world population over historical times together with a U.S. Census Bureau projection of the future growth of the world population. Source: U.S. Census Bureau.

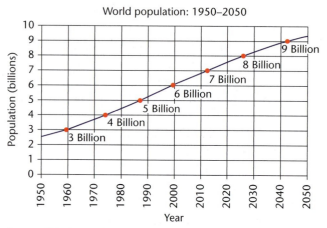

Source: U.S. Census Bureau, International Data Base, 2008 First Update

FIGURE 1-2 Composite photograph of the world at night showing light from the surface such as coming street lights, malls, traffic, offices, and homes. Source: NASA.

Prediction of Population Catastrophe and Famine

Thomas Malthus wrote in 1798 "An Essay on the Principle of Population." In this he stated that "the power of population is infinitely greater that the power of the earth to produce subsidence for man. Population, when unchecked, increases in a geometrical ratio. Subsistence increases only in an arithmetical ratio."

Extent of Hunger in the World

Today, there are over 800 million people who are hungry and do not get enough to eat. These people live predominantly in Africa and Asia, and, to a lesser extent, Latin America.

Urbanization

There is tremendous growth of cities in the world. Using satellites, the National Aeronautics and Space Administration (NASA) has produced pictures of the world at night showing the light coming from human activities. Figure 1-2 illustrates the "night glow" of the planet and its people.

The largest urban metropolitan areas in the world are presently the following: Tokyo (Japan), with 35 million people; Mexico City (Mexico), 19 million; Mumbai, formerly Bombay (India), 19 million; New York (United States), 19 million; São Paolo (Brazil), 19 million; Delhi (India), 15 million; Shanghai (China), 15 million; Kolkata, formerly Calcutta (India), 14 million; Jakarta (Indonesia), 13 million; and Buenos Aires (Argentina), with 13 million people. It is surprising that none of the great metropolitan areas or cities of Europe makes the top 10. This is coupled with the presence of only one U.S. metropolitan area, and reflects European and U.S. cities being overtaken by the rapidly growing cities in developing countries. Urbanization is occurring at a rate unprecedented in human history. The vast majority of the increase in the world's population is occurring both in the developing countries and in their cities.

Let us look at the situation globally. In 1950, only one city in the world, New York, had a population greater than 10 million. In 2006, 21 urban areas have a population greater than 10 million. Another surprise is that despite China being the most populous country in the world, it has only one metropolitan area in the top 10. This situation in China was caused by relatively slow metropolitan growth and a focus on rural areas until about 20 years ago. This has now changed.

It is estimated that the world population in urban areas will grow from 2.9 billion in 2000 to 5.0 billion by the year 2030, creating what some have called "megacities" (see fig. 1-3). This growth of cities is an opportunity for economic growth but a logistic and infrastructural "nightmare." The growth is occurring in developing countries. How will these countries afford the essential infrastructure of clean water, sewers, schools, housing, transportation for people, food and other essentials, health care, etc.?

QUOTATIONS ABOUT CITY LIFE

"A city is a place where there is no need to wait for next week to get the answer to a question, to taste the food of any country, to find new voices to listen to and familiar ones to listen to again."

Anthropologist Margaret Mead (1901–1978)

"Clearly, then, the city is not a concrete jungle but a human zoo."

British zoologist and ethnologist Desmond Morris (1928–)

FIGURE 1-3 Overcrowded city in the developing world. Reproduced by permission from Ho Philip. © 2010 by Shutterstock.com.

ISSUES *for discussion*

In any discussion, whether in a class or not, it is important to be respectful of other points of view and considerate of others. Develop your arguments based on information that is known and that you have researched. If the facts do not fit your position, you might consider shifting your position. Avoid attacking people individually (the *ad hominem* attack, such as "You stink," or "Only a moron would believe that").

WORLD POPULATION GROWTH

1. How do you feel about current world population growth?
2. Is it a problem?
3. Should countries attempt to control their populations through legal and political policy channels?
4. Which if any of these approaches is ethical or practical?
5. Should countries attempt to slow or stop urbanization?
6. What policies are ethical or practical?
7. How should a country prioritize government spending between the developing urban areas and rural areas?

Population of the United States

The United States has a population 302 million people and is growing at 0.9%; about six tenths of this increase is the greater number of births than deaths, and four tenths is due to immigration. The rate of population growth was much greater early in the history of the United States and the 13 colonies. Between 1660 and 1880, the U.S. population had a doubling rate of about 24 years or a rate of population growth of 3%. For instance, in 1650, the population of the 13 original colonies that went on to form the United States was about 50,000. The population of the original colonies was about 250,000 in 1700. In 1760, the population of the original colonies was about 1.7 million. The population of the United States was about 48 million in 1880. (For more information on U.S. population growth, see the Central Intelligence Agency's Web site https://www.cia .gov/library/publications/the-world-factbook.)

The population of the United States is not evenly distributed across the country, as can be readily seen if you fly across the country at night or as shown in Figure 1-2.

Concepts

Green field development is when a housing development, mall, or factory is constructed on agricultural or other land that has not previously been developed. The advantage is that there are no issues of contamination.

Brown field development is when a new development is occurring on land previously used for factories, etc., and where there is significant contamination of the soil with chemical toxicants.

ISSUES *for discussion*

POPULATION IN THE UNITED STATES

1. How do you feel about population growth in this country?
2. It is a problem?
3. Should the United States attempt to control its population?
4. What policies might be used?
5. Which of these approaches is ethical or practical?
6. Should immigration be restricted or halted?

Urban areas are continuing to grow in the United States. New residential subdivisions are developed using larger lots, surrounded by mega-malls, office parks, restaurants, and new roads removing areas that were once prime agricultural land. Among the reasons for green field development are that it is less costly than development in a city, where the soil may be contaminated with toxicants. Another reason is that people "vote with their feet," moving to the new subdivisions, shopping in the new stores, eating at the new restaurants, and working in the office parks (see fig. 1-4).

Some call this "urban sprawl" and are concerned about the loss of farmland. Some have argued for restricting growth, no growth, or "smart growth." Smart growth aims to focus growth into planned communities with town centers and high population densities. This allows walking or bicycling to stores or work or other activities and a greater use of public transit. This is similar to "new urbanism," with the concept of communities to "live, work and play" (see fig. 1-5).

ISSUES *for discussion*

URBAN SPRAWL AND URBAN DEVELOPMENT

1. How do you feel about using agricultural land for development? Is it a problem?
2. Do communities that restrict growth prosper?
3. Does restriction of growth lead to increased prices for housing and, therefore, people with low incomes having to commute?
4. What is urban sprawl?
5. Should farmers and/or developers be allowed to develop on green fields?
6. Should there be a subsidy for farmers who commit to keep land in production?
7. Do you frequent malls and restaurants in new developments?

FIGURE 1-4 Green field development of housing, business parks, and retail uses a large amount of agricultural land and requires considerable driving. Courtesy of Getty Images.

THE EXTENT OF HUNGER AND INADEQUATE NUTRITION IN THE WORLD

The United Nations (UN) Food and Agriculture Organization (FAO) estimates that the present (as of year 2002) number of undernourished people is 840 million. Most of the people in the world who receive inadequate nutrition live in the developing world, as can be seen in Figure 1-6. Geographical areas containing undernourished people are the developed world (11 million); economies in transition (30 million); and developing countries (799 million), including East Asia (China and North Korea, 128 million), Southeast Asia (64 million), South Asia (the Indian subcontinent, 315 million), North Africa and the Near East (40 million), Sub-Saharan Africa (196 million), Latin America and the Caribbean (55 million), and Oceania, which includes the islands of the Pacific Ocean (1 million).

It is encouraging that the number and proportion of people who are undernourished are decreasing in some regions, particularly East Asia. However, the number either is not decreasing or is, in fact, increasing in other regions, e.g., Sub-Saharan Africa.

Associated with hunger, undernutrition, and malnutrition are a number of further issues, including low-birth-weight babies (and higher infant mortality), poor nutrition of young children that impairs neural and cognitive development, hungry children who are inattentive and do not learn well, and inadequate nutrition leading to reduced growth and stunting.

800 Million People Go Hungry

The first goal of Millennium Development Goals (based on the UN Millennium Declaration 2000) is to eradicate extreme poverty and hunger. In 1990, it was estimated that 815 million people in the world are hungry and get inadequate nutrition. One of the specific targets of the Millennium Development Goals is to cut the proportion of people in the world who suffer from hunger or have inadequate nutrition by 50% by the year 2015.

In 1990, 28.3% of all people in the world lived on less than $1 per year. Extreme poverty is an important issue, and one of the specific targets of the Millennium Development Goals is to cut the proportion of people in the world living on less than $1 per year by 50% by 2015. Other goals include the following:

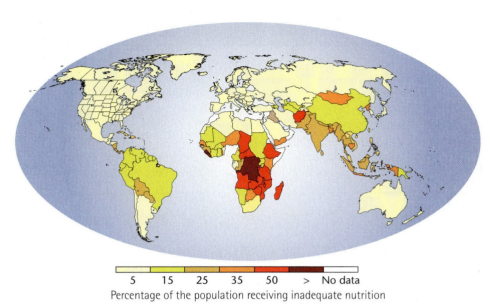

FIGURE 1-6 World map showing the counties where people do not get adequate nutrition. Source: United Nations Food and Agriculture Organization (FAO).

5   15   25   35   50   >   No data

Percentage of the population receiving inadequate nutrition

- To achieve universal primary education
- To promote gender equality and empower women
- To reduce child mortality
- To improve maternal health
- To combat HIV/AIDS, malaria, and other diseases
- To ensure environmental sustainability
- To develop a global partnership for development

Hunger Can Be a Cause of Poverty

About 45% of children in India are malnourished (being underweight). The FAO calculates that the effects of stunting, and iron and iodine deficiency alone, are reducing the productivity of adults (due to cognitive impairment and other developmental issues) in a number of Asian countries by about 3% of the gross domestic product. Not only is hunger due to poverty, but hunger is also a cause of poverty.

Improving Nutrition and Health Facilitates Development

American economic historian and Nobel Prize winner Robert Fogel has estimated that improved nutrition and health are the basis of about half the economic growth of Britain and France in the 18th and 19th centuries.

Food Security

The mission of the FAO of the UN is helping to build a world without hunger (food security). The FAO of the UN provides time-series and cross-sectional data relating to food and agriculture for some 200 countries (http://faostat.fao.org).

The FAO defines food security as follows: "Food security exists when all people, at all times, have physical, social and economic access to sufficient, safe and nutritious food to meet their dietary needs and food preferences for an active and healthy life." According to the FAO, about 40% of the world's population raises its own food.

ISSUES *for discussion*

HUNGER AND UNDERNUTRITION

1. How do you personally feel about world hunger and people getting inadequate nutrition?
2. Do you consider it any of your concern?
3. Do you feel powerless to alleviate hunger?
4. What, if anything, should the United States and international agencies be doing?
5. Are there actions people as individuals can take to reduce hunger in the world?

| Definition

The term *green revolution* contrasts with violent revolutions but, nonetheless, has had a dramatic effect on peoples' lives in the developing world. It stems from research initially in Mexico and India, resulting in wheat and rice with greatly improved yields. The distribution of this improved genetics resulted in greatly increased production, reducing hunger by increasing the availability of food and at a lower price.

THE IMPORTANCE OF AGRICULTURE

There has been tremendous progress in the ability of the planet (or more specifically the world's farmers) to feed the growing population with large increases in total agricultural production and production efficiency. The animal industry has made great strides in many countries with major increases in production and consumption of poultry meat, egg, beef, milk, pork, and other meats. China and India are making tremendous progress in improving the nutrition of their population.

Improvements in agricultural production in China are because of the following:

- The "green revolution," using improved genetics for wheat and rice
- Increased agricultural productivity due to improved seeds, fertilizer, pesticides, irrigation, and "know how"
- Opening of the Chinese economy with farmers having the following: a profit motive, access to markets, and availability of technology
- Overall economic development (8% annual growth in gross domestic product for nearly 30 years) such that consumers can buy agricultural products, particularly meat

Beginning in the early 20th century, there were research-based advances in the production of cereals with better agronomic practices, increased use of fertilizer and irrigation, and, most importantly, greatly improved genetics of the seed. The coming of hybrid corn together with genetic selection has increased yields of corn. The advent of agricultural biotechnology with genetically modified seed is reducing losses because of weeds and other pests, such as the European corn borer.

In the United States, the average yield for corn increased tremendously in the 20th century. The 1860–1920s' yield was 26.0 bushels per acre (1.2 metric tons per hectare [t/ha]). In 1940, 39.1 bushels per acre (1.8 t/ha) were produced. A total of 161 bushels per acre (7.4 t/ha) was grown in 1990. In 2000, 187 bushels per acre (8.6 t/ha) were yielded. A total of 202 bushels per acre (9.3 t/ha) was produced in 2004.

Globally, there have also been increases in the yield of maize and corn, with, for instance, a 63% increase in yield in Brazil between 1990 and 2005. There were also marked increases in the yield for wheat, rice, and other cereals. The production of corn shows marked increases because of both increased yield and land area being used (see Table 1-2).

UNITS AND CONVERSIONS

1 bushel = 0.0254 metric ton

1 metric ton = 39.4 bushels

Hectare (10,000 m²) = 2.47 acres

Acre (4,840 square yards) = 0.40 hectare

TABLE 1-2 Maize production (million metric tons)

	1990	2005
United States	201.5	282.3
China	97.2	139.5
Brazil	21.3	35.1
Argentina	5.4	20.4
India	9.0	14.7
France	9.4	13.8
South Africa	9.2	11.7
Romania	6.8	10.4
Canada	7.1	9.5
Egypt	4.8	7.7
Ukraine	a	7.1
Spain	3.0	4.1
Germany	4.1	4.1
Russian Federation	a	3.2

[a]No data for 1990 are reported for the former Soviet Union, which included Russia and the Ukraine.

The Green Revolution and the World Food Prize

The Green Revolution brought genetically improved high-yielding dwarf wheat and rice to the developing world, particularly Asia and Latin America. This, together with improved agronomic practices (fertilizer, irrigation, etc.), resulted in a 6-fold increase in grain production and countries like China becoming net exporters of grain. The research leading to these improved seeds was performed at the International Maize and Wheat Improvement Center (*Centro Internacional de Mejoramiento de Maíz y Trigo*) near Mexico City and other International Research Centers, with strong financial support from the U.S. Agency for Economic Development, and both the Rockefeller and Ford Foundations.

Norman Borlaug was born in 1914 in Cresco, Iowa. He is hailed as the "Father of the Green Revolution." After earning a Ph.D. at the University of Minnesota, he led the research at the International Maize and Wheat Improvement Center, known as CIMMYT (based on the acronym for its name in Spanish: Centro Internacional de Mejoramiento de Maíz y Trigo), in Mexico that resulted in high-yielding, disease-resistant dwarf wheat. This wheat not only had high yield, but also much reduced losses from breakage or bending of the stem. Borlaug worked diligently, persistently, and ultimately successfully to persuade leaders of developing countries to adopt the new varieties of wheat and later other cereals, including rice. He has championed the application of science, including biotechnology, to solving humankind's pressing need for adequate food and the end of hunger. In recognition of his work, Borlaug received the Nobel Peace Prize in 1970, and later the Presidential Medal of Freedom and the Congressional Gold Medal (see fig. 1-7).

Norman Borlaug founded the World Food Prize to give recognition to food and agricultural pioneers throughout the world (see fig. 1-8). Among the prizewinners are the following:

- Modadugu Gupta for work on aquaculture
- Yuan Longping and Monty Jones for the development of hybrid rice

FIGURE 1–7 Norman Borlaug received the Congressional Gold Medal, United States, in 2006. Image shows Dr. Borlaug with Speaker of the House Nancy Pelosi, President George W. Bush, and Senate Majority Leader Harry Reid. Courtesy of Justin Cremer, World Food Prize.

THE WORLD FOOD PRIZE

FIGURE 1–8 Logo of the World Food Prize. Courtesy of Justin Cremer, World Food Prize.

- Walter Plowright for developing a vaccine for Rinderpest or cattle plague, which has such a devastating effect on livestock production in many developing countries
- He Kang, the then Chinese Minister for Agriculture, whose leadership and adoption of new policies have moved China so far toward food security,
- M.S. Swaminathan, the architect of India's "Green Revolution"

Corn Usage

Today, grain from corn is used in food (e.g., tortillas, corn flake, corn chips), and is a major ingredient in livestock feed, particularly for pigs, poultry, and feedlot

---

ISSUES *for discussion*

THE GREEN REVOLUTION

1. What do you know about the Green Revolution?
2. What are the advantages?
3. What are the benefits to the farmer?
4. Are these benefits to all farmers?
5. Are there increased costs to farmers?
6. What are the benefits to urban dwellers?
7. Does the Green Revolution solve the population problem?
8. Does the Green Revolution solve hunger?
9. Are there adverse effects of increased use of chemical fertilizers and/or pesticides?
10. What are consequences of increased irrigation?
11. With increased production, there is the potential to mechanize. What does this do to rural employment and the drift to the cities?
12. Is agricultural biotechnology a second Green Revolution?

cattle, and as the raw material to produce ethanol to add to gasoline. Corn silage is also used extensively as a diet for cattle.

In 2004–2005, usage of corn in the United States was 298 million metric tons in production, 52 million metric tons in exports, 154 million metric tons in feed use, and 70 million metric tons in other uses (as ethanol for an alternative/additive for gasoline, and as human food) (U.S. Department of Agriculture Foreign Agricultural Service Database; http://www.fas.usda.gov).

The increased availability of grain coupled with increased purchasing power also leads to huge increases in the production and consumption of meat, eggs, and milk products (see Table 1-3).

TABLE 1-3 Comparison of global production of beef, pork, and poultry meat (in million metric tons) for 1995 and 2005

COMMODITY	1995	2005
Pork	80.1	102.8
Poultry meat	54.2	82.8
Beef	54.2	60.1

Source: Data from FAO.

Importance of Animal Agriculture in the Developing World

Animal agriculture is very important in the developing world. Livestock makes a significant contribution to the poor in developing countries, with estimates that 60% of income comes from livestock. Livestock provides a route out of the poverty trap. Livestock provides a source of cash (milk and egg money plus hides). There is high-quality nutrition derived from the food that the animals (milk, meat, and eggs) provide. For instance, animal products provide energy (see Table 1-4); proteins with a high biological value, having optimal amino acid composition (see Table 1-4); minerals such as calcium, iron, phosphorus, zinc, magnesium, and manganese; and vitamins such as thiamine (vitamin B_1), riboflavin (vitamin B_2), niacin, pyridoxine (vitamin B_6), and vitamin B_{12}.

TABLE 1-4 The importance of animal products to human nutrition

REGION	PERCENTAGE ENERGY (CALORIES) FROM ANIMAL PRODUCTS	PERCENTAGE PROTEIN FROM ANIMAL PRODUCTS
Developing countries	11%	26%
Developed countries	27%	56%
World	16%	36%

Source: Data from FAO.

Livestock provides power for plowing. They also provide manure for fertilizing the soil. Moreover, dried cattle dung can be used as a fuel in Asia and Africa. Livestock contributes significantly to the economic vitality in rural communities in the United States and other developed countries by being a value-added product from plant agriculture, and by generating income and jobs by production and further processing.

Definition

The word *corn* is synonymous with maize in most, if not all, of the United States, but in England, *corn* means "wheat." This apparent conflict reflects the meaning of the corn, that is, the most prevalent cereal or grain produced in a specific region.

REVIEW QUESTIONS

1. What is the population in the world?

2. How fast is the world's population growing?

3. When is the world's population projected to reach 10 billion people?

4. How many people have been added to the world's population between 1960 and 2007?

5. What is a reasonable estimate of the average calories that people across the planet consume?

6. What is the population of the United States?

7. What is the rate of increase of the U.S. population?

8. What is the relative contribution of birth rate and immigration to the increasing population in the United States?

9. How has the U.S. population changed over its history?

10. What is the rule of 72?

11. As world populations increase, is the increased number of people living in the developed world or the developing world? Are they in cities or rural areas? What are the consequences?

12. Who was Thomas Malthus, and what was his significant contribution?

13. Was Thomas Malthus wrong in his conclusions?

14. What is the extent of hunger and inadequate nutrition in the world?

15. Where do people suffering from hunger and inadequate nutrition live in the world?

16. What are the consequences of hunger and inadequate nutrition?

17. What are the Millennium Development Goals of the United Nations (UN)?

18. What is the mission of the Food and Agriculture Organization (FAO) of the UN?

19. What is "food security"?

20. How has China greatly reduced the number of its people suffering from hunger and inadequate nutrition?

21. What is the "Green Revolution," and what was the contribution of Norman Borlaug to it?

22. Is the consumption of livestock products increasing in the world? If yes, why?

23. Why have the yields of corn, wheat, and rice increased?

24. How does this affect livestock production?

25. Why is livestock production important in rural communities in the United States?

26. Why is livestock production important in the developing world?

27. What are the critically important contributions of livestock products to human nutrition?

REFERENCES AND FURTHER READING

Diamond, J. (1997). *Guns, germs, and steel: The fates of human societies*. New York: W. W. Norton.

Malthus, T. (1798). *An essay on the principle of population: An essay on the principle of population, as it affects the future improvement of society with remarks on the speculations of Mr. Godwin, M. Condorcet, and other writers*. Retrieved July 19, 2009, from http://www.ac.wwu.edu/~stephan/malthus/malthus.0.html

NASA Earth at night. Retrieved October 21, 2009, from http://earthobservatory.nasa.gov/IOTD/view.php?id=896

The road ahead: FAO and the Millennium Development Goals 2009. Retrieved July 18, 2009, from www.fao.org

Scanes, C. G., & Wilham, R. L. (2005). Agriculture—A short history (from domestication to present). In J. A. Miranowski & C. G. Scanes (Eds.), *Perspectives in world food and agriculture* (Vol. 2) (pp. 273–294). Ames, IA: Blackwell.

U.S. Census Bureau. Retrieved July 19, 2009, from http://www.census.gov

U.S. Census Bureau International Data Base (IDB). Retrieved July 19, 2009, from http://www.census.gov/ipc/www/idb/

United Nations Web sites such as esa.un.org or www.un.org/popin/

United Nations (1999). *The world at six billion, Table 1*. "World Population From Year 0 to Stabilization" (p. 5). Retrieved July 12, 2009, from http://www.un.org/esa/population/publications/sixbillion/sixbilpart1.pdf

Von Kaufmann, R. R., & Fitzhugh, H. (2004). The importance of livestock for the world's poor. In C. G. Scanes & J. A. Miranowski (Eds.), *Perspectives in world food and agriculture* (pp. 137–159). Ames, IA: Blackwell.

The world's largest cities and urban areas in 2006: Urban areas ranked 1 to 100. Retrieved October 21, 2009, from http://www.citymayors.com/statistics/urban_2006_1.html

CHAPTER 2

Origin and Development of Agriculture

OBJECTIVES

This chapter will consider the following:

- An introduction to the domestication of animals and plants
- Where, when, and why plants (cereals, legumes, fruits, and nuts) were domesticated
- Changes in the productivity of cereal production
- Where, when, and why animals were domesticated
- Domestication in reverse, such as the case of feral animals

ORIGIN OF AGRICULTURE

Definition

Neolithic time is the "New Stone Age."

During the hunter-gatherer stage of human development, people gathered grain from various wild grasses, fruits, and roots; hunted large and small wild animals; and fished. The Neolithic revolution encompassed a shift from nomadic hunting and gathering to planting and staying in one place (agriculture). An intermediary step would be gathering wild grain and staying in one area for the winter, then moving on to another area in the spring or summer months.

The shift from hunter gathering to agriculture allowed, or perhaps caused, the population to grow. The population density could expand by 10- to 100-fold. More children meant more help to grow plants and to take care of animals. It was advantageous to have more children. In contrast for hunter-gatherers, children had to be carried and made nomadic movement more difficult. There was a premium on small families.

Staying in one place contributed to permanent settlements. Some of these settlements grew into towns or cities. As agriculture developed with surplus food, people could adopt specialized jobs outside of agriculture (e.g., wood, stone, or metal work).

There was also a need to protect the land, animals, and stored grain. It is easy to imagine that stored grain or domestic animals would be a tempting target for neighboring hunter-gatherers. This need to protect stored agricultural products (or a desire to take other people's products or land) led to the development of the first military. The costs associated with soldiers, their equipment, and leaders led to the advent of taxation, requiring written language and number systems for record keeping. Consequently, the higher density population of the farming settlements compared with hunter-gatherer groups gave the former an obvious military advantage. Society also became more hierarchical with a division of labor between farmers and soldiers who would also run the society as an "elite" together with artisans and officials to serve the elite.

With specialization of human activity, there came occupations such as metal or stone workers. The development of agriculture is summarized in Figure 2-1.

> **QUOTATION**
>
> ABOUT THE IMPACT OF AGRICULTURE ON HUMAN CIVILIZATION
>
> *"When tillage begins, other arts follow. The farmers, therefore, are the founders of civilization."*
>
> Daniel Webster (1782–1852), American statesman

 ## WHERE AND WHEN DID ANIMAL AND PLANT DOMESTICATION FIRST OCCUR?

Much of what we eat or wear today is the result of Neolithic (see sidebar for definition) domestication of animals and plants in the "Fertile Crescent" (see fig. 2.2). It is no wonder that this region is referred to as the "Cradle of Civilization." The Fertile Crescent is the area of the Middle East around the rivers Tigris, Euphrates, and Nile (present-day Iraq, Syria, Lebanon, Israel, Palestine, and Jordan, parts of Turkey and Iran, and stretching into Egypt). Development along the Nile occurred somewhat later. Plants and animals that were domesticated during the Neolithic times in the Fertile Crescent included cereal (such as wheat and barley), pulses (such as lentils, peas, chickpeas, and bitter vetch), flax (used to make linen), cattle, sheep, goats, pigs, geese, and cats.

There were other "cradles of civilization," including East Asia (the area of present-day China along the Yangtze and Yellow Rivers) (see fig. 2-2); South Asia/the Indian subcontinent, particularly the areas along the Ganges and Indus Rivers; and Meso- and South America (see fig. 2-3).

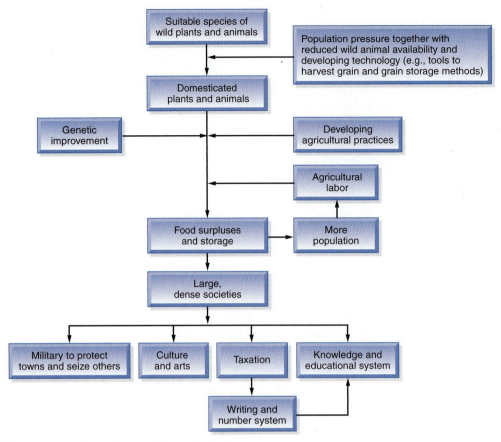

FIGURE 2–1 Development of agriculture and the onset of culture.

FIGURE 2–2 Agriculture was first developed along rivers in tropical and semitropical areas such as the Middle East (the Fertile Crescent along the rivers Tigris, Euphrates, and Nile), the Indian subcontinent (along the Indus), and East Asia (along the Yangtze and Yellow/Huang He Rivers).

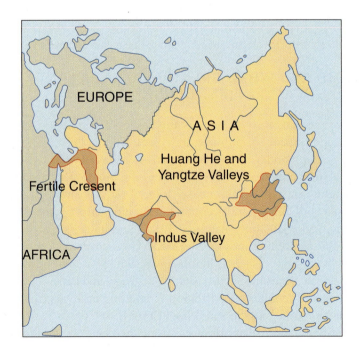

FIGURE 2-3 Agriculture developed independently in the New World. Positions of the cradles of civilization in the New World include the Aztec, Maya, and Inca civilizations.

Plants and animals that were domesticated in the Neolithic times in East Asia (present-day China) included rice, pigs, chickens, ducks, and geese.

Plants and animals that were domesticated in the Neolithic times in the Indian subcontinent included rice and *Bos indicus* cattle. Plants and animals that were domesticated in the Neolithic times in different areas of the Americas included potatoes (Andes in South America), turkeys (Mesoamerica), llamas and alpacas (pre-Incan Andes in South America), Muscovy ducks (South America), wild rice (North America), and guinea pigs (South America).

 DOMESTICATION OF PLANTS

Cereals (wheat, rice, corn, barley) were the first plants to be domesticated. In addition, as part of the Neolithic revolution, people domesticated pulses, including lentils, peas, chickpeas, and bitter vetch. The initial domestication of fruits and vegetables occurred later in the Neolithic or into the Bronze Age. Soybeans (*Glycine max*) were domesticated in Northeast China or the Korean peninsula 2,500–3,000 years ago. Other significant crops include potatoes (domesticated in South America Andes region), sorghum (Sub-Saharan Africa), sunflower (Eastern North America, i.e., the present-day Eastern United States), and oats (Western Europe).

Cereals

All cereals are domesticated grasses. During domestication, people choose to plant seeds from individual wild plants selected for certain characteristics.

Seeds were chosen for the amount of seed (yield per plant) and its ready recovery (e.g., threshing characteristics). The location and timeline of the domestication of cereals are summarized in Table 2-1.

TABLE 2-1 The wheres and whens of the domestication of cereals

CEREAL	WHERE DOMESTICATED	WHEN DOMESTICATED (BC)
Wheat	Northeast Fertile Crescent	8000
Rice	East Asia and possibly also the Indian subcontinent	8800
Barley	Fertile Crescent	7000
Maize	Mesoamerica	5000–6000

The domestication and development of wheat involved three processes:

1. Increases in chromosome number or ploidy.
2. Spontaneous hybridization (crosses) between different but closely related species of grasses.
3. Selection for mutations for favorable traits, for example nonfragile rachis. This prevents scattering of the seeds in the field together with a free-threshing character.

The most cultivated wheats in the world are bread or common wheat (by far the most common), durum or macaroni wheat (*Triticum turgidum*) (tetraploid), einkorn (*Triticum monococcum monococcum*) (diploid), emmer (*Triticum dicoccon*) (tetraploid), and spelt (*Triticum aestivum spelta*) (hexaploid).

Bread or common wheat (*Triticum aestivum aestivum*) is hexaploid with 42 chromosomes. This is the result of the spontaneous hybridization (about 9,000 years ago) of the tetraploid wild wheat crossing with the diploid goat grass *Aegilops squarrosa*, which is also referred to as *Aegilops speltoides* with 14 chromosomes (see Table 2-2). In turn, tetraploid wild wheat is the result of a cross between two diploid grasses (*Triticum uratus* and *Triticum speltoides*), each with 14 chromosomes.

TABLE 2-2 The hybridization that produced bread or common wheat

WILD WHEAT		GOAT GRASS		BREAD OR COMMON WHEAT
T. turgidum	+	*Aegilops tauschii*	⇒	*T. aestivum aestivum*
Genome AABB		Genome DD		Genome AABBDD

Domestic barley (*Hordeum vulgare*) was domesticated from wild barley (*H. vulgare* ssp. *spontaneum*) in the Fertile Crescent about 9,000 years ago. Based on archeological investigations, the earliest domesticated barley had two-rowed spikes as does the wild progenitor. Cultivation of the superior six-rowed barley started about 8,000 years ago and spread rapidly. There is a single recessive allele responsible for barley being six rowed. Barley has been used since at least the time of ancient Egypt not only for bread making, but also for beer.

Corn (*Zea mays*) was domesticated from teosinte about 7,000–8,000 years ago in Central America with crosses between teosinte and Tripsacum. Corn is today one of the major cereals. Once domesticated, corn spread throughout Central America and then to North America before the era of Christopher Columbus (500–1,000 years ago). It was grown extensively and became a major staple of the diet.

Starch Digestion and Human Genomics

The domestication of grain with its high starch content has been accompanied by changes in the genome of human populations that consume high amounts of grain or potatoes. The saliva contains the enzyme amylase that initiates the digestion of starch. This enzyme is encoded by the *AMY1* gene. Interestingly, there are more copies of the gene in populations that consume high amounts of starch-containing foods, such as Americans, Europeans, and Asians, compared with human hunter-gatherer populations.

ISSUES *for discussion*

CONSEQUENCES OF GRAIN DOMESTICATION AND PRODUCTION

1. What were the consequences of the Neolithic revolution and the shift toward agriculture?
2. Both wheat and corn are the result of crosses between different species of plants that do not normally cross. Should they be considered as genetically engineered or transgenetic?
3. What is the effect of increased production of cereals on the livestock industry?
4. What is the effect of increased production of cereals on the price a farmer receives for the grain?
5. What would the consequence be if we moved to a greater use of cereals for industrial or energy use, or a shift toward more vegetarian diets?

Other Domesticated Plants

Although cereals provide much of human food either directly or after passage through livestock, there are many other domesticated plants. Table 2-3 provides a list of some of these domesticated plants and the approximate time of their domestication.

 ## ANIMAL DOMESTICATION

Domestication of Dogs and Cats

Dogs are the result of the domestication of the grey wolf, which was the first animal to be domesticated. The earliest remains of domesticated dogs date from about 15,000 years ago in Europe and the Middle East. Genetic evidence can be interpreted as either multiple founding or domestication events in Eurasia about

TABLE 2-3 Date of domestication of selected plants, excluding cereals

PLANT	YEAR
Fruits	
Grapes	3000 BC
Melon	2000 BC
Peach	2000 BC
Apple	1000 BC
Banana	1000 BC
Kiwi fruit	1970 (AD)
Legumes	
Chickpea	6500 BC
Bean	6500 BC
Vegetables	
Potato	3300 BC
Lettuce	3000 BC
Onion	2200 BC
Cabbage	400 BC
Carrot	AD 60
Nuts	
Almond	3000 BC
Walnut	1500 BC

Based on data summarized by Duarte et al.

100,000 years ago or, more likely, a single founding from as little as three females in East Asia about 15,000 years ago. The first domesticated dogs assisted hunting and provided protection for the hunter-gatherer communities.

The first Americans brought dogs with them when they crossed the Bering land bridge to North America at the end of the last Ice Age (about 14,000 BC).

Cats are the result of domestication of the Near Eastern wildcat (*Felis silvestris lybica*). This occurred in the Fertile Crescent by 7500 BC, at the time when agricultural villages were developing, with cats reducing losses of stored grain from rodents. It is thought that there were only about five founders.

Domestication of Livestock and Poultry

Around 9000 BC, sheep were domesticated by some of the new Neolithic agricultural communities in the Fertile Crescent. This was followed by the domestication of the goat in about 8000 BC and cattle in about 5800 BC in the Fertile Crescent. Independently, Asian cattle were domesticated in the Indian subcontinent in about 2500 BC. Sheep, goats, and cattle provided meat and milk, converting indigestible plant materials to high-quality protein foods for the burgeoning Neolithic agricultural communities (see Table 2-4). Sheep, goats, and cattle were raised not only by farmers, but also by nomadic people. Livestock, including cattle, sheep, goats, pigs, buffalo, and chickens, is raised today in the developing world under different systems, including pastoral, with nomadic groups on the native range

TABLE 2-4 The wheres and whens of animal domestication

ANIMAL	WHERE DOMESTICATED	WHEN DOMESTICATED (BC)
Cattle (European breeds) *Bos taurus*	Near East, i.e., around present-day Turkey	5800
Cattle (Indian breeds) *Bos indicus*	Indian subcontinent	2500
Pigs	Two sites: East Asia (area of present-day China) Fertile Crescent in the Middle East	>7000 >7000
Sheep	Fertile Crescent	>9000
Goats	Fertile Crescent	8000
Chickens	Northeast area of present-day China along the Yellow River	5500

and grasslands; mixed crop and livestock, with small-scale family or village farms, tenant farmers, and sharecroppers; mixed crop and livestock (small-scale family or village farms); landless peasants; and industrial, with large-scale confinement systems used in the United States for poultry and pig production.

Pigs were domesticated independently in both present-day China and the Fertile Crescent around 7000 BC (see Table 2-4). Chickens were domesticated by 5500 BC in what is now Northeast China. It is thought that captive jungle fowl (*Gallus gallus*) from Southeast Asia provide the ancestral stock. Both pigs and chickens are excellent scavengers and provide an excellent source of meat with chickens also providing eggs.

Domestication of Horses

Around 4000 BC, horses were domesticated by nomadic people in central Eurasia (the Steppes) (see Table 2-5). It is thought that they took tamed horses that next

> **INTERESTING FACTOID**
>
> What is the difference between taming a wild animal and domestication? According to Jared Diamond in his book *Guns, Germs, and Steel: The Fates of Human Societies*, in domestication, there must be breeding in captivity and selection such that the animal becomes "something more useful to man."

TABLE 2-5 The wheres and whens of companion/working animal domestication

ANIMAL OR PLANT	WHERE DOMESTICATED	WHEN DOMESTICATED
Dogs	East Asia or possibly multiple domestications in other loci in Eurasia	>15,000 and possibly 50,000 years ago
Cats	Fertile Crescent	7500 BC
Horses	Eurasian Steppes, with multiple domestications probable	4000 BC
Donkeys	At least two domestications in Northeast Africa	3000 BC
Dromedary camels	Arabian peninsula	1500 BC
Bactrian camels	Central Asia	1500 BC
Llama and alpacas	South America	3000 BC

Note: Cattle have been used extensively as draft animals in the past.

step to domestication and began selecting for attributes useful to them. These were used for food, hunting, and transportation.

Other Domesticated Mammals

Other domesticated mammals included camels, llamas and alpacas, donkeys, reindeer, water buffalo, and guinea pigs. More details are included on the domestication of domestic animals when each is considered in Chapters 3, 5, and 6 on the individual livestock species, or on horses or companion animals (dogs and cats), respectively (see Table 2-5).

Feral Animals or Domestication in Reverse

Feral animals are those whose ancestors had been domesticated but have reverted and now live in a wild state. Feral animals breed in the wild and have relatively stable populations. This is different from a pet parrot or dog being lost or escaping.

Examples of feral animals include the following:

- Soay sheep on the Island of St. Kilda off the Scottish coast
- Dingoes (dogs) in Australia
- Feral cats that can be found in both rural and urban areas; feral cats can interbreed with the European wildcat (*F. silvestris*).
- Feral horses, including mustangs of the American West and the Chincoteague ponies from an island off the coast of Maryland. Horses were originally brought to the Americas by European settlers.
- Feral donkeys, including burros of the American West. Donkeys were originally brought to the Americas by European settlers.
- Wild pigs in the southern United States are feral. Pigs were originally brought to the Americas by European settlers.

Feral animals can have a dramatic effect on the ecosystem as they compete against native animals, or they eat native plants and prey on animals.

REVIEW QUESTIONS

1. Where geographically did agriculture begin?

2. What is the Neolithic time?

3. When did agriculture begin?

4. What is the Fertile Crescent?

5. How did the advent of agriculture affect population? Why?

6. What was the relationship between the development of agriculture, and the development of cities, learning, and the military?

7. Where, when, and why were specific plants (cereals, legumes, fruits, and nuts) domesticated?

8. What were the changes to wheat and corn after domestication?

9. What is ploidy?

10. Where, when, and why were the major companion/working animals (cats, dogs, horses, donkeys, camels, dromedaries, llamas, alpacas) and livestock species (cattle, sheep, goats, pigs, chickens) domesticated?

11. What is the difference between captive animals and domesticated animals?

12. What is a feral animal? Name three species of feral animal.

REFERENCES AND FURTHER READING

Beja-Pereira, A., England, P. R., Ferrand, N., Jordan, S., Bakhiet, A. O., Abdalla, M. A., et al. (2004). African origins of the domestic donkey. *Science, 304*, 1781.

Diamond, J. (1997). *Guns, germs, and steel: The fates of human societies*. New York: W. W. Norton.

Driscoll, C. A., Menotti-Raymond, M., Roca, A. L., Hupe, K., Johnson, W. E., Geffen, E., et al. (2007). The Near Eastern origin of cat domestication. *Science, 317*, 519–523.

Duarte, C. M., Marbá, N., & Holmer, M. (2007). Ecology. Rapid domestication of marine species. *Science, 316*, 382–383.

Keegan, J. (1993). *A history of warfare*. New York: Vintage Books.

Komatsuda, T., Pourkheirandish, M., He, C., Azhaguvel, P., Kanamori, H., Perovic, D., et al. (2007). Six-rowed barley originated from a mutation in a homeodomain-leucine zipper I-class homeobox gene. *Proceedings of the National Academy of Sciences of the United States of America, 104*, 1424–1429.

Savolainen, P., Zhang, Y. P., Luo, J., Lundeberg, J., & Leitner, T. (2002). Genetic evidence for an East Asian origin of domestic dogs. *Science, 298*, 1610–1613.

Scanes, C. G., & Wilham, R. L. (2005). Agriculture–A short history (from domestication to present). In J. A. Miranowski & C. G. Scanes (Eds.), *Perspectives in world food and agriculture* (Vol. 2) (pp. 273–294). Ames, IA: Blackwell.

CHAPTER 3

Global Overview of the Livestock Industry

OBJECTIVES

This chapter will consider the following:

- An overview on the importance of animal production globally
- The importance of cattle to global animal production
- The importance of pigs to global animal production
- The importance of poultry to global animal production

AN OVERVIEW OF THE IMPORTANCE OF GLOBAL ANIMAL PRODUCTION

Livestock and poultry production is a very important part of agriculture in the United States and throughout the world (see fig. 3-1). People depend on animal production and processing for their livelihood, for food (particularly high-quality protein), and for leather and wool.

According to the Food and Agriculture Organization (FAO) of the United Nations, the average per person consumption of animal products in the world is meat (based on carcass weight), 84 lb (38 kg) per year; milk, 101 lb (46 kg) per year; eggs, 18 lb (8 kg) per year; and fish, 44 lb (20 kg) per year.

Consumption of meat in the United States is higher, with meat consumption (based on carcass weight) being over 264 lb per capita per year, and much lower in Sub-Saharan Africa, where it is frequently lower than 22 lb per capita per year. In developing countries, increasing consumption of animal products improves pregnancy outcome together with both growth and cognition in infants and children. It is often said that livestock is competing with people for cereal grains. However, less than 35% of grains are consumed by livestock globally, and the percentage continues to decline as shifts toward the more efficient systems continue.

Globally, animal agriculture is of growing importance. The FAO states that livestock production accounts for 37% of world agriculture (compared with 63% as crop production). Both are growing in size at 2% per year. The increases are disproportionate in the developing world, where output is increasing by more than 3% per year. The increases in the production of meat between 1980 and 2004 globally are shown in Table 3-1.

UNITS AND CONVERSIONS

One metric ton equals 1 million g or 2,205 lb.

One U.S. ton equals 2,000 lb.

One imperial or United Kingdom ton equals 2,200 lb.

Animal Production in Different Countries

The top three countries for meat production are the United States, People's Republic of China, and Brazil (see Tables 3-1 and 3-2). Production is increasing in all three, but the rate of increase is far greater in China and Brazil. This is due to the availability of grain and capital for intensive animal production. Meat production as a percentage of global production is declining in Western European countries

FIGURE 3-1 Composite view of the earth from space in July showing areas where there are high densities of green plants growing and, therefore, where animals are likely to be raised. Source: NASA.

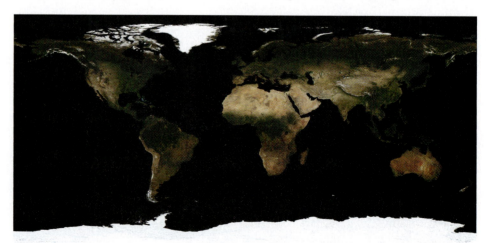

TABLE 3-1 World production of meat (in million metric tons) with comparisons of the United States with China and Brazil

	1980	1990	2000	2004
World	136	178	235	260
China	14.6	30.4	63.0	74.4
United States	24.3	28.5	37.6	38.7
Brazil	5.2	8.2	15.3	20.0

TABLE 3-2 Relative size of meat production in countries producing at least 1% of the global production of meat in 2004

COUNTRY	SHARE OF WORLD PRODUCTION OF MEAT (%)			
	1980	1990	2000	2004
China	10.7	17.1	26.8	28.6
United States	17.9	16.0	16.0	14.9
Brazil	3.8	4.6	6.5	7.7
Germany	5.1	3.9	2.7	2.6
France	4.0	3.2	2.8	2.4
India	1.9	2.2	2.2	2.3
Spain	1.9	1.9	2.1	2.1
Mexico	1.9	1.6	1.9	1.9
Russian Federation	[a]	[a]	1.9	1.9
Canada	1.8	1.6	1.7	1.8
Argentina	2.7	2.0	1.7	1.6
Italy	2.6	2.2	1.8	1.6
Australia	2.0	1.7	1.6	1.4
Poland	2.0	1.6	1.2	1.3
United Kingdom	2.2	1.9	1.5	1.3
Japan	2.2	1.9	1.3	1.2

[a]*Figures for the former Soviet Union are omitted here and elsewhere in this chapter.*

such as Germany, France, Italy, and the United Kingdom, together with the Asian advanced-industrial country of Japan. The basis of the decline is the higher costs of production, including high labor, and land and feed costs, together with the expense of addressing stringent government regulations.

Why is animal agriculture increasing? The simple answer is that production of meat, eggs, milk, and aquaculture products (finfish and shellfish) is increasing to meet consumer demand.

By and large, people like animal products, regardless of where they live in the world. Not only have the number of people increased, but also as the global economy has grown, they have had much greater ability to afford to buy food and, particularly, animal products. There is a strong correlation between per capita income (per capita gross domestic product) in a country and the consumption of meat, milk, and eggs.

There are good nutritional reasons to eat animal products, including high-quality protein, with optimal or close-to-optimal amino acid balance; iron; zinc; calcium; and vitamins such as niacin and vitamin B_{12}.

Definitions

What are shellfish and finfish? Shellfish are mollusks (such as mussels and squid) and crustaceans such as shrimp, crabs, crawfish, and lobsters. Finfish are vertebrate fish with a backbone, eyes, a mouth, gills, and fins.

Animal products in the average U.S. diet provide 35% of the energy, 68% of the protein, 78% of the calcium, 39% of the iron, 42% of the vitamin A, and 37–98% of the B vitamins. The proportion of protein and other nutrients coming from animal products is much lower in developing countries. The percent protein coming from animal products has been estimated at 28% in China and 15% in India, although with higher incomes and increased production, these are certainly underestimates. The percent protein coming from animal products in Sub-Saharan Africa is about 20%.

 ## PRODUCTION OF MAJOR ANIMAL PRODUCTS

This section will cover the global production of products from cattle, pigs, and poultry—the major agricultural livestock in the United States and the world. Other species or groups of species such as horses, sheep and goats, alternate agricultural animals, and aquaculture species are considered in Chapters 5, 9, 10, and 12, respectively. Today beef, pork, and poultry meat each represent almost a third of meat produced and consumed globally.

Beef and Dairy Production

There has been a major increase in global production of meat from cattle (beef together with veal) (see Table 3-3) and milk (see Table 3-4). This reflects increases in all production systems. Extensive production systems are range- or grass-fed cattle and the pastoral systems of developing countries. The land available for expansion of production in extensive systems is limited. Intensive systems use cereals being fed and/or supplements. Mixed-crop systems are less intensive systems, including cattle eating the residue after the harvesting of cereals.

Beef production is growing, with global production up 11% between 1990 and 2005 (see Table 3-3 and fig. 3-2). This increase is marked but is less than that observed with either pork or poultry meat. Perhaps surprisingly, China is exhibiting a very major increase in the production of beef (89%).

Milk production for the 10 highest producing nations is summarized in Table 3-4. These figures show the total milk production from cattle and water buffalo.

TABLE 3-3 World production of beef (cattle meat) in million metric tons

NO.	COUNTRY	1990	2005
1	United States	10.5	11.3
2	Brazil	4.1	7.8
3	China	1.3	7.1
4	Argentina	3.0	3.0
5	India	2.4	3.0
6	Australia	1.7	2.3
7	France	2.2	1.9
8	Mexico	1.3	1.8
9	Russian Federation	[a]	1.8
10	Canada	1.1	1.7

TABLE 3-4 Comparison of milk production in the top milk-producing countries shown in million metric tons

RANKING IN 2005	COUNTRY	PRODUCTION IN 1995	PRODUCTION IN 2005
1	India	65.4	95.6
2	United States	70.4	80.3
3	Russian Federation	39.3	31.1
4	Pakistan	19.0	29.7
5	China	9.5	29.4
6	Germany	28.6	28.5
7	France	29.1	26.1
8	Brazil	17.1	25.5
9	United Kingdom	14.8	14.6
10	New Zealand	9.3	14.5

Data from the Food and Agriculture Organization of the United Nations (FAO).

The United States was the leading milk-producing nation for virtually all the 20th century but was overtaken by India in 1997. The United States is still the number-one producer of milk from cattle. Milk production systems for North America and Europe are considered in Chapter 7.

India is now the world's leading milk-producing nation (Table 3-4). Production of milk in India showed a modest increase between 1950 (17 million metric tons) and 1970 (21 million metric tons). Between 1970 and 2005, production of milk from both water buffalo and cattle has increased by 4.6-fold. This is due at least in part to the successes of a series of cooperatives that market the milk.

FIGURE 3-2 Cattle are tremendously important across the world, providing milk, meat, hides, manure (fertilizing soil and used for heat or cooking in some societies), and traditionally locomotive power pulling plows and other equipment. Reproduced by permission from Pichugin Dmitry. © 2010 by Shutterstock.com.

Production is typified by very small herds of two or three stalled animals fed crop residue such as wheat straw, rice straw, and corn stalks, together with grass and other vegetation collected from along roadsides. The feed is supplemented using a nitrogen source such as urea. The increase in milk production and availability has enabled per-person consumption to double to more than a cup per day. This is critically important to children and infants, together with pregnant and lactating women in whom needs for protein are high, and protein deficiencies are not uncommon.

Milk production is rapidly increasing in China (3.1-fold in 10 years), Pakistan (56% in 10 years), and Brazil (49% in 10 years). However, production is static or declining in Western Europe and the Russian Federation (see Table 3-4).

Pork Production

There have been major increases in pork production across the world (see Table 3-5). This reflects increases in all production systems, including pastoral systems of developing countries, mixed-crop systems, and an intensive industrial model of cereals or other high-energy components such as sweet potatoes in Vietnam being fed together with protein sources.

TABLE 3-5 World production of pork (pig meat) in million metric tons

NO.	COUNTRY	1990	2005
1	China	24.0	51.2
2	United States	7.0	9.4
3	Germany	4.5	4.5
4	Brazil	1.0	3.1
4	Spain	1.8	3.1
6	Canada	1.2	2.6
7	France	1.7	2.3
7	Vietnam	0.7	2.3
9	Denmark	1.2	2.1
10	Poland	1.9	2.0

Data from FAO.

With the shift toward the intensive system, fewer farmers are producing pigs, regardless of whether in the developed or developing world.

Traditionally, pigs were scavengers in villages and farmsteads. This is still the pattern in some developing countries. In Western countries, pigs became more concentrated in many farms, particularly in the corn belt (see fig. 3-3). The situation today shows a strong trend toward large highly efficient pig-producing operations with integration with breeders and/or meat processors. Economists refer to this as the "industrialized model."

As indicated in Table 3-5, there are large increases occurring in pig production, particularly in China and Southeast Asia. There are also increases in two of the major corn-producing countries, the United States and Brazil. In contrast, production is relatively static in Europe.

FIGURE 3-3 Pigs are important agricultural animals in the United States and many countries of the world. Courtesy of the U.S. Department of Agriculture. Photo courtesy of Ken Hammond.

The development of pig and poultry production in developing countries is frequently occurring in densely populated areas near cities. This enables ready access to market and little transportation costs. However, there are adverse effects such as pollutants and pathogens released into populated areas, and the destruction of fragile ecosystems such as wetlands.

Poultry and Egg Production

Globally, the production of the primary poultry products (meat and eggs) has been increasing rapidly, as summarized in Tables 3-6 and 3-7. Among the largest increases in the production of chicken were those in Vietnam (136%), with production increasing from 0.14 million metric tons in 1995 to 0.32 million metric tons in 2005; India (217%); China (67%); and Brazil (112%). The increase in chicken production in the United States was 38% over the 10-year period (see fig. 3-4). This reflects consumption based, in turn, on consumer preference for these high-quality products and the relatively low price that has been a consequence of the efficiency of production. Over a 10-year period between 1995 and 2005, there were the following percent increases (based on carcass weights) globally: duck (67%); chicken (53%); goose (53%); chicken eggs (39%); other eggs (27%), predominantly duck and goose; and turkey (13%).

TABLE 3-6 Global production of chicken in million metric tons

NO.	COUNTRY	1990	2005
1	United States	8.7	15.9
2	China	2.7	10.2
3	Brazil	2.4	8.7
4	Mexico	0.8	2.4
5	India	0.3	1.9
6	United Kingdom	0.8	1.4
7	Japan	1.5	1.3
8	Russian Federation	a	1.3
9	France	1.1	1.2
10	Spain	0.8	1.0
11	Argentina	0.3	0.8

Data from FAO.

FIGURE 3-4 Poultry are a very important part of animal agriculture in the United States and globally. Reproduced by permission from R. © 2010 by Shutterstock.com.

TABLE 3-7 Comparison of egg production in the top egg-producing countries shown in million metric tons

RANKING IN 2005	COUNTRY	PRODUCTION IN 1995	PRODUCTION IN 2005
1	China	17.1	28.6
2	United States	4.4	5.3
3	India	1.5	2.5
4	Japan	2.5	2.5
5	Mexico	1.2	2.3
6	Russian Federation	1.9	2.1
7	Brazil	1.4	1.7

Data from FAO.

Globally, total egg production is 78% by weight that of poultry meat production, with China by far the major producing nation (see Table 3-8). Egg production is growing rapidly (39% over 10 years), with Asian countries (e.g., China and India) having high rates of increase of, respectively, 42% and 67%. Despite the increased concerns of welfare, the aggregate production of chicken eggs in the European Union has increased by 6% from 9.4 million metric tons in 1995 to 10.0 million metric tons in 2005.

TABLE 3-8 Comparison of global and Chinese production of chicken and other eggs (in million metric tons) for 2005: Comparison with 1995

COMMODITY/COUNTRY	1995	2005
Chicken eggs		
World	42.8	59.7
China	17.1	24.3
Other eggs (predominantly duck and goose eggs)		
World	4.0	5.1
China	3.4	4.3

Data from FAO.

ISSUES *for discussion*

GLOBAL LIVESTOCK PRODUCTION

1. Why is China the number-one producer of meat?
2. What has contributed to the increased production and consumption?
3. Why is India the number-one producer of milk?
4. What has contributed to the increased production and consumption?
5. How do you view the United States as being no longer the number-one producer of meat, milk, and eggs?
6. What are your views on how livestock is produced globally?
7. What are your views on how livestock is produced in the United States?
8. What do animal products contribute to human well-being?
9. What are the costs of animal production?
10. Why is animal production declining or static in Europe?

REVIEW QUESTIONS

1. Has the production of meat, milk, and eggs increased globally?
2. How does the consumption of animal products improve the well-being of people?
3. What are the major countries producing beef?
4. What are the major countries producing milk?
5. Why has milk production increased in India?
6. What are the consequences?
7. What are the major countries producing pork?
8. What are the major countries producing chicken meat?
9. What is the major country producing chicken eggs?
10. What is the major country producing duck and goose eggs?
11. Where are the increases in animal production occurring?
12. Where is animal production decreasing or static?
13. What are the major production systems for cattle?
14. What are the major production systems for pigs?

REFERENCES AND FURTHER READING

Global Animal Production

Brown, L. R. (2003). *Outgrowing the Earth: The food security challenge in an age of falling water tables and rising temperatures.* New York: W. W. Norton.

Food and Agriculture Organization of the United Nations (FAO). (2007). *Core production data.* Retrieved January 25, 2007, from http://faostat.fao.org/site/340/default.aspx

Food and Agriculture Organization of the United Nations (FAO). (2007). *FAO statistical yearbook 2007–2008.* Retrieved July 18, 2007, from http://www.fao.org/economic/ess/publications-studies/statistical-yearbook/fao-statistical-yearbook-2007-2008/en/

Food and Agriculture Organization of the United Nations (FAO). (2007). *ProdSTAT: Livestock (primary and processed).* Retrieved January 25, 2007, from http://faostat.fao.org/site/569/DesktopDefault.aspx?PageID=569Chicken Meat

Speedy, A. W. (2003). Global production and consumption of animal source feeds. *The Journal of Nutrition, 133,* 4048S–4053S.

Importance of Animal Production in the Developing World

Allen, L. H. (2003). Interventions for micronutrient deficiency control in developing countries: Past, present and future. *The Journal of Nutrition*, *133*, 3875S–3878S.

Hambidge, K. M., & Krebs, N. F. (2007). Zinc deficiency: A special challenge. *The Journal of Nutrition*, *137*, 1101–1105.

Von Kaufmann, R. R., & Fitzhugh, H. (2004). The importance of livestock for the world's poor. In C. G. Scanes & J. A. Miranowski (Eds.), *Perspectives in world food and agriculture* (pp. 137–159). Ames, IA: Blackwell.

CHAPTER 4

Careers in Animal Science

OBJECTIVES

This chapter will consider the following:

- Overview of careers in animal science, including consideration of the different sectors, such as production agriculture, the agricultural service sector, business and finance, the dairy industry, the feed industry, government, the horse industry, the meat industry, and so on, where animal science graduates can find exciting and rewarding careers
- Options for professional school
- Opportunities for graduate school
- Résumé building
- Applying for jobs and internships, including the written application and the interview
- Advantages of internships

OVERVIEW OF CAREERS IN ANIMAL SCIENCE

Students in animal sciences gain a strong grounding in the sciences, an appreciation of animal behavior, breeding, genetics, microbiology, management, nutrition, physiology, reproduction, muscle biology, and the animal industry. In addition, such traits as critical thinking and international awareness are gained, and these are valued by potential employers. Other valued abilities are communication skills (written and oral), and the ability to work with other people and in teams.

Graduates of an animal science program accept positions in diverse and challenging careers. Potential employers can be found in such areas as agricultural service sector, animal breeding, animal health (vaccines and pharmaceuticals), biotechnology, dairy industry, education, exotic animals, corporate farms and ranches, finance, feed manufacturers, food (including meat) industries, government, horse industry, laboratory animal care, pharmaceutical companies, research, technical support (e.g., veterinary technicians), and zoos/historical farms/animal exhibitors such as theme parks. Salaries vary by career, but those in sales may have commission pay or bonuses on top of salary, together with a car and cell phone provided by the employer. A graduate in animal science may use his or her degree as a general education in the same way that liberal arts graduates use theirs. Examples of some careers for graduates in animal science are considered in detail below.

> ### SKILLS THAT EMPLOYERS ARE LOOKING FOR
>
> 1. Critical thinking
> 2. Problem solving
> 3. Good oral and written communication
> 4. Ability to work in teams
> 5. International awareness
> 6. Knowledge of academic field

Animal Production

There are careers in which an animal science graduate will work directly with animals, including farmer, rancher, broiler manager, hatchery manager, animal health supervisor, animal welfare specialist, contract producer, livestock herdsperson, dairy herdsman, livestock manager feedlot operators, horse unit manager, and kennel manager.

Agricultural Service Sector

Careers in agricultural sales and marketing include nutrition consultant, livestock product specialist, urban pet food distributor, animal/livestock marketer, salesperson for feed or feed ingredients, equipment, semen/genetics and animal health products, international market development, and sales specialist. Animal scientists have interesting careers in agricultural organizations such as the national and state beef, dairy, pork, and poultry organizations (e.g., National Cattlemen's Association, National Pork Producers Council, National Chicken Council, and the National Dairy Herd Improvement Association). The responsibilities of such positions are to serve the farmer members, to promote their respective products, to educate the public about the products and production systems, scientists, and to work in government agencies and with legislators to protect the interests of agriculture.

Positions in agricultural and other animal communications are available. Major animal journals employ some animal scientists as feature article writers, advertising development specialists, or marketers.

Animal Breeding, Genetics, and DNA Testing Industries

Increasingly, animal breeding is based on the "marriage" of population genetics and molecular genetics. Moreover, with livestock, horses, and companion animals, there is increasingly DNA testing to ensure that the correct breeding occurred. A further area of growth is the production of drugs using transgenic animals. All these exciting developments are providing job opportunities for animal scientists.

Animal Health, Biotechnology, and the Pharmaceutical Industry

There has been significant growth in the biotechnology and pharmaceutical industries with the development of new drugs, vaccines, and diagnostic tests. Pharmacologic and biotechnical advances are the result of extensive research and testing laboratories. An analogous situation exists for the animal health industry. There are many opportunities for animal scientists, including laboratory animal care, laboratory technician, research scientist, product safety and quality assurance, manager, and sales.

Business and Finance

Careers for animal scientists in banking, insurance, and real estate companies include agricultural loan officer, financial analyst, and financial representative. Moreover, animal scientists can work directly with farmers as agricultural finance advisors or small business management advisors.

Dairy Industry

Positions in the dairy industry include herdsperson, dairy manager, field representative for dairy cooperative/breeder, animal health, nutrition consultant, artificial inseminator, milk quality assurance/control, and new product research.

Education

Animal scientists teach at several different levels, including high school (e.g., vocational agriculture), youth in 4-H programs, community colleges, 4-year colleges, and research universities.

Feed Industry

Positions for animal scientists include such areas as sales and service, animal nutritionist, manager of research, feed mill manager, feed company manager, pet food research and manufacturing, and feed quality control (control officer and laboratory technician).

Government

There is a considerable scope for animal scientists being recruited into positions in the government. Possibilities are for the U.S. Department of Agriculture, including livestock marketing, livestock forecasting, reporting, environmental regulation, animal health, disease control, animal and meat inspection (e.g., Animal Plant Health Inspection Service), production credit officer, and public information, together with positions with state departments of agriculture.

There are opportunities to work in the field of international agriculture through international assistance projects such as those under the U.S. Agency for International Development or the World Bank. In addition, there are employment possibilities in the U.S. Department of Agriculture's Foreign Agricultural Service. This can be based in Washington, D.C., in a state in the United States, or internationally as agricultural attachés in U.S. embassies. Regardless of location, the Foreign Agricultural Service helps open markets to U.S. agricultural products.

Horse Industry

Positions in the equine industry include breeding farm management, equipment sales and service, farrier, farm or stable manager, nutrition consultant, pharmaceutical sales, riding instructor, or trainer. There are also careers related to racehorse management or training.

Meat Industry

Meat packers, processors, and related industries recruit animal scientists particularly with a meat science background for positions in management, product and process development, new foods development, purchasing (livestock buying for meat processor), quality control/assurance, food safety, technical and consumer services, advertising, merchandising, and sales.

Table 4-1 summarizes some of the careers in animal science. There are possibilities to work with livestock, horses, dogs and cats, exotic animals, farmers, and ranchers; work in the animal nutrition, feed, and other animal-related industries; lead or participate in a research program in such areas as biomedical research and genomics; conduct research in areas from animal diseases to nutrition; or be a communicator, teacher, or medical professional (e.g., veterinary technician, veterinarian, or medical doctor).

College is an ideal time to add to your accomplishments in the following ways: in the classroom; in extracurricular activities, which is an excellent way to expand your résumé and show your organizational talents and skills; working with a faculty member doing research, which is a great way to expand consciousness and confidence; and in outside employment related to your area of interest.

During your time as a student, you will have learned how to study, to get tasks accomplished, and to work with very different people. You should become increasingly aware of your own strengths and weaknesses during your college years. In seeking a career, it is best to build on your strengths and in areas where you are highly motivated. Having enough money is important, particularly if you do not have enough, but being happy and fulfilled in a position is also critically important.

OPTIONS FOR PROFESSIONAL SCHOOL

Many students who take an undergraduate degree in animal science have a burning desire to go to veterinary school and be a veterinarian. This is a laudable goal, but not everyone will be accepted to veterinary school. Grades are one of the most important considerations. Another is that you have to apply to be accepted.

What else is relevant to the application process? Specifically, what do they look for other than grades? Your university may have a college of veterinary medicine that

---

MATCHING YOU WITH THE RIGHT OPPORTUNITY

When considering any career, it is important to try to match your strengths and talents with the opportunities that are or may become available. Your strengths may include the ability to do one or more of the following:

- Work well with other people in teams
- Organize groups
- Plan methodically
- Motivate others
- Market
- Work with numbers
- Write
- Understand science
- Communicate successfully
- Be creative

TABLE 4–1 Careers in animal science: What is possible

Agriculture communications	International assistance/development specialist
Animal genetics	Laboratory animal science
Animal nutrition	Management
Animal welfare	Military
Aquaculture	Pharmaceutical industry
Artificial inseminator	Pork production
Biomedical careers	Poultry production
Biotechnology	Processing
Careers with dogs	Produce or merchandize niche products
Careers with exotic animals	Professor
Careers with horses	Public relations
Cattleman	Research technician
Education: kindergarten through 12th grade	Sales
Embryo transfer	Service
Export market	Start your own company
Extension agent or specialist in the United States or internationally	Research scientist
Dairy specialist	Toxicologist
Food industry	Veterinarian
Food safety	Veterinary assistant
Genomics	Veterinary technician
Government inspectors	Vocational education

requires a separate application. This application process usually occurs during the junior or senior year of undergraduate studies. If your state does not have a veterinary college, it may have contractual relationships with colleges of veterinary medicine either in state or out of state that can facilitate the application process and result in lower tuition costs. Although your state may have a college of veterinary medicine or have contracts with colleges of veterinary medicine in other states, you are not prevented from applying to other schools. However, if successful, you will have to pay out-of-state tuition.

Students with an animal science major frequently take classes that are required or recommended for admittance to colleges of veterinary medicine. Classes usually include subjects like anatomy, physiology, and biochemistry. It may be that during your undergraduate career, your goals change. A strong degree in animal science with basic science classes may allow students to pursue a medical degree, degree in dentistry, or graduate degree in nursing or the health sciences. Another professional degree that animal science students may consider is law school.

 ## OPPORTUNITIES FOR GRADUATE PROGRAMS

For many careers, a graduate degree is very useful and, in some, essential. For instance, if you are considering a career in business, a master's degree in business administration is excellent training, particularly if you want to be a "high flyer" (someone who moves up—often rapidly—within an organization). For a career in research, education, and much of government, a graduate degree is either required or recommended. A master's of science degree is recommended for many people considering a career in the sciences such as a laboratory technician, extension agent, educator (e.g., at a 2-year college), nutritionist, or geneticist. For someone wanting a career in research, a doctorate (doctor of philosophy) is usually required and essential if you want to lead your own laboratory or team, regardless of whether in a university, industry, or federal government laboratory, such as the U.S. Department of Agriculture's Agricultural Research Service, or other agencies such as the Centers for Disease Control or National Institutes of Health. A graduate degree may also be required for some regulatory agencies such as the Food and Drug Administration.

Advice for When Considering a Potential Graduate Program and School

The choice of a graduate program and graduate school comes down to first exploring what is available, the quality and reputation of the program and school, your preference to where you will live for 2 (for a master's degree) to about 5 years (for a doctorate), and what financial package is offered. Assistantships with a modest salary and tuition covered are frequently available for master's students and are almost a matter of course for doctoral students. The quality and reputation of the program and school are the most important parameters because you will be known by where you did your graduate degree for your entire career. Coupled to this is the reputation of your advisor/major professor, that is, the person you do research with. You will be known as that person's academic "son or daughter" for your career. Working with someone with an outstanding reputation will stand you in good stead; however, it is also important to have a research advisor/supervisor you can work with. Many graduate programs require applicants to take the Graduate Record Examination and submit scores and a grade point average as part of the application. Do not hesitate to contact someone you want to do research with in addition to submitting the formal application. Figure 4-1 provides an example of the close working relationship between faculty and students.

HOW TO BUILD A RÉSUMÉ

College is a time when you can hone your talents and develop in ways that you may not have considered. This could include taking a leadership role in an animal science and other student organizations such as Block and Bridle, running an activity, learning about a different culture in study abroad, assisting people and organizations by service learning, and conducting research with a faculty member. Not only are new skills learned, but also once you have a series of accomplishments under your belt, your confidence increases and you have something interesting to put into your résumé and talk about at the job interview.

APPLYING FOR JOBS AND INTERNSHIPS

Check with your school's career counseling center or your advisor to get a list of the companies that offer internships and jobs to graduates in your area of study. Once you have a list of potential employers, you need to conduct some research on the company to gain an initial understanding of what it might want and how you can assist it. The importance of internships should not be underemphasized and is considered later in this chapter.

FIGURE 4–1 Microbiologist James Mecham and student research aide Jenny Dockham examine autoradiographic film. Courtesy of the Agricultural Research Service/U.S. Department of Agriculture. Photo by Scott Bauer.

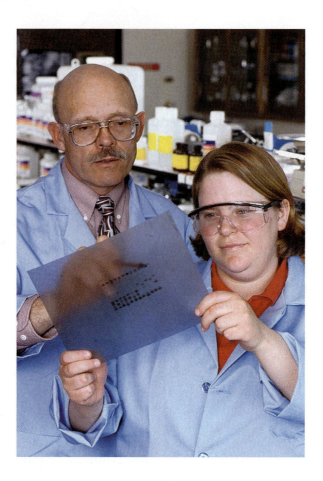

Preparing the Application

This consists of your résumé customized for the specific job and individual employer, together with a cover letter explaining why you are well suited to the position, paying particular attention to your skills that are transferable to the position. It is recommended not to just rely on "spell check" when preparing an application and the cover letter because you may have the wrong word. If the application is to be sent electronically, it may be better in the form of an e-mail with the résumé attached as a document file.

Preparing for the Interview

You want to prepare for an interview such that you demonstrate knowledge about the prospective employer. Make sure you understand the issues facing it, and be able to articulate how you can help it. Your first step to understanding the company is to learn as much as you can about your potential employer by visiting its Web site, reading its annual report, and "Googling" the employer's name together with the interviewer's name. It can also be helpful to talk to your network of contacts who know the prospective employer either through work or other interactions.

It is critical to have thought through the interview. How could you be an asset to the employer? What are your strengths that you want the interviewer to see?

What are your transferable skills? You need to customize your strengths to the employer's needs, and think in advance of the likely questions, preparing draft answers. Again, check with your college's career center for interview preparation opportunities. If these opportunities are not available, you might want to practice the interview with a friend who can assume the part of the employer.

It may seem obvious, but it is important to dress in clean, business-appropriate clothes. It is also essential to be on time, so check to see how long it is going to take to get to the exact location of the interview. Another critical action is to go to your Web page or social network space (e.g., on Facebook or MySpace), and remove any information and pictures that could be embarrassing, using what would embarrass your grandparents or parents as a guide. Frequently, human resources departments will check such sites.

Interviewing

Throughout the interview, you want to give the impression that you are a professional, would be suitable for the job, and would be an asset to the employer.

It is great to be memorable, but do not be remembered as the butt of jokes for many years. The last thing you want the interviewer to think is that you are weird! You need to be confident and assertive, but do not offend the interviewer.

The first essential is showing up on time. Allow yourself a time margin such that if anything goes wrong, you can still be on time for the interview. On the other hand, you should not be too early either. You should be no more than 10 minutes early for the appointment. If you have extra time, find a coffee shop nearby to relax and mentally prepare yourself.

Your appearance is the first impression that you make. This applies to an interview at the employer or to campus interviews. Not only should you be dressed appropriately and professionally for an interview, but also your clothes should be clean and pressed and your shoes polished. Another reason to be early is for a trip to the bathroom. This is recommended for the obvious reasons but also to check your hair and that you do not have anything between your teeth or on your face and that your fingernails are clean. I would suggest that tattoos and facial piercings not be visible (see fig. 4-2).

> ## QUOTATION
> ### THAT COULD BE APPLIED TO A JOB INTERVIEW
>
> *"We have nothing to fear but fear itself."*
>
> Franklin Delano Roosevelt (1882–1945), 32nd president of the United States, statesman, and leader

FIGURE 4-2 How not to be successful in interviewing for a job. *A,* Chewing gum. Reproduced by permission from Emiliano Rodriguez. © 2010 by Shutterstock.com. *B,* Eating messy foods. Reproduced by permission from m. emerson photographic. © 2010 by Shutterstock.com. *C,* Not paying attention. Courtesy of Getty Images.

TIPS *for interviewing*

- Timeliness. Be punctual!
- Turn off your cell phone before entering the building.
- Look professional with appropriate clothing and grooming (see fig. 4-3). Have clean shoes and fingernails, and do not wear facial jewelry.
- Use a firm handshake at the beginning and end of the interview. If your hands are clammy, wash and dry them.
- Do not slouch in your chair.
- Have a professional disposition and posture.
- Provide short, honest answers to questions.
- Be prepared, but present yourself as spontaneous, as opposed to giving memorized answers.
- Send a thank-you letter to the interviewer; beforehand, be sure to check for the person's correct job title, and the correct spelling of his or her name.

Greet the interviewer and shake his or her hand firmly. Sit down and start the back and forth of the interview (see fig. 4-4).

Answering Questions and What Questions to Expect

The interviewer can ask you virtually anything, although there are some questions that should not be asked, such as your age or marital status. The most obvious first questions are "Tell me about yourself," and "Why are you applying for this position?" These are questions for which you should have an answer

FIGURE 4-3 How to be successful in interviewing for a job. Show up early and dressed appropriately. Reproduced by permission from Carlos E. Santa Maria. © 2010 by Shutterstock.com.

FIGURE 4-4 An interview is like a race. The swiftest person wins. You need to be prepared, thoughtful, and eager to win. © Dmitriy Shironosov, 2011. Used under license from Shutterstock.com.

DOS AND DON'TS *for interviewing*

THE "DOS"

- Practice in front of a mirror.
- Dress appropriately; "clothes make the man/woman."
- Use a firm handshake.
- Sit up straight.
- Make eye contact. If you cannot look someone in the eye, you might try looking 1–2 inches above the eyes.
- Show respect for yourself and the person interviewing you; for example, turn off your cell phone.
- Keep your responses clear and to the point.
- Answer questions honestly in a way that makes you look as good as possible, but do not volunteer candid comments that may "torpedo" the interview.
- Smile, and show your interest in the company and the position.

THE "DON'TS"

- Do not have a limp handshake.
- Do not criticize your family, your university, previous employers, and, particularly, the company conducting the interview.
- Do not look bored, such as checking your watch as if you have somewhere better to be.
- Do not chew gum or tobacco products.
- Do not smoke.
- Do not pick or touch your face or other parts of your body.
- Do not discuss compensation or benefits unless the interviewer brings them up.
- Do not bring someone with you.

ADVICE FOR INTERVIEWING

During the interview, remember the adage "Sit up, speak up, and then shut up!" Project confidence, and speak powerfully but with brevity.

you have thought about and planned well before the interview. This is your opportunity to make a great first impression. In both cases, you want to give the interviewer the impression that you have knowledge of the prospective employer and that your talents meet its needs. Call attention to your accomplishments and future direction. Customize your answers based on your research on the prospective employer. Be as specific and succinct as possible. Be honest, but do not volunteer information that might be disadvantageous to you. For a difficult question, you might want to pause before responding or ask the interview to clarify the question. These approaches give you more time to carefully consider your response.

Remember, an interviewer may ask you an "off-the-wall" question to see how you will react, or because they are not an experienced interviewer. Another technique that an interviewer might use is "silence" after you have finished your answer in an attempt to get you to volunteer information unhelpful to you. A response to silence might be to ask the interviewer if there are areas where he would like you to elaborate. Watch the interviewer for nonverbal cues. Is he getting impatient? Does he like you? If he does, it is advisable to be more cautious lest you present yourself as overconfident.

A list of some of the other likely questions includes the following:

* Give an example of a difficult situation and how you dealt with it.
* Give an example of a difficult person and how you dealt with him or her.
* If you get this job, where will you be in 5 years? What will you have achieved? (Saying you will be on the golf course is not likely to get you hired!)
* Give an example of a goal you have set for yourself, or someone has set for you, and how you achieved it.
* How do you manage time?
* What are your strengths and weaknesses?
* How do you deal with pressure?
* Describe a typical workweek (if you were an intern or have other related work experience).

As important as your responses are to all the other questions is the likely last question: "Do you have any questions?" It is not a good idea to say "No." This is the last opportunity to present yourself to the interviewer. Questions you might ask include, "What are your expectations for the person who is hired?" and "What are the career paths in this department?" Summarize your strengths and how they would fit with the job opportunity, and then ask if there are further questions of you.

The interviewer may provide you with the next steps in the interview process. As you leave the interview, shake the interviewer's hand and thank him or her. Follow up on the interview with a short letter. This gives you another opportunity to communicate your strengths and suitability for the position.

Frequently, the interview process involves a short interview of between 30 and 90 minutes. The top tier of candidates then may get a longer interview, including a meal. If you are flying to an interview, you should bring interview clothing in your carry-on. Checked bags can and do go missing.

The meal is often the final test in the interview. I know of candidates who lost the job or internship because of problems with the interview meal. For tips on an interview meal, see the boxed text on the next page.

Internships

Internships offer a tremendous advantage to you and can lead to a full-time position. It is an advantage to an employer because internships bring intelligent and enthusiastic young people to a company (see fig. 4-5). The company also can select the people it wants to hire, that is, someone who fits the company culture and brings the talents that the company wants or has the potential that the company recognizes. One way to put this is there is a "good fit."

An internship is a time for the potential employer to know whether the intern would make a good, or even great, employee. It is also the time when the intern can evaluate whether the job and company is something that he or she would like long term. This could also be viewed as a disadvantage of an internship. Perhaps we should consider an internship like a 3-, 6-, or even 12-month interview.

There are many advantages of internships:

- Experience.
- Building a résumé.
- Networking.
- Having people who know you well, who can write a good letter of recommendation.
- Providing something to talk about at an interview, as well as a competitive edge over other candidates for the same job.
- You can get to know a prospective employer and the work environment, and the employer gets to know you.

FIGURE 4–5 Internships provide opportunities to improve communication skills, and to work in teams. © Dmitriy Shironosov, 2011. Used under license from Shutterstock.com.

QUOTATIONS

ON HOW TO MAKE AN INTERNSHIP OR JOB SUCCESSFUL

"I'm a great believer in luck, and I find that the harder I work, the more I have of it."

Thomas Jefferson (1743–1826), third president of the United States

"Promotion should not be more important than accomplishment, or avoiding instability or more important than taking the right risk."

Peter Drucker (1909–2005), author and consultant on management who was born Austrian, then became a naturalized American

DOS AND DON'TS *of a meal at an interview*

THE "DOS"

- Appear professional at all times.
- Be punctual.
- Turn off your cell phone.
- Use a firm handshake at the beginning and end of the meal.
- Dress professionally, and remember it is better to be overdressed than underdressed.
- Have good table manners.
- Smile.
- Appear interested in the conversation.
- Let the other person talk.
- Use your napkin.
- Be polite at all times; for example, thank the server.
- Use "sir" and "ma'am" when appropriate.
- Send a thank-you note after the meal.

THE "DON'TS"

- Do not eat garlic.
- Do not drink alcoholic drinks; if wine is offered, slowly sip the wine, and drink more water.
- Do not smoke.
- Do not chew gum or smokeless tobacco products.
- Do not eat messy foods, such as pasta with sauce, ribs, or large messy sandwiches.
- Do not make jokes, particularly of a sexual, religious, political, or racial/ethnic nature.
- Do not use colloquialisms or profanity, even if the interviewer does.
- Do not slouch.
- Do not talk with your mouth full.
- Do not pick your teeth or any other body part.

ISSUES *for discussion*

1. What are the reasons to consider a specific career?
2. How do extracurricular activities help in getting the job you want, if they do?
3. How do research experiences and/or international experiences help in getting the job you want, if they do?
4. The advantages and disadvantages of internships.
5. The advantages and disadvantages of graduate or professional school.
6. Ethics and résumé preparation.
7. Ethics and interviewing.
8. Why be as positive as possible at an interview?
9. What aspects of body language appear open, honest, and friendly?
10. Why should you not lie or embellish your record or résumé?

REVIEW QUESTIONS

1. What are traits or characteristics of graduates that employers particularly value?

2. Give examples of careers for animal science graduates in the following sectors:
   - Animal production
   - Agricultural service sector
   - Animal health, biotechnology, and pharmaceutical industries
   - Animal breeding, genetics, and DNA-testing industries
   - Business and finance
   - Dairy industry
   - Feed industry
   - Food safety
   - Government
   - Horse industry
   - Meat industry

3. What professional degrees can follow a degree in animal science?

4. What graduate degrees can follow a degree in animal science?

5. What are the advantages of a professional or graduate degree?

6. How can you build a résumé?

7. What do you do to prepare yourself for writing an application for a job or internship?

8. When should you prepare yourself for writing an application for a job or internship?

9. Why should you prepare yourself for writing an application for a job or internship?

10. What do you do to prepare yourself for an interview for a job or internship?

11. When should you prepare yourself for an interview for a job or internship?

12. Why should you prepare yourself for an interview for a job or internship?

13. What makes you a winning candidate at an interview?

14. What time should you arrive at an interview?

15. What are behaviors to avoid at an interview?

16. What should you do/avoid doing at a job interview meal?

17. What are the advantages of an internship?

REFERENCES AND FURTHER READING

American Society for Animal Science. Retrieved July 17, 2009, from http://
www.asas.org/career.htm

Department of Animal Science, University of Connecticut. *Careers opportunities
in animal science.* Retrieved July 14, 2009, from http://www.canr.uconn.
edu/ansci/handbook/career.htm

Iowa State University. *Careers for animal and dairy science graduates*. Retrieved July 14, 2009, from http://www.ans.iastate.edu/stud/ugrad/career/careers.html

Michigan State University. *Careers in animal science*. Retrieved July 14, 2009, from http://www.canr.msu.edu/canrhome/career_aniSci.htm

Rutgers Career Services. *Career opportunities for majors in animal science*. Retrieved July 14, 2009, from http://careerservices.rutgers.edu/mh/animal_science.shtml

Simms, R., & Lane, C. D., Jr. *Careers in animal science*. Retrieved July 14, 2009, from http://www.ansci.cornell.edu/extension/beef/beefu4.pdf

University of Wyoming. *Careers in animal science*. Retrieved July 17, 2009, from http://www.uwyo.edu/anisci/info.asp?p=8111

CHAPTER 5

Horses

OBJECTIVES

This chapter will consider the following:

- Introduction to horses
- Classification and evolution of horses
- Domestication of horses
- History of the use of horses
- Definitions and terminology
- Horse reproduction
- Horse genetics
- Horse nutrition
- Horse housing

Objectives continue on the next page.

INTRODUCTION

Horses and ponies are domesticated members of the species *Equus caballus*. Throughout Eurasia, Northern Africa, and in the Americas after the arrival of Europeans, there has been a very close relationship between people and horses. Given this, it is not surprising that the horse is one of the animals on the Chinese calendar and zodiac.

The difference between a horse and a pony is normally size, with a pony being below 14.2 hands in height at the shoulders or withers (58 in or 147 cm). There are exceptions to this because some breeds, such as Welsh ponies, are called ponies regardless of their height, whereas the breed of very small horses is called miniature horses. The external anatomy of a horse, giving the names of specific features, is shown in Figure 5-1.

CLASSIFICATION AND EVOLUTION

Horses are in the order Perissodactyla, which contains the odd-toed ungulates (plant eating/grazing mammals with an odd number of toes). Other Equidae (the family to which horses belong) include the donkey or ass (discussed later in this chapter), the zebra (*Equus hippotigris*), the Asian wild ass or onager (*Equus hemionus*), and the wild horse, that is, Przewalski's horse (*Equus ferus*). The ancestors of today's horses diverged from those of donkeys and zebras about 2.4 million years ago.

The fossil record provides a clear picture of the evolution of the horse. This is now being supplemented by comparative genomics of existing species together with DNA from ancient bones. The evolution of horses appears to be a gradual increase in size from the earliest of the Equidae—*Hyracotherium*. This was previously called eohippus or dawn horse. It was about 16 in (40 cm) high at the shoulders, or approximately the size of a small dog, living about 55 million years ago in the Eocene period. There were a series of small Equidae species. The ancestors of horses were then medium-sized Equidae, such as *Mesohippus*, about 40 million years ago (the Oliogocene period), and *Miohippus*, about 24 million years ago. These stood about 24 in (60 cm) at the shoulders. The increase in size was again evident with *Merychippus* (17 million years ago), standing

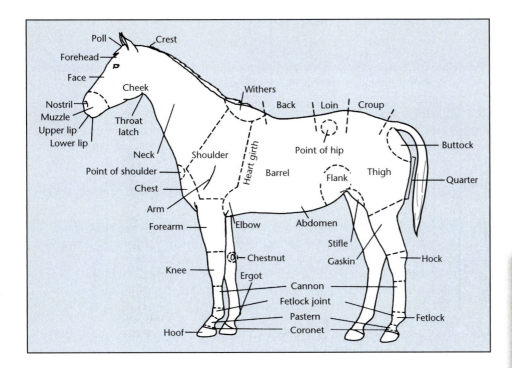

FIGURE 5-1 External anatomy of a horse. Courtesy of Appaloosa Horse Club, Inc., Idaho.

Definition

A *hand* is an archaic measure of length being the width of a man's hand. It is used to measure the height of horses: one hand (h or hh) = 4 in or 10.16 cm.

CLASSIFICATION

Living things are classified as kingdom (e.g., animal), phylum, class, order, family, genus, and species. The classification of horses is phylum, Chordata; class, Mammalia; order, Perissodactyla; family, Equidae; genus, *Equus*; and species, *E. caballus*.

NOMENCLATURE AND TERMINOLOGY

A *pony* is a small breed of horse. The process of giving birth in horses is called *foaling*. A newborn horse is a *foal*. A young horse up to its first birthday is a *weanling*. A young horse after its first birthday is a *yearling*. A young female horse 3 years old or less is a *filly*. A young male horse 3 years old or less is a *colt*. An adult female horse (4 years old or greater) is a *mare*. An adult male horse (4 years old or greater) is a *stallion* or *horse*.

40 in (10 hands or 1 m), and the first species of the genus *Equus* (4 million years ago), standing 53 in (13.2 hands or 1.33 m) and being about the size of a pony. Much of the evolutionary history of the Equidae was in North America, with wild equids present until about 12,000 years ago. Fossils of *Hyracotherium* have been found in Wyoming, and those of *Mesohippus* in Colorado, Nebraska, and the Dakotas; and *Miohippus* together with *Merychippus* were widely distributed in North America. Some Equidae of the genus *Equus* migrated to South America. Others migrated to Asia about 2.6 million years ago and there diverged to the ancestors of today's horses and the ass/zebra branch. It should be emphasized that the evolution of the Equidae was not a linear change; there have been many branches and twigs from the root, and with the exception of the genus *Equus*, all have become extinct.

DOMESTICATION

Horses (*E. caballus*) were domesticated about 6,000 years ago (4000 BC). It is probable that the first domesticated horses were used both for food (meat, milk, and blood) and transportation. Initially, they could be rode bareback or yoked to early chariots. Men with horses were much more successful hunting large game, and the population of the group of early people could then increase. Horses revolutionized warfare. It is easy to see a transition for using early chariots for hunting with a bow to using it to attack another group of people.

HORSE REPRODUCTION

Horses are long-day breeders (breeding in the spring such that the foals are born in the spring) and are seasonally polyestrous. The length of gestation is

A castrated male horse of any age is a *gelding*. The process of castration is called *gelding*. The mother of a horse is called the *dam*. The father of a horse is called the *sire*. A group of horses is a *herd*. *Equine* is an adjective referring to horses and other equids; it can also be used as a noun with similar meaning.

HISTORY OF THE USE OF HORSES

- 4000 B.C. - Horses have been suggested in playing a critical role in the spread of the proto Indo-European language (this lead to the dominance of Indo-European languages across Europe and South and West Asia).
- 2000 B.C. - Horses were used to pull chariots (see fig. 5-2)
- 860 B.C.–1890 A.D. - Horses were used for calvary. Heavy calvary was used to charge the enemy, and light calvary was used to fire arrows or slingshot at the enemy (see fig. 5-3*A*).
- 432–453 A.D. - Attila the Hun dominated much of Europe using bowmen on horseback (light cavalry).
- 500 A.D. - The stirrup was used extensively for the first time.
- 800 A.D. - The Chinese horse collar was developed such that the horse could be used for plowing. Up until then plowing was largely restricted to oxen and also some donkeys.
- 1206–1240 - The Mongol Empire was established by Genghis Khan and expanded into Europe largely because of mounted troops (light cavalry).
- 1493 - The first domesticated horses were brought to the Americas on Columbus's second voyage.
- 1519 - Cortéz brought horses as part of the force to conquer the Aztecs in the area that is now Mexico.
- 1532 - Horses played a critical role in the conquest of the Incas by the Spanish.
- 1539 - De Soto brought horses to what is now the United States as part of an expedition.
- 1540 - Feral horses (mustangs) began to spread through Meso-America and into North America and the Great Plains. They were successfully redomesticated by America Indians and used in hunting bison and in warfare. General Custer's defeat in 1876 was a significant event that showed how horses affected the history of the West.
- 1860-1865 - The U.S. Civil War was one of the last major wars with a cavalry (see fig. 5-3*B*).
- 1866–1890 - Cattle drives occured from Texas to Kansas railhead cow-towns such as Ellsworth, Dodge City, Newton, Wichita, and Abilene on the Chisholm Trail, the Goodnight Trail, and other trails. Cowboys rode ahead on horses while flanking the herds of Longhorn cattle.

FIGURE 5-2 Chariots were first developed about 2000 BC and used successfully by the Ancient Egyptians. Courtesy of Getty Images.

FIGURE 5–3 *A* and *B*, Horses were used for cavalry. Heavy cavalry charged the enemy while light cavalry rode close to the enemy and used bows. *A* reproduced by permission from Willierossin and Jim Parkin. *B* © 2010 by Shutterstock.com.

336 days. There are problems with breeding horses. First, drugs used illegally in racing may interfere with reproduction. Second, the birth date of horses is set as January 1 by many breed associations in the Northern Hemisphere. There is an obvious reason, therefore, to have foals born as early as possible in the year so that they have the maximum growth possible before their first birthday when they become yearlings. Therefore, breeders are attempting to breed horses outside their normal breeding season (see fig. 5-4).

 ## HORSES AND STRESS

Blood concentrations of the stress hormone cortisol are increased in horses during transportation and treadmill exercise. A large percentage of performance horses develop gastric ulcers, which may be stress related. Many of these horses are treated with omeprazole. Horses are also susceptible to problems with glucose balance. As with people, glucose tolerance tests are performed. Severe stress is thought to change the nutritional requirements of horses (see the section on horse nutrition later in this chapter).

 ## HORSE GENETICS

Horses have 64 chromosomes. Maps of the genome of the horse have been reported, as is detailed in Chapter 15. The mare provides not only half the chromosomes of the offspring, but also mitochondrial DNA. There is evidence that part of the mitochondrial DNA (the D loop) affects horse racing performance.

FIGURE 5-4 Mare and foal. Courtesy of Rutgers University Center of Excellence in Equine Science.

HORSE NUTRITION

Meeting the nutritional requirements of horses is essential for optimal performance. In horses, there is a breakdown of cellulose by microorganisms in the hind gut (cecum and colon). Horses require protein, vitamins, minerals, and water. In a maintenance situation, horses can receive sufficient nutrition by being fed hay or grazing pastures. Examples of good grass for pastures for horses include orchard grass, reed canary grass, and bromegrass. In addition, legumes such as alfalfa provide more protein. However, for performance, pregnancy, lactation, and growth, there also needs to be concentrates. Concentrated feeds contain cereals such as oats, corn, or wheat as an energy source.

Feeds are optimized for protein content, with the National Research Council recommendation varying with the age of the horses. Concentrates also supply sufficient vitamins and minerals, particularly calcium and phosphate. Horses suffering from severe stress (trauma, sepsis, and severe burns) need a higher percentage of protein (12–16% on a dry-matter basis) in their diet.

 ## HORSE HOUSING

The well-being of horses requires shelter to protect them against adverse weather conditions, such as cold rains, sleet, snow, and extremely low temperatures. A horse barn does not need to be elaborate, but it can be. The structure should be constructed on well-drained land and provide a place to store feed and tack under dry conditions. Figures 5-5 and 5-6 provide schematic views of simple horse barns.

Comfortable quarters for four horses are provided by this 20' × 60' barn. Four 12' × 12' box stalls, as well as a combination tack and feed room, open to an 8' covered way.

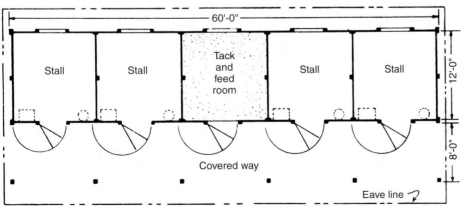

FIGURE 5-5 A four-stall horse barn. The box stalls are 12 × 12 ft. A combination tack and feed room is included. There is an 8-ft covered way on the front. Courtesy of Clemson University.

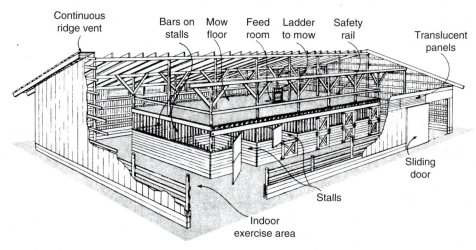

The clay or wood floor in the stalls should be covered with wood shavings, sawdust, or straw to cushion the horses' hooves and for bedding. It is important to clean the stalls to remove soiled bedding.

SIZES OF HORSES AND PONIES

Ponies: < 14.2 hands (14 hands 2 in), 300–900 lb (136–408 kg)

Light horses, including riding and racing: 14.2–17 hands

Heavy or draft horses: > 8 hands

 BREEDS OF HORSES

The five most popular breeds of horses in the United States are the quarter horse, Thoroughbred, standardbred, Appaloosa, and Arabian. These five breeds represent three quarters of all registrations in the United States.

Quarter Horse

The quarter horse is the number-one horse in the United States (see fig. 5-7). In the colonial era, horse racing was becoming a common sport. The races were

a quarter of a mile or less, often along a straight road. What was desirable was a sprinter (quarter horses can achieve speeds of 55 mph), and the term *quarter miler* was used to describe it. The foundation stock was English Thoroughbreds and the native Chickasaw horses, which were derived from the horses brought by the Spanish. Additional blood lines include Arabian, Morgan, and standard-bred. The quarter horse was extensively used during the United States's westward expansion by the pioneers and on the western ranches. The breed is also called "America's horse" and is a light horse breed.

There are many colors of quarter horses. Quarter horses are used for riding, showing, racing, and stock horses. The breed association is the American Quarter Horse Association, with over 3 million quarter horses registered.

Thoroughbred

The Thoroughbred breed, a "hot blood" (highly spirited breed) and a light horse, is a racehorse (see fig. 5-8). In the United States, they race on the flat not only in, for instance, the races of the U.S. Triple Crown (Kentucky Derby, Preakness, and Belmont Stakes) but also in the races of the Triple Crown of Ireland or England. In the United Kingdom, Thoroughbreds are also raced with jumping over hurdles or steeplechasing. The breed has also been successfully used in dressage, for example by the U.S. Olympic team.

The Thoroughbred breed was developed beginning in the late 17th and early 18th centuries in England, breeding English mares with three imported Arabian

FIGURE 5-8 Thoroughbred horse. Courtesy of Agriscience and Fundamentals and Applications.

stallions (Darley Arabian, Godolphin Arabian, and Byerly Turk). The first Thoroughbred was brought to North America in 1730.

Thoroughbreds are 15.2–17 hands in height and are usually bay, brown/chestnut, or black but can be roan or gray, or even white. The face and legs may have white markings. Thoroughbreds are registered with The Jockey Club in the United States and Canada, and with Weatherbys in the United Kingdom.

Standardbred

The standardbred is a light horse breed (see fig. 5-9). It originated in the United States as a harness-racing horse with the term *standardbred* first used in 1879. The foundation stock for the breed is Narragansett pacers, Canadian pacers, the English Thoroughbred, Norfolk Trotters, hackneys, and Morgans.

Standardbreds are between 14.1 and 17 hands in height. Standardbred horses are mainly bay, black, brown, or chestnut, with other colors, including gray and roan. The breed registry is the United States Trotting Association. Trotting races include the Hambletonian and the Yonkers Trot, and a major pacing event is the Meadowlands Pace.

Appaloosa

The Appaloosa has a distinctive look because of the presence of dark or mottling spots in the coat (see fig. 5-10). The breed originated in the northwest of North

FIGURE 5–9 Standardbred horse. Courtesy of the United States Trotting Association.

FIGURE 5-10 Appaloosa pony. Reproduced by permission from Eline Spek. © 2010 by Shutterstock.com.

America, probably from the descendants of light horses brought to the Americas by the Spanish or possibly horses from Russian traders in the northwest. The Nez Perce Indian tribe in the 1800s developed the breed by selective breeding. They were initially called Palouse horses after the Palouse River.

The Appaloosa is used for pleasure riding, showing, racing, and parades, and as a stock horse. The Appaloosa Horse Club registers Appaloosa horses based on there being two Appaloosa parents or one Appaloosa parent plus a parent from an approved breed.

Arabian

The Arabian breed is one of the oldest breeds, going back perhaps as much as 4,500 years in Arabia. It is small to medium in size and traditionally 14.1 to 15.1 hands, but with breeding programs in Europe and North America, Arabians are now between 15 and 16 hands. Arabians range in weight from 850 to 1,100 lb (385.6–499 kg) and are mainly bay, gray, or chestnut, with a few white or black in color (see fig. 5-11).

The Arabian is a versatile breed of light horses being used for light cavalry, pleasure riding, dressage, racing, showing, and long-distance racing, and as a stock horse. The breed associations are the Arabian Horse Registry of America and the International Arabian Horse Association.

Other Breeds

Other breeds will be covered briefly in order of size.

Miniature Horses

Miniature horses are, not surprisingly, defined by their height (see fig. 5-12). The two registries for miniature horses, namely the American Miniature Horse Association and the American Miniature Horse Registry, define miniature horses a little differently. The American Miniature Horse Association defines miniature horses as not more than 34 in at the withers (at the last hair of the mane). The American Miniature Horse Registry has two divisions: the "A" division for horses 34 in (82 cm) and under, and the "B" division for horses 34–38 in (82–91 cm).

FIGURE 5-13 *A*, Shetland pony. Reproduced by permission from Eline Spek. © 2010 by Shutterstock.com. *B*, Welsh pony. Reproduced by permission from Zuzule. © 2010 by Shutterstock.com.

Ponies

Ponies are excellent for children learning to ride. There are a number of breeds of ponies, including Shetland ponies (see fig. 5-13*A*), which are <10.3 or 10.3–11.1 hands; and Welsh ponies (see fig. 5-13*B*), which are <12.2 or 12.2–14 hands.

Tennessee Walking Horses and Other Light Horse Breeds

The Tennessee walking horse is another light horse breed found in middle Tennessee. It is said to be good for all ages and levels of experience. The Tennessee Walking Horse Breeders and Exhibitors' Association requires that the verification of the parentage of all foals be verified through genetic fingerprinting of DNA. Other popular light horses are pintos or paints and Morgans (see fig. 5-14).

Clydesdales

Clydesdales are draft or heavy horses more than 18 hands in height (see fig. 5-15). In addition, they have been used as drum horses, that is, horses that carry large military drums. Clydesdales weigh about 176 lb (80 kg) at birth, and adults weigh about 1 ton (2,000 lb). It is thought that Clydesdales resulted from crossing local mares in Southwest Scotland with large English or Flemish war horses. Other draft horses or cold bloods (placid or quiet breeds) include the Percheron, Shire, Belgian Draft, and German draft breeds.

The Anheuser-Busch company (the brewing company that makes Budweiser) uses Clydesdales as emblems of the company. It owns a number of Clydesdales, and these can be viewed by the public near the Anheuser-Busch headquarters in St. Louis, Missouri. These are often seen in television commercials, particularly around Christmas, pulling a sleigh or carriage of beer.

FIGURE 5-14 *A*, Morgan. Reproduced by permission from B. Speckart. © 2010 by Shutterstock.com. *B*, American paint. Courtesy of the American Paint Horse Association.

FIGURE 5-15 Anheuser-Busch Clydesdales. Reproduced by permission from Michelle Marsan. © 2010 by Shutterstock.com.

 ## LOCOMOTION OR GAITS IN HORSES

The natural gaits for a horse are walking, trotting, cantering, and galloping. These represent a hierarchy in the speed attainable, with the fastest gait being the gallop. For instance, with transition from cantering to galloping, there is an increase in speed.

Walking

The walk has one foot off the ground at a time with a sequence: left hind leg, left front leg, right hind leg, and right front leg, in a regular one, two, three, four beat. This is viewed as one through three, with one foot off the ground and three on the ground. The average speed is 4 mph.

Trotting

The trot is a two-beat diagonal gait; the left front and right rear legs move in unison, as do the right front and left rear. Two legs are in the air, not in contact with the ground, at the same time. This is viewed as two-two, with two feet on the ground and two off the ground. Speeds of 8 mph are common when using trotting as the working gait for a horse. This gait is used in standardbred racing, and considerable speeds are possible. A slow trot is called a jog.

Cantering

The canter is a three-beat gait. There are three legs moving forward in the air, not in contact with the ground, at the same time, with either of the back legs propelling the horse forward.

Galloping

The gallop is a four-beat gait. The gait used in horse races (except harness racing) is the gallop because it is the fastest gait. The fastest galloping speed in horse recorded history is in a sprint by an American quarter horse, with a speed of close to 55 mph.

 USAGE OF HORSES

People in the United States and throughout the world enjoy horses. This interest includes the following:

- Horse racing (e.g., Thoroughbreds racing on the flat [see fig. 5-16], standardbreds in trotting/pacing races, and quarter horse races)
- Horse shows (annually 14,000 sanctioned horse shows in the United States), including dressage competitions (see fig. 5-17)
- Rodeos
- Riding for pleasure, including trail riding (see fig. 5-18)
- Endurance riding (see the section on endurance rides later in this chapter)
- Using horses as working animals (such as for herding cattle or for police work)
- Carriage horses ranging from the Budweiser Clydesdales to romantic carriage rides in historical cities

Economic and Societal Impact of Horses

There are estimated to be about 3.6 million horses and ponies in the United States (data from 2005). Horses have a significant impact socially and economically. For

> **INTERESTING FACTOID**
>
> The maximum speed of a quarter horse is 55 mph. The maximum speed of the fastest land animal, the cheetah, is 70 mph (110 km/h).

instance, in the United States, it was estimated by the American Horse Council in the early 2000s that the following impacts had occurred:

- Horse racing has an impact of almost $35 billion to the U.S. economy, employing 441,000 people.
- 4.3 million Americans ride for pleasure, with 3 million pleasure horses.

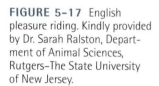

FIGURE 5-17 English pleasure riding. Kindly provided by Dr. Sarah Ralston, Department of Animal Sciences, Rutgers–The State University of New Jersey.

FIGURE 5-18 Trail riding. Kindly provided by and featuring Dr. Sarah Ralston, Department of Animal Sciences, Rutgers–The State University of New Jersey.

- The pleasure horse industry has an economic impact of almost $30 billion, employing over 300,000 people.
- About 250,000 children are active in 4-H equine programs and pony clubs.

The eating of horse meat is not widely practiced in the United States and many other English-speaking countries. Indeed, there is almost a taboo against it. However, global consumption of horse meat is about 1 million metric tons per year.

Horse Racing

Horses have been raced for thousands of years. About 2,000 years ago, the Romans raced chariots around a ring circuit or circus.

Thoroughbred and Quarter Horse Racing

Thoroughbred horses are raced between five-eighths to 1½ miles in a circular route(s), with there being a length of straight track before the finish. As might be expected with the gallop being the fastest gait, Thoroughbred racing uses horses galloping. A jockey rides the horse. Thoroughbred horse races are of such a length that the horse must be paced and not gallop at full speed throughout the race. In contrast, quarter horse races are shorter (one-quarter mile), and the horses gallop at maximal speed for the entire race.

Standardbred or Harness Racing

Standardbred racing is governed by United States Trotting Association (http://www.ustrotting.com). Standardbred racing is synonymous with harness racing, with the horse pulling a two-wheeled streamlined vehicle called a *sulky* or *race bike* carrying a driver (see fig. 5-19). In training, the horse pulls a jog cart that is

FIGURE 5–19 Equipment for pacing races. Courtesy of United States Trotting Association, Ohio.

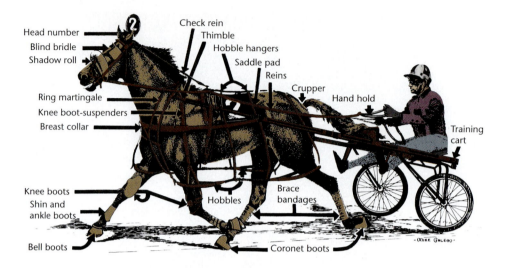

heavier than the sulky. The horse is attached to the sulky by a harness that also affixes the hobbles and enables the driver to steer the horse.

There are two gaits with standardbred racing: the trot and the pace. Standardbred horses are trained and race only as trotters or pacers, as illustrated in Figures 5-20 and 5-21. Trotters race only as trotters, and pacers only race as pacers. Standardbreds start racing as 2 or 3 year olds.

FIGURE 5–20 Standardbred pacer. Courtesy of United States Trotting Association, Ohio.

FIGURE 5-21 Standardbred trotter. Courtesy of United States Trotting Association, Ohio.

Drug Use and Abuse in Racehorses

As might be expected from cases with human athletes, drugs are used by equine athletes, regardless of whether Thoroughbreds, standardbreds, or quarter horses. Some drugs are legal for racehorses such as furosemide (Lasix) and phenylbutazone (Bute) below established levels. Lasix prevents bleeding, whereas Bute is an analgesic reducing pain and acts as a nonsteroidal anti-inflammatory drug. In addition, acepromazine is used as a tranquillizer for horses in the ring.

To protect the integrity of the racing industry and the confidence of the public, horses are routinely tested for a large number of drugs that could influence their performance in a race. The Association of Racing Commissioners International has published the Uniform Classification Guidelines for Foreign Substances and Recommended Penalties and Model Rule (see the boxed text below with drug classifications).

Illegal uses of drugs for horses include central nervous stimulants and depressants, anabolic steroids, the beta-adrenergic agonist clenbuterol, and erythropoietin (EPO).

Recombinant human EPO is used in the treatment of anemia in people. In addition, EPO and related darbepoetin have been used for "blood doping" in human and equine athletes, as has chemically stabilized hemoglobin. EPO has been used/abused to increase red blood cell production and circulating concentrations of red blood cells (packed cell volume) and hemoglobin with consequent improved

HORSE-RACING TERMINOLOGY

A *jockey* is a paid rider of a horse in a race.

A *maiden* (male or female) is a horse that has not yet won a race.

A *trainer* is a person paid to train racehorses (Thoroughbreds, quarter horses, and standardbreds).

oxygen-carrying capacity. This leads to improved aerobic performance in human and equine athletes. In horses, EPO has been demonstrated in controlled studies to increase packed cell volume, hemoglobin, aerobic capacity, and the speed at maximum aerobic capacity. To deter the abuse of EPO in the horse-racing industry, sensitive and reliable detection methods have been developed.

CLASSIFICATION OF DRUGS (ASSOCIATION OF RACING COMMISSIONERS INTERNATIONAL, JULY 2007)

Class 1 contains drugs having the highest potential to stimulate or depress performance and that have no generally accepted medical use in the racing horse. Examples include opiates, amphetamines, all Drug Enforcement Administration Schedule I substances (see http://www.usdoj.gov/dea/pubs/scheduling.html), and many Drug Enforcement Administration Schedule II substances. The penalty recommendations are 1–5 years suspension, a $5,000 fine, and the loss of prize money (the *purse*).

Class 2 drugs have a high potential to affect performance and either are not generally accepted as therapeutic agents in racing horses or are therapeutic agents that have a high potential for abuse. Examples include psychotropic drugs and other drugs that affect the nervous system and/or cardiovascular system (stimulants such as caffeine, EPO, and somatotropin, and depressants such as barbiturates), injectable local anesthetics, and nerve blockers such as snake venom. The penalty recommendations are 6 months to 1 year suspension, a $1,500–2,500 fine, and loss of the purse.

Class 3 contains drugs that may have a generally accepted medical use in the racing horse but have less potential to affect performance than drugs in Class 2. Examples include bronchodilators and other drugs with primary effects on the autonomic nervous system (e.g., acepromazine), procaine, antihistamines with sedative properties, and the high-ceiling diuretics. The penalty recommendations are 60 days to 6 months suspension, up to a $1,500 fine, and loss of the purse.

Class 4 drugs have less potential to affect performance than those in Class 3. Drugs in this class include less potent diuretics; anabolic steroids; corticosteroids; antihistamines and skeletal muscle relaxants without prominent central nervous system effects; expectorants and mucolytics; hemostatics; cardiac glycosides and antiarrhythmics; topical anesthetics; antidiarrheals, mild analgesics; and nonsteroidal anti-inflammatory drugs such as phenylbutazone, at concentrations greater than established limits. The penalty recommendations are 15–60 days' suspension, up to a $1,000 fine, and loss of the prize money.

Class 5 contains drugs for which concentration limits have been established by the racing jurisdictions, such as dimethylsulfoxide, together with drugs with localized actions only, such as antiulcer drugs, and certain antiallergic drugs. The anticoagulant drugs are also included. The penalty recommendations are 0–15 days suspension with the possible loss of the prize money and/or fine.

Drugs that are "Non-classified" are antibiotics, anthelmintics, sulfonamides, and vitamins.

OTHER USES OF HORSES

Endurance Rides

These are races over a trail of 50–150 miles—for example, 160 km—and take place over 1–3 days. The terrain may also be part of the challenge, and the ride may include a short steeplechase event. People ride either to compete or just to finish.

Precautions are taken to ensure the health of the horse. Qualified veterinarians check the horses before, during, and after the event. In a manner similar to athletes drinking Gatorade to combat water deficit, horses are frequently provided with a salt/sugar solution to maintain hydration. It is critical that both the horse and rider receive extensive training before undertaking an endurance ride. Arabians are often thought of as the ideal endurance horse

Riding for the Disabled

Beginning in Europe in the 1950s and later in North America, people with disabilities have had quality of life enhanced by riding horses. Riding horses improves physical and mental health. Improved physical health is evident by greater muscle tone and coordination. Improved mental health is seen by the smile of the person and the increased confidence. Today about 30,000 people with various disabilities enjoy therapeutic riding. Organizations that promote riding for the disabled include the North American Riding for the Handicapped Association, promoting the benefit of riding horses for people with physical, emotional, or learning disabilities.

Crowd Control

Horses have been used by police and military for crowd control.

DONKEYS

Donkeys or asses have been used extensively in agriculture and for transportation. The animal is referred to in the Bible with, for instance, Jesus riding into Jerusalem on an ass. The traditional name is *ass*, but due to obvious connotations, many people feel more comfortable calling them *donkeys*.

Domestication of Donkeys

Donkeys (*E. asinus*) were domesticated on multiple occasions from both the African wild ass (*Equus africanis africanis*) and the Nubian wild ass (*Equus africanis somaliensis*).

MULE

Mules are crosses or hybrids between male donkeys and female horses. Mules were used in the past extensively in the United States in farming to pull plows. There were about 2.5 million mules in the United States around 1900–1910.

> **CLASSIFICATION**
>
> The classification of donkeys is phylum, Chordata; class, Mammalia; order, Perissodactyla; family, Equidae; genus, *Equus;* and species, *Equus asinus.*

> **NOMENCLATURE AND TERMINOLOGY**
>
> A male donkey is called a *Jack* and a female donkey a *Jenny* or *jennet.* A male mule is a *John,* and a female a *Molly.*

Today, the number has declined by over 90%. It is not uncommon to see mules, particularly in the southern United States or when traveling internationally. The country with the most mules is China. The countries with the highest number of mules per square mile or hectare are Mexico, Columbia, and Morocco.

Mules have many of the traits of horses and donkeys. They can be used to pull a plow or cart, be ridden, or be used as pack animals, that is, carrying goods on their backs. They have hybrid vigor exhibiting hardiness, and resisting sun, heat, and disease. They are said to be stubborn, but this reflects that they will not work past their endurance. They are sure footed, which is advantageous when they are carrying people or goods over mountainous terrain.

Mules can be either male or female but are normally sterile, with either very low fertility or complete infertility. The male has a full male reproductive system with ducts, glands, penis, and testes. Spermatogenesis is either very low or absent. The female has the full female reproductive track and external genitalia, ovaries, and estrus cycles, but no ovum. There are documented cases of female mules crossed to a horse, giving birth to a foal. The infertility of mules can be explained by the difference in chromosome numbers between horses and donkeys. In more detail, it is thought that there was repositioning of centromeres in the donkey chromosomes some time after their divergence from the horse ancestors and during the 2.4 million years of evolutionary separation between horses and donkeys.

A hinny is a cross or hybrid between a male horse and a female donkey. Hinnies can be either male or female but are normally infertile. It is more difficult to produce hinnies than mules. Other equid hybrids are possible, including the zebronkey, which is a cross between a zebra and a donkey; the ancestors of asses and zebras diverged 0.9 million years ago.

CONSEQUENCES OF THE DECLINE IN HORSE AND MULE NUMBERS

At the beginning of the 20th century (1900–1906), there were between 14 and 19 million horses and 2.1–3.4 million mules. Today, the numbers have declined by about 80%. The decline in horse and mule numbers in the United States has freed up land previously being used for their forage and pasture needs. Today, the numbers of horses for recreation have increased. This is keeping green field for pasture in rural areas and outer suburbs.

INTERESTING FACTOID

Horses have 64 chromosomes, donkeys have 62, and mules (and hinnies) have 63.

REVIEW QUESTIONS

1. What are the names for adult female and adult male horses?

2. What is a foal?

3. What is the process of foaling?

4. What is a weanling horse?

5. What is a yearling horse?

6. What is a filly?

7. What is a colt?

8. What is a horse that has been castrated called?

9. What is signified by the terms *dam* and *sire* of a horse?

10. What is the name of the species of the horse?

11. What are other Equidae or equid species?

12. What is the evolutionary history of the horse, and where did most of it occur?

13. Where and why were horses domesticated?

14. What were the major uses of horses through history? What were some of the major events and changes?

15. What is the impact of horses on the U.S. economy?

16. How many people are employed in the U.S. horse industry?

17. How many horses are there in the United States?

18. How were horses important to the development of the American West?

19. When do horses naturally breed? Why is this important?

20. How many chromosomes does the horse have?

21. In addition to chromosomes, genetically what does the dam also contribute?

22. What is a horse called that has not won a race?

23. What are the characteristics of the following: Arabians, Thoroughbreds, standardbreds, quarter horses, and Appaloosas?

24. What are the ancestries of the following: Thoroughbreds, standardbreds, quarter horses, and Appaloosas?

25. Name five additional breeds of horses and their characteristics.

26. What are Thoroughbred horses used for?

27. What is the difference between a quarter horse race and that for a Thoroughbred in the United States?

28. What gait is used by a Thoroughbred or quarter horse in a race?

29. What types of racing use standardbred horses?

30. What is the name for a rider in a horse race?

31. Give examples of major horse races in the United States.

32. What are the natural gaits for a horse, and how do they differ?

33. In trotting and pacing, what is the person called who is controlling the horse?

34. In trotting and pacing, what is the name of the device being pulled by the horse?

35. In trotting and pacing, what are the purposes of the harness and hobbles?

36. What are the gaits used by trotters and pacers, and how do they differ?

37. Why are many drugs prohibited from being used in racehorses?

38. Name five drugs that are prohibited from being used in racehorses.

39. What drugs can be used in racehorses?

40. What is the significance of riding for the disabled?

41. What is the difference between a donkey (ass) and a horse?

42. Where were donkeys domesticated?

43. How many chromosomes does a donkey have?

44. What are the terms for an adult female and adult male donkey and for these in mules?

45. What is a mule?

46. What are the characteristics/advantages of a mule?

47. Why has the number of mules declined in the United States?

REFERENCES AND FURTHER READING

American Horse Council. *National economic impact of the U.S. horse industry*. Retrieved July 17, 2009, from http://www.horsecouncil.org/nationaleconomics.php

Diamond, J. (1997). *Guns, germs, and steel: The fates of human societies*. New York: W. W. Norton.

Florida Museum of Natural History. *Fossil horse cybermuseum*. Retrieved July 14, 2009, from http://www.flmnh.ufl.edu/natsci/vertpaleo/fhc/firstCM.htm

Keegan, J. (1993). *A history of warfare*. New York: Vintage Books.

Kraft, N., & Rucker, K. *Harness racing 101: An educational primer*. Retrieved July 14, 2009, from http://www.ustrotting.com/pdf/HarnessRacing101.pdf

McKeever K. H., Agans, J. M., Geiser, S., Lorimer, P. J., & Maylin, G. A. (2006). Low dose exogenous erythropoietin elicits an ergogenic effect in standardbred horses. *Equine Veterinary Journal. Supplement*, *2006*(36), 233–238.

The TalkOrigins Archive. *Fossil horses FAQs*. Retrieved July 14, 2009, from http://www.talkorigins.org/faqs/horses/

CHAPTER 6

Companion Animals

OBJECTIVES

This chapter will consider the following:

- Cats. The discussion will include their numbers in the United States, classification and domestication, definitions and terminology, cats in folklore, their uses, reproduction, genetics, breeds, and nutrition, including obesity.

- Dogs. The discussion will include their numbers in the United States; classification and domestication; definitions and terminology; their uses, from hunting to herding, sniffing, guarding, and service; reproduction; genetics; breeds; and nutrition, including obesity.

Objectives continue on the next page.

OBJECTIVES, *continued*

- Ferrets
- Rodents
- Pet birds
- Monkeys

(Rabbits are considered in Chapter 10.)

 CATS

Introduction

There are about 85 million cats in the United States, with the American Veterinary Medical Association estimating 82 million cats and the Humane Society of the United States estimating 88 million. About 34% of households have at least one cat. Of owned cats, 84% are spayed or neutered. In addition, there are barn cats and the like. Cats develop close relationships with people, are affectionate (enjoy being stroked), and make excellent companion animals (see figs. 6-1 and 6-2).

Classification

The species of the domestic cat is referred to as *Felis catus*. An alternative, and scientifically more correct, classification has the domestic cat as *Felis silvestris catus* because the cat was domesticated from the African cat (*Felis silvestris lybica*). The cat is a member of the genus *Felis,* which also includes the wildcat (*Felis silvestris*) and the cat family. Other members of the cat family (family Felidae) include the lion, tiger, leopard, jaguar, bobcat, cougar, lynx, European or African wildcat, cougar, and extinct saber-toothed tiger

Domestication

The domestic cat is thought to have as its ancestor the Near East/African cat (*F. silvestris lybica*; alternatively *Felis lybica)* with domestication occurring in the Fertile Crescent. The cat was venerated by the ancient Egyptians, beginning about 5,000 years ago. However, domestication now appears to have been earlier. Recently, archaeologists unearthed a burial on Cyprus of a woman and a cat together from 9,500 years ago. This provides evidence that cats were domesticated at that time. Wildcats are not native to Cyprus, and, therefore, the cat or its ancestors had to be transported to Cyprus by boat.

Cats were probably originally domesticated to reduce the problem of mice and rats infesting stored grain.

Feral cats are domesticated cats that can successfully revert to the wild. Populations of feral cats can

FIGURE 6-1 Alert and well-cared-for cat showing whiskers. Courtesy of PD Photos.

NOMENCLATURE

An adult female cat is a *queen*, and an adult male is a *tom*. A castrated male is a *gib*. The mother of a litter is a *dam*, and the father is the *sire*. Young cats are *kittens*. The process of giving birth is called *kittening*. *Feline* is an adjective referring to cats.

FIGURE 6-2 Cats are very popular companion animals. Reproduced by permission from Vlasov Pavel. © 2010 by Shutterstock.com.

CLASSIFICATION

The classification of cats is phylum, Chordata; subphylum, Vertebrata; class, Mammalia; order, Carnivora; family, Felidae; genus, *Felis*; and species, *F. catus*.

threaten wildlife populations, including wild birds. Feral cats can interbreed with the European wildcat (*F. silvestris*).

Cats in History and Folklore

Examples of cats in history and folklore include the following:

- The ancient Egyptians venerated cats.
- Cats have proven excellent pets and useful in reducing rodent infestations. One of the great mistakes of history was the order to kill the cats of London at the beginning of the Great Plague because of the wrong-headed belief that cats spread plague. Without cats, there was an increase in the rat population. The bubonic plague was being spread by fleas on the rats!
- Cats have been linked to witchcraft. Despite their ability to kill rodents, at times in history, cats have not been well accepted.

Therapy Cats

The presence of a docile cat that likes to be stroked can improve the quality of lives of nursing home residents. This approach of having companion animals wandering has been very successful.

Cat Reproduction

The ejaculate volume in cats is 1.5 mL (15 billion/mL) (*see also* Chapter 16). Cats are seasonally polyestrous and are induced or reflex ovulators (ovulating only after mating). The length of gestation is about 60 days. Kittens are born in an immature state (scientifically called *altricial*) with limited capacity to see and move around.

Cat Genetics

So much of what we see in a cat is determined by its genetics. Cats have 38 chromosomes (19 pairs). A research group from Texas A&M University in 2007 published a map of the domestic cat genome. Comparative genomics suggests that during evolution, the cat genome has undergone primarily intrachromosomal rearrangements.

Cats can be either shorthaired or longhaired. This difference in the fur or coat is controlled by a single gene locus. In this case, the shorthair genotype is dominant (L), and the longhair genotype (l) is recessive. Thus, cats need to be homozygous for the longhair genotype (ll) with the gene on two chromosomes. A series of genes control the color of cats.

There are also examples of genetic abnormalities in cats. For instance, the Manx cat lacks a tail. This is due to an autosomal dominant trait. All Manx cats are heterozygous for the trait, that is, they have one copy of the mutated gene. No Manx cats have two copies (homozygous) because it is lethal, and they die before birth. Other simple genetic traits include the Scottish Fold with a unique ear fold and the Munchkin cat, which has a form of dwarfism with short legs. Both of these are autosomal dominant traits. Another such genetic trait in cats is polydactyly or

Definition

Altricial refers to an animal born in a very immature development state, such as a cat or rat, with limited senses and ability to move independently. Parent care is critical for altricial species. The converse of altricial is *precocial*. Precocial refers to when an animal is born and can function well independently, such as a calf or foal.

having extra digits. A cat with polydactyly may have six or seven toes or more. This is a relatively common genetic trait in cats, particularly in the United States.

Cat Nutrition

Cats are carnivores and can live on a diet that is entirely meat. Domestic cats often receive a diet with high levels of carbohydrate, fat, and protein, often in amounts far higher than required for the nutrition of the cat. This leads to obesity.

Obesity in Cats

About one third of all domestic cats in the United States are overweight or obese. Obesity can be readily identified by the body mass index (based on weight and size/length) or the girth (measuring circumference around the abdomen). There are long-term health consequences of feline obesity, including diabetes mellitus, lower urinary tract disease, hepatic lipidosis, and lameness due to arthritis.

Removal of the gonads or "gonadectomy" (spaying females and castrating males) is one factor leading to obesity. Researchers have demonstrated in controlled studies that the amount of adipose tissue increases after gonadectomy when the cats have free access to food (are fed *ad libitum*). What seems to be happening is that cats are eating more after removal of their gonads, but there is no change in the basal rate of metabolism. There is, however, also much less activity and, therefore, less energy expenditure. After castration, male cats fight less and roam less. The only approaches available to address the problem are reducing the amount of food provided, giving smaller meals, or reducing the fat/caloric content of the food coupled with attempting to increase activity.

Breeds

The Cat Fanciers' Association was established in 1906. This organization recognizes and registers cats in 41 pedigreed breeds (see fig. 6-3), which are shown in Table 6-1.

Abyssinian

American Curl

American Shorthair

American Wirehair

FIGURE 6-3 Breeds of cats. Except for the Manx, all photos are by Isabelle Francais.

Balinese · Bombay · British shorthair · Burmese

Chartreux · Colorpoint Shorhair · Comish Rex · Devon Rex

Egyptian Mau · Exotic · Havana Brown · Japanese Bobtail

Javanese · Korat · LaPerm · Maine Coon

FIGURE 6-3 (Continued)

Manx

Norwegian Forest Cat

Ociat

Oriental

Persian

RagDoll

Russian Blue

Scottish Fold

Selkirk Rex

Siamese

Siberian

Singapura

Somali

Sphynx

Tonkinese

Turkish Angora

Turkish Van

FIGURE 6-3 (Continued)

TABLE 6-1   Pedigreed breeds of cats

Abyssinian	LaPerm
American Bobtail	Maine Coon
American Curl	Manx
American Shorthair	Norwegian Forest Cat
American Wirehair	Ocicat
Balinese	Oriental
Birman	Persian
Bombay	RagaMuffin
British Shorthair	Ragdoll
Burmese	Russian Blue
Chartreux	Scottish Fold
Colorpoint Shorthair	Selkirk Rex
Cornish Rex	Siamese
Devon Rex	Siberian
Egyptian Mau	Singapura
European Burmese	Somali
Exotic	Sphynx
Havana Brown	Tonkinese
Japanese Bobtail	Turkish Angora
Javanese	Turkish Van
Korat	

DOGS

Importance of Dogs in the United States

There are estimated to be about 73 million dogs in the United States, with 39% of households having at least one dog (see fig. 6-4). Dogs are loyal and affectionate, and make excellent companion animals. Given the love of people for dogs, it is not surprising that the dog is one of the animals on the Chinese calendar and zodiac.

More than 70% of owned dogs are spayed or neutered. This leaves 30% of dogs that can and do breed. Many dogs are abandoned for reasons from the trivial such as moving home, to changes in family circumstances, unplanned puppies, and safety issues such as biting. Abandoned dogs are most commonly placed in shelters until they can be adopted, or else they are euthanized by either gas or lethal injection. It is estimated that up to 2 million dogs are killed each year in the United States.

Because of humans' close relationship with dogs, there are many expressions in the English language about dogs, including "good dog" (see sidebar for interesting quotation), "That dog won't hunt," "He does have a dog in the fight," "He's an old

FIGURE 6-4 Well-cared-for dog. Courtesy of PD Photos.

NOMENCLATURE

An adult female dog is a *bitch*, and an adult male is a *dog*. The mother of litter is a *dam*, and the father is the *sire*. Young dogs (<1 year old) are *pups* or *puppies*. The process of giving birth is called *whelping*. *Canine* is an adjective referring to dogs.

dog," "son of a bitch," "You can't teach an old dog new tricks," "He's her lapdog," people putting "dogged" resistance, "Dog is man's best friend," "If you lie down with dogs, you get up with fleas" (proverb), "Every dog has his day," and "to chain the dog of war."

Classification

Dogs are traditionally classified as being in the species *C. familiaris*. However, because dogs were domesticated from gray wolves (*Canis lupus*), they could be placed in a subspecies of gray wolves: *C. lupus familiaris*. Dogs, like wolves, jackals, and coyotes, are members of the *Canis* genus and the Canidae family. Interestingly, interbreeding of all members of the *Canis* genus, including wolves, jackals, and coyotes, can produce fertile hybrids.

Domestication

The domestic dog was domesticated from ancestors of today's gray wolf, probably at the end of the last Ice Age (or glaciation). All breeds of dogs are descended from tamed gray wolf ancestors (see fig. 6-5), with there being no contribution from other canids such as jackals, coyotes, or Ethiopian wolves.

When were dogs domesticated? Based on archaeological research in the Middle East and Europe, dogs were domesticated at least 15,000 years ago by hunter-gatherers before the advent of agriculture. It is also possible that dogs

CLASSIFICATION

The classification of dogs is phylum, Chordata; subphylum, Vertebrata; class, Mammalia; order, Carnivora; family, Canidae; genus *Canis*; and species, *Canis familiaris* (better *Canis lupus familiaris*).

FIGURE 6–5 Gray wolf in the wild. Courtesy U.S. Fish and Wildlife. Photo by William C. Campbell.

were domesticated several times, or, to put it another way, there were multiple domestication "events."

Where were dogs domesticated? Based on genomic comparisons between dogs and wolves from different populations across Eurasia, it seems that domestication occurred in East Asia. How were dogs domesticated? It is easy to imagine that there was taming of orphaned wolf pups by people, perhaps by children. Wolf packs would live in close proximity to groups of hunter-gatherers scavenging for the waste. There would then be selection for reduced aggressiveness or tameness.

Dogs in North America

Domestic dogs were brought with the first people to settle in North America. Dogs accompanied the first Americans as they moved to the continent some 14,000 years ago across the Bering Strait. They were used in the Americas to pull carts and sleds.

Dingos

These wild dogs were the only placental mammalian animals in Australia until the coming of Europeans. The Australian Aborigines arrived there some 50,000 years ago, crossing the Torres Straits from what is now Papua/New Guinea on rafts or dugouts. However, they did not bring the dingos. Dingos are thought to be the feral descendants of a small population of domesticated dogs either traded with the Australian Aborigines or abandoned in a shipwreck on a single occasion about

5,000 years ago. The first archaeological evidence of dingos is from 3,500 years ago. The Australian Aborigines adopted the dingo as a companion animal and for hunting.

Working Dogs

Dogs are intelligent, have an excellent sense of smell, are loyal, and are readily trained. There are multiple types of dogs that serve people:

- Herding dogs
- Hunting dogs
- Guardian dogs
- Racing dogs
- Sled dogs
- Fighting dogs (illegal in the United States)
- Sniffer dogs
- Police dogs
- Military or war dogs
- Service or assistance dogs (such as guide dogs)
- Therapy dogs

Herding Dogs

Perhaps the best example of a herding dog is a sheepdog (see fig. 6-6). When trained, this breed will move sheep from pasture to pasture, alone or with a human companion. Examples of sheepdogs include the collies, German shepherd,

FIGURE 6-6 Animal scientist Dr. Sandy Velleman enjoying sheep herding with her dog. Image kindly provided by Dr. Velleman.

Shetland sheepdog or Sheltie, and Old English sheepdog. Herding dogs are also used with cattle, reindeer, and poultry.

Hunting Dogs

Dogs have played a role in hunting, probably since they were domesticated. Examples of the role of dogs in hunting include the following:

- Gundogs or bird dogs
  - Flushing out the prey animals such as quails and pheasants (e.g., spaniels)
  - Pointing to where the prey animals are (e.g., English Pointers or German Shorthaired Pointers)
  - Retrieving the shot animal prey, including ducks, often from lakes and rivers (e.g., golden retrievers, Labrador retrievers, Irish setters)
- Foxhounds: Fox hunting involves a group of people on horseback following a pack of foxhounds chasing a fox (because of their sense of smell). The chase ends when the fox is run to earth. In the United States, foxhunts may pursue coyotes or bobcats instead of foxes. Foxhunting has been very contentious in the United Kingdom due to welfare concerns about the hounds' "kill" of the fox.
- Dogs chase a raccoon until either they catch it or it climbs a tree.
- Hounds with their tracking ability are used to hunt larger game in the United States, including mountain lions.

Guardian Dogs (Watchdogs, Guard Dogs, and Attack Dogs)

Guard dogs and watchdogs protect property, people, and livestock. The latter includes protecting sheep from wolves. The watchdog will bark at the sight or smell of an intruder, whereas the well-trained guard dog will either restrain or attack the intruder. Examples of guard and watchdogs include rottweilers, German shepherds, and Doberman pinschers. There are cases in which guard or attack dogs attack a supposed intruder, killing a child or innocent adult. Guard dogs have traditionally been used to guard sheep. Examples include German shepherds and Great Pyrenees (also known as the Pyrenean Mountain dog).

Racing Dogs

Greyhounds are said to be the second fastest mammal running, the fastest being the cheetah. It is perhaps not surprising that greyhounds have been used for racing. The greyhounds chase a lure, an artificial rabbit, around a track. In some states in the United States, there is greyhound racing with on-track betting. Greyhound racing also occurs in Western Europe and Australia. There are groups that foster the adoption or rescue of greyhounds that are too old to competitively race or that cannot race for other reasons.

Another example of dog racing is the Iditarod Trail Sled Dog Race across snow-covered Alaska. In the race, teams of 12–16 dogs pull a sled with a musher. Teams compete to go more than 1,150 miles from Anchorage to Nome. This can take between 10 and 17 days. At each checkpoint, there are veterinary inspections of the dogs. There have been fatalities of both human and canine members of the teams.

Fighting Dogs

Dogs have been bred and trained to fight other dogs. This is conducted in a pit and, therefore, explains the name of one of the breeds: the pit bull (the American pit bull terrier, American Staffordshire terrier, and Staffordshire bull terrier). Dogfighting is illegal throughout the United States because of concerns of severe cruelty. There are also dog–pig fights. These are also known as hog–dog fights, hog–dog rodeos, hog dogging, hawg dawgin', or hog baiting. Again, pit bulls are used. The legality of this activity is questionable at best.

Sniffer Dogs

Sniffer dogs are being used by law enforcement, the U.S. Department of Homeland Security, the military, the U.S. Department of Agriculture, and others based on their acute sense of smell and ability to be trained. They become very proficient at their task. They are also referred to as *detector dogs* or *canine detectors*.

These dogs are trained to detect the following:

- Illicit drugs (narcotics)
- Explosives and weapons
- Cadavers
- People
- Other odors

Drug detection dogs are used to detect illegal drugs (cocaine, heroin, methamphetamine, cannabis/marijuana, or opium) at border crossings by customs officials or by police. In addition, they are used by some employers to enforce a drug-free work environment. Beagles are frequently used as drug detection dogs because of their friendly nature therefore reducing the risk of frightening people, such as plane passengers (see Figure 6-7). The drug detection dog detects the presence of drugs without a search warrant.

FIGURE 6-7 U.S. Department of Agriculture official Ilka Mathis and Comet of the Beagle Brigade conduct a "field interview" after Comet alerted on a bag at Dulles International Airport, Virginia. Courtesy of U.S. Department of Agriculture. Photo by Ken Hammond.

Dogs that detect explosives and weapons are known as bomb dogs, bomb-sniffing dogs, explosives detection dogs, or weapons detection dogs. Labradors are an example of a breed employed by police, Homeland Security, and the military to detect explosives and weapons at public places such as airports, railroad stations, and sporting events. Sirius was a bomb dog killed in the September 11, 2001, terrorist attack.

Cadaver dogs or human remains detection dogs detect the odors from decomposing human remains (whole corpse or fragments). They can differentiate these from the odors of decomposing animals and live people or other odors that might distract them. Cadaver dogs are used when a person disappears, and either murder or suicide is suspected. They work with the police with both trained to work as a team such that forensic evidence is not compromised. They are also used with natural disasters such as earthquakes and hurricanes together with fires. They were used at ground zero after September 11.

There are a variety of ways that dogs are used to detect people. Search and rescue dogs are used to find an individual or any survivor of a natural or man-made disaster. Their training involves ensuring that they are both air scenting and trailing. The New York Police Department search and rescue dogs sniffed through the rubble at ground zero. A specialized example of a search and rescue dog is the avalanche dogs, which find survivors of an avalanche.

A trailing dog does what the name implies—it trails a person. Perhaps the best example of a trailing dog is the bloodhound.

Dogs are used to combat the illicit trade in agricultural products and wildlife that can bring new diseases into the United States. The U.S. Department of Agriculture's Animal and Plant Health Inspection Service use "sniffer" dogs to detect prohibited agricultural products (e.g., fruits and meat) coming into the United States and, therefore, to protect U.S. agriculture from plant and animal diseases. At airports, the Animal and Plant Health Inspection Service uses the "beagle brigade" to sniff out illicit agricultural products (see fig. 6-7). In addition, there are other detector dog programs using larger dogs at main international mail distribution centers. Dogs have also been used to detect pirated compact discs/digital video discs.

Police Dogs

Police dogs are also known as police K9 (sounds like *Canine*). German shepherds are a frequently used breed of police dogs. They are used as sniffer dogs, and to apprehend a criminal suspect in a chase and hold the situation. The dog works closely with a police officer handler. A number of police dogs have been killed in the line of duty. It is not surprising that killing a police dog is a felony in some states in the United States.

Military or War Dogs

There have been military uses of dogs for millennia. The ancient Egyptians, the classic Greeks and Romans, and armies in the Middle Ages used attack dogs to menace the infantry and to bite the horses. Such uses of dogs have continued up to the present.

In World War I, dogs were used as sentries and messengers. In World War II, the U.S. Marine Corps and U.S. Army used war dogs. The most common breeds were Doberman pinschers, German shepherds, and Labrador retrievers. These were used as sentries (guard dog), messengers, and scouts. More recently, dogs have been used as sniffers to detect mines and improvised explosive devices (IEDs). Dogs are also used as mascots.

There is a war dog memorial in Guam commemorating the dogs that served with the marines in World War II. A replica has been donated to the University of Tennessee. The memorial consists of a sculpture of a Doberman pinscher.

Service or Assistance Dogs

Dogs and other animals are being used to assist people with disabilities. Examples of service or assistance animals are the following:

- Guide dogs assisting people with blindness
- Guide horses assisting people with blindness
- Hearing or signal dogs assisting people with deafness
- Seizure response dog
- Psychiatric service dogs
- Mobility assistance dogs
- Assistance monkeys for quadriplegics and people with major disabilities that restrict movement

The best example of a service or assistance animal is the guide dog for people who are blind or have severe visual impairment. Dogs were first used to assist blinded World War I veterans. These assist people to stay on the sidewalk, avoid obstacles, and generally work with the person. Although the term *seeing eye dogs* is often used as a generic term for guide dogs, Seeing Eye, Inc., is a guide dog school in New Jersey. Among the breeds of dogs trained successfully as guide dogs are German shepherds, Labrador retrievers, golden retrievers, and Labradoodles (crosses between Labradors and poodles). Guide miniature horses are now being trained, and these have the advantage of living longer than guide dogs.

Another example of an assistance dog is a seizure response dog. These are trained to summon assistance when a person with epilepsy is having a seizure. Alternatively, some dogs may detect an impending seizure and signal to the person the need to take medication. There are also psychiatric service dogs to provide support to people with severe problems. Mobility assistance dogs may assist people who have problems with mobility by opening doors or directly assisting the rehabilitation returning to walking after injury. They can be referred to as "living canes" or "walker dogs" (see fig. 6-8). Capuchin monkeys have been trained to assisted people with disabilities such as quadriplegics and people with major disabilities that restrict movement. The monkeys need to be socialized before they can be trained.

Service dogs are being trained to help people with Alzheimer's disease. These dogs can assist the person when confused when given a command like "home." They have not only a service role but also a therapy role because contact with the dog may relieve depression and soothe loneliness.

Therapy Dogs

Animal-assisted therapy is different from service dogs. These therapy dogs are not specifically trained; rather, they are domestic pets that are docile and friendly, and, therefore, interact well with people. The presence of a dog in a nursing home can improve the quality of lives of the residents. Stroking the animal provides a warm tactile interaction and a soothing effect, and reduces a sense of isolation. This physical contact is something that is often lacking in an elderly person's life. Moreover, the act of stroking helps the person physically because it strengthens the muscles of the hand and arm.

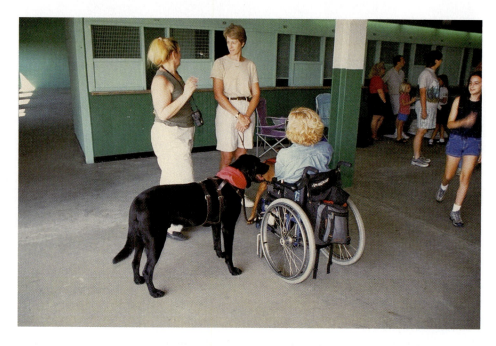

These therapy animals either are residents in a nursing home or are brought by volunteers. Different groups (e.g., Therapy Pets and the American Red Cross) have established visitation programs that bring domestic pets to nursing homes regularly to give residents a chance to interact with the animals. Many of the residents had pets when they lived independently.

Dog Genetics

Dogs show tremendous variation in size, coat (color, texture, length of the hair), and behavior based on their genetics. Dogs have 78 chromosomes (39 pairs).

In 2004, the sequence of the genome was reported from a boxer dog. About 94% of the dog genes are in the same order as in the human and mouse genomes. In view of dogs' sense of smell, efforts are underway, focusing attention on sequencing and understanding the olfactory related genome.

Breed parentage can be identified in dogs by DNA analysis. Genetic testing can be useful where there are disputes on parentage, or forensic needs, or to test for genetic diseases. A company (Mars Veterinary [United States] in partnership with Waltham Centre for Pet Nutrition in the United Kingdom) is now offering mixed-breed DNA analysis to establish within a reasonable certainty the breed parentage of an individual dog. The test, performed with a veterinarian, can distinguish between more than 130 American Kennel Club (AKC)–recognized breeds. It is accurate at 95% for mixed breeds and 99% for purebred dogs. In another example, the Canine Heritage Breed Test from MetaMorphix, Inc., also is used to determine the breed(s) of an individual dog. Owners send a swab sample from the inside of the dog's cheek, and the company compares the DNA with analyses of up to 100 different breeds.

Why would you want to test your dog for breed background?

1. You have a natural curiosity about the animal you love.
2. You want to understand the dog's veterinary needs because many genetic diseases, or genetic predisposition to diseases, are related to breed.

> **INTERESTING FACTOID**
>
> There is a 40-fold variation in the size of dogs across different breeds.

A series of genes control the color of dogs. A single allele in one gene (the *IGF-I* gene encoding the growth factor, insulin-like growth factor I) is a major determinant of size in many small breeds of dogs. A single nucleotide polymorphism in the *IGF-I* gene leads to small size in dogs.

Genetic abnormalities are found in dogs. The accumulation of adverse mutations occurs with the lack of careful breeding in a number of high-demand breeds.

Dog Breeds

It is estimated that there are between 350 and 400 different breeds of dogs. The oldest dog breeds may be the Pharaoh hound and Ibizan hound; these look like the pictures of the dogs in ancient Egyptian tombs from 5,000 years ago. Most modern breeds of dogs have been around for less than 400 years and have been genetically isolated since the mid-19th century by the "Breed Barrier Rule." A dog may only be registered if both the dam and sire are also registered. This selective breeding ensures genetically distinct breeds. The genetic relatedness of dog breeds has been studied. Of the 350–400 different breeds of dogs, 150 breeds are recognized by the AKC.

Groups of dogs (according to the AKC) include sporting (spaniel, retriever), working, terrier, toy, nonsporting, herding, and miscellaneous.

An index of the most popular breeds of dogs can be seen by the registration data of the AKC in Table 6-2 (also see fig. 6-9)

> **INTERESTING FACTOID**
>
> Using DNA typing and sequencing, the breeds of dogs can be readily determined. Breeds have been genetically isolated since the mid-19th century by the "Breed Barrier Rule." A dog may only be registered if both parents are registered. Mitochondrial DNA sequencing was used to explore the domestication of dogs from the gray wolf.

TABLE 6-2 AKC registrations of dogs

2006 RANK	BREED
1	Retrievers (Labrador)
2	Yorkshire terriers
3	German shepherd dogs
4	Retrievers (golden)
5	Beagles
6	Dachshunds
7	Boxers
8	Poodles
9	Shih tzu
10	Miniature schnauzers
11	Chihuahuas
12	Bulldogs
13	Pugs
14	Pomeranians
15	Boston terriers
16	Spaniels (Cocker)
17	Rottweilers
18	Maltese
19	Pointers (German shorthaired)
20	Shetland sheepdogs

Retrievers (Labrador) Yorkshire terriers German shepherd dogs Retrievers (golden)

Beagles Dachshunds Boxers Poodles

Shih tzu Miniature schnauzers Chihuahuas Bulldogs

Pugs Pomeranians Boston terriers Spaniels (Cocker)

FIGURE 6-9 Breeds of dogs. Photos by Isabelle Francais.

Rottweilers   Maltese   Pointers (German shorthaired)   Shetland sheepdogs

FIGURE 6-9 (Continued)

Dog Reproduction

Dogs are monoestrous. One estrus is followed if not pregnant by a period of anestrus or reproductive quiescence for about 6 months. Pregnancy or gestation lasts for 52–58 days. There are about six puppies per litter

The ejaculate volume in dogs is 10 mL (3 billion spermatozoa/mL). Dog semen also contains protein (0.6 mg/mL); nutrients for spermatozoa metabolism, including fructose (0.6 mg/mL); and electrolytes such as sodium, calcium, and chloride (see also fig. 16-1 in Chapter 16). Artificial insemination can be performed and the resulting puppies registered by the AKC, provided that certain conditions are met to ensure correct pedigree identification.

- When using fresh semen, both the sire and dam must be present when the semen is collected and during the insemination of the bitch.
- When fresh extended semen is used, a licensed veterinarian must collect it, dilute it, and inseminate the bitch. The sire must be AKC-DNA certified.
- When using frozen semen, the collection of semen for the artificial breeding must be reported to the AKC with DNA certification.

Dog Cloning

The first dog cloned was Snuppy in 2005 at Seoul National University in Korea. Semen from this dog has been used to inseminate female cloned dogs, and healthy puppies have been produced.

Dog Nutrition

Dogs, like their gray wolf ancestors, are largely carnivorous and require a diet containing protein, energy as carbohydrate and fat, together with a balance of minerals and vitamins. There are many commercial dog foods available that will meet the dog's dietary requirements. Dogs should be fed twice daily with attention to ensuring that excessive food is not available. Dog owners are concerned not only with the health of the dog, but also with dog food that does not lead to high amounts of either flatulence or feces.

Milk replacer formulas based on cow's milk and egg yolks are frequently recommended for use in neonatal puppies. These formulas may not be as good for fostering the growth and development of the puppies. The formulas are lower in protein, calories, calcium, and phosphorus than bitch's milk. In addition, the cholesterol content is greater.

Obesity in Dogs

Obese dogs have excess body fat and weigh 20% above the ideal. They have a thick layer of fat over the ribs, spine, tail base, and abdomen. The incidence of obesity in dogs is between 20% and 40% of dogs in the United States. Reduced life span, cardiovascular disease, diabetes mellitus, and arthritis are associated with obesity.

Obesity is caused by overfeeding, particularly of high-calorie snacks/treats, coupled with inactivity. Weight reduction programs in dogs reduce the daily caloric intake together with increased activity such that the dogs lose 1% of their body weight per week. Examples of low-caloric dog foods include Iams ProActive Health Weight Control and Purina Fit & Trim. As with people, dogs on a weight reduction program may be stressed from the hunger. Recently, the Food and Drug Administration approved a drug, dirlotapide, specifically for canine weight reduction.

ISSUES *for discussion*

DOGS IN HUMAN SOCIETY

1. Should insurance companies charge high household premiums to owners of certain breeds (rottweilers, German shepherds, Doberman pinschers, Akitas, pit bulls, or chows)?
2. Should certain breeds (rottweilers, German shepherds, Doberman pinschers, Akitas, pit bulls, or chows) be banned?
3. How can we reduce the number of dogs euthanized by shelters (estimated as much as 2 million per year)?
4. Should dogs or cats be used in animal experimentation?
5. Should dogs or cats be used as meat or fur?
6. Should dogs be used as entertainment such as in circuses, in greyhound races, or in the Iditarod Trail Sled Dog Race?
7. What should the penalties be for dogfighting (illegal in all states in the United States)?
8. If dogfighting is to be illegal, should the breeding of fighting dogs also be illegal?
9. Are there other issues with dogfighting, such as should it be a crime (misdemeanor or felony) to be a spectator?
10. What should the regulations be (if any) for breeding dogs and cats?
11. Should there be regulations for dogs or cats requiring them to be on leashes in urban and suburban settings?
12. Is it ethical to hunt with dogs?
13. If it is ethical, are all forms of hunting morally acceptable?
14. Is dog racing morally acceptable? Explain your position.
15. What are the issues of having a watchdog or guard dog?

FERRETS

In addition to cats and dogs, another domesticated carnivore is the ferret (see fig. 6-10). The American Veterinary Medical Association (AVMA) U.S. pet ownership survey estimated that there were over 1 million ferrets as pets in the United States in 2007. Owning ferrets in California as companion animals is not legal. Ferrets are domesticated European polecats. It is thought that ferrets were domesticated as early as 2400 BC. Ferrets can be used for rabbit hunting.

Ferret Reproduction

The ferret is an induced ovulator. The gestation length is 40–44 days, and the litter size is 8–10 kits.

RODENTS

There has been a growing interest in keeping rodents as pets. Examples of rodents as pets include gerbils, hamsters, guinea pigs or cavies, chinchilla, rats, and mice (see fig. 6-11). The AVMA reports there are about 2 million pet rodents in the United States.

Pet rodents are frequently kept in cages or aquaria with saw dust, wood shavings, or shredded newspaper as bedding. They can become friendly after taming. A significant number of people enjoy having pet rodents, partly because they are easy to take care of. Some pet rodents will show aggression and bite their owner. For instance,

CLASSIFICATION

The classification of ferrets is class, Mammalia; order, Carnivora; family, Mustelidae; genus, *Mustela*; and species, *Mustela putorius*.

TERMINOLOGY

A male ferret is a *hob*, and a female is a *jill* (if spayed, a *sprite*). Baby ferrets are called *kits*.

A   B   C

D   E

FIGURE 6–11 About 2 million rodents are kept as pets in the United States. Examples include: gerbils (*A*); hamsters (*B*); guinea pigs (*C*); rats (*D*); and mice (*E*). *A–E* reproduced by permission from Eric Isselee, Peter Igel, Sascha Burkard, Oleg Kozlov and Sophy Kozlov, and Sylvaine Thomas and 2happy, respectively. © 2010 by Shutterstock.com.

male rats and pregnant or lactating females may exhibit aggressive and biting behavior. There are shows where different rodent species are exhibited by fanciers.

Coloration of Rodents

As with other mammals, rodents can have either of both pigments in their fur; these pigments are black (eumelanin) or yellow (pheomelanin). Each hair in the fur can have no pigment or varying one or the other, or both. Moreover, in at least some species, such as the gerbil, there can be banding of color along the hair follicle. This is called agouti. There can also be patches of color, with rats being hooded or caped. Color is genetically controlled with genes for each pigment, agouti, dilute (decreasing the amount of the pigment), and no color/albino.

Rodent Nutrition

There are commercially available feeds for the different rodent species that meet their requirements.

Pet Rodents and Human Health

Pet rats and other rodents may carry diseases. Examples include the lymphocytic choriomeningitis virus and bacteria such as *Campylobacter* and *Salmonella* with

INTERESTING FACTOID

Many famous cartoon animals are rodents, including Mickey Mouse (created by Walt Disney with the movie debut in 1928) and Remy, a Norway rat who is the leading character in the 2007 Pixar movie *Ratatouille*.

multidrug-resistant *Salmonella* serovar Typhimurium (isolated from hamsters). These cause illness in people, including diarrhea, fever, vomiting, and abdominal cramps. The Centers for Disease Control recommends that "pregnant women and people with weakened immune systems should not bring a new pet rodent into their household." Moreover, people should be careful to thoroughly wash their hands after handling their pet rodents or their bedding, and never to kiss their pet rodents.

Laboratory Animals

The most widely used animals in research are rats and mice, with hamsters the third most commonly used animal species. Examples of strains of rats used in the laboratory include the Sprague Dawley outbred strain of white (albino) brown rats, Wistar, and Long Evans.

Rats and Mice

Pet rats or fancy rats are usually derived from the brown rat, and some are derived from the black rat. Rats have been bred in captivity for about 150 years. During rat and mice reproduction, the estrous cycle length is 4–5 days, gestation length is 21–23 days, and the litter size of the mouse is 4–12 and for rats is 6–12 pups.

Hamsters

There are several species of hamsters that are pets. These include the golden or Syrian hamster, Chinese hamster, and dwarf hamsters of the genus *Phodopus*. Hamsters are rodents. The hamster is widely used as a laboratory animal. These are also pets, with the AVMA estimating that there about 0.8 million pet hamsters in the United States

The golden or Syrian hamster originates in Southeast Europe and Asia Minor (Turkey, Syria, etc.), and was domesticated in 1930. In their natural environment, they live in tunnels. This provides a cooler temperature and higher humidity than the prevailing arid conditions. The Chinese hamster originates from the desert areas of Northern China and Mongolia.

Guinea Pigs or Cavies

Guinea pigs were domesticated in the Andes in South America. Domestication occurred well before the arrival of the first European settlers. Guinea pigs are kept as pets. The AVMA estimates that there about 0.6 million pet guinea pigs in the United States. During their time of reproduction, the estrous cycle is 15–17 days, gestation is 68 days, and the litter size is two to five.

The word *guinea pig* is used in everyday language for experimental animals or subjects due to the extensive use of this species in the first half of the 20th century. Guinea pigs are used as research models for human diseases, such as cardiovascular disease, high-risk pregnancies, and toxic effects of environmental contaminants.

Unlike most animals, guinea pigs require vitamin C in their diet. Otherwise, they come down with vitamin C deficiency (scurvy).

CLASSIFICATION

The classification of rodents is phylum, Chordata; subphylum, Vertebrata; class, Mammalia; and order, Rodentia. For the mouse it is order, Rodentia; family, Muridae; genus, *Mus*; and species, *Mus musculus*. For the rat it is order, Rodentia; family, Muridae; genus, *Rattus*; and species, *Rattus norvegicus* (for the brown, or Norway, rat) or *Rattus rattus* (for the black rat).

For the golden or Syrian hamster it is order, Rodentia; family, Cricetidae; genus, *Mesocricetus*; and species, *Mesocricetus auratus*. For the Chinese hamster it is order, Rodentia; family, Cricetidae; genus, *Cricetus*; and species, *Cricetus griseus*.

For the guinea pig it is order, Rodentia; family, Caviidae; genus, *Cavia*; and species *Cavia porcellus*.

TERMINOLOGY

The male adult rat is a *buck*, the female is a *doe*, and young rats are *pups*.

Gerbils

The most common pet gerbil is the Mongolian gerbil. The first gerbils were brought to the United States in 1954. Although widespread as pets, it is illegal to have pet gerbils in California. During their time of reproduction, the estrous cycle is 4–5 days, and gestation is 16 days.

BIRDS

It is perhaps surprising to find out that there are so many pet birds in the United States. The AVMA estimates that there about 11.2 million pet birds, with 4% of households having at least one. Important groups of pet birds include parrots, parakeets, and budgerigars. Other companion birds include chickens (particularly bantam and other heritage breeds), and such passerines as canaries and zebra finches.

Parrots

There are many species of parrots that can be pets (see fig. 6-12). The parrot species are all in the order Psittaciformes, and all have four toes on each foot. Parrots are typified by their ability to learn and speak words; their ability to develop a linkage with a single person or with people in general; their consumption of fruits, nuts, and other plant products; and their requiring more attention than dogs or cats.

Examples of parrots include cockatoos (21 species), macaws (19 species), New World parrots, African greys, Amazons, lovebirds, and parakeets. There are many species of parakeets. These are small or medium-sized parrots with long tails. The budgerigar parakeet is a small parrot often kept as companion animals.

FIGURE 6-12 Caique Parrots are one of the many species of birds kept as pets. There are over 11 million pet birds in the United States. Shown here is a White-Bellied Caique (*Pionites leucogaster*) and a Black-Headed Caique (*Pionites melanocephala*). Courtesy Carla Carpenter. Source: American Veterinary Medical Association.

Budgerigars (Budgies)

Budgerigars (or budgerigar parakeets) are the most common species of parrot that is kept as pets. They have been bred in captivity for 150 years. People frequently consider budgerigars as interesting pets. These birds can learn to say words and interact with people. The origin of the word (etymology) *budgerigars* is "good eating" in some aboriginal languages.

Budgerigars have been bred to have different colors, including yellow, blue, white (nonalbino), violet, olive, or white (albino). There are more than 30 color mutations in budgerigars, enabling a great many more combinations. The traits include the absence, dilution, or overproduction of pigment.

In the United States, the Department of Agriculture through the Animal and Plant Health Inspection Service has established rules and a quarantine period for importing pet birds (nonpoultry) and commercial birds. The rationale for this is that pet birds could bring various avian diseases to the United States. Parrots from South America are thought to have been the source for an outbreak of Newcastle disease in Southern California in the 1970s, with costs of $56 million.

To protect the United States from bird diseases, there are a series of requirements that have to be met.

1. Pet birds can only enter the United States through specific ports of entry.
2. The birds are then required to have 30 days of quarantine in a U.S. Department of Agriculture animal import facility, with the pet owner paying the cost for boarding the bird.
3. The bird is then tested for a series of diseases paid for by the pet owner.

> **CLASSIFICATION**
>
> The classification of budgerigars is phylum, Chordata; subphylum, Vertebrata; class, Aves; order, Psittaciformes; family, Psittacidae; genus,: *Melopsittacus*; and species, *Melopsittacus undulatus.*

 ## MONKEYS

The Allied Effort to Save Other Primates estimates that there are about 15,000 monkeys kept as pets in the United States. Monkeys are a tremendous responsibility for their owners. Moreover, they can carry zoonotic diseases that infect people after, for example, a bite from a monkey.

It is illegal to own monkeys as pets in a number of states, including New York, Georgia, Maryland, Pennsylvania, Kentucky, Louisiana, and Minnesota, together with all of New England and California. There are efforts to ban ownership of monkeys in other states. Other states require a permit, including Michigan, South Dakota, Oregon, Idaho, Oklahoma, and Delaware.

 ## COMPANION REPTILES AND AMPHIBIANS

Lizards, snakes, turtles, frogs, other reptiles, and amphibians are kept as companion animals, although they will not be considered in any detail. The AVMA U.S. pet ownership survey estimates that there are 2.0 million turtles (see fig. 6-13), 586,000 snakes (see fig. 6-14), and 1.08 million lizards (see fig. 6-15) as companion animals in the United States. There are also numerous ornamental fish kept as companion animals (see fig. 6-16).

FIGURE 6–13 Turtle. Reproduced by permission from ANP. © 2010 by Shutterstock.com.

FIGURE 6–14 Snake. Reproduced by permission from Brad Thompson. © 2010 by Shutterstock.com.

FIGURE 6–15 Iguana. Reproduced by permission from Sergey Goruppa. © 2010 by Shutterstock.com.

REVIEW QUESTIONS

1. How many domesticated cats are there in the United States?

2. What is the proportion of U.S. households that has a cat?

3. What is the name for an adult female cat?

4. What is the name for an adult male cat?

5. What is the species name for the cat?

6. What was the species that was domesticated to produce the cat?

7. Where and when were cats domesticated?

8. Why were cats domesticated?

9. What are other members of the cat family?

10. Cats are reflex ovulators. What does that mean?

11. Kittens are born in an altricial state. What does that mean, and what are the consequences?

12. Do cats have a role as a working animal? If yes, how?

13. How many chromosomes does a cat have?

14. Have cats been cloned?

15. Give examples of genetic abnormalities in cats.

16. Give five examples of breeds of cats.

17. Are cats and dogs carnivores (both nutritionally and as members of the order Carnivora)?

18. How many domesticated dogs are there in the United States?

19. What is the proportion of U.S. households that has a dog?

20. What is the name for an adult female dog?

21. What is the name for an adult male dog?

22. What is the species name for the dog?

23. What was the species that was domesticated to produce the dog?

24. Where, when, and why were dogs domesticated?

25. Did the ancestors of the American Indians who arrived in North America over 10,000 years ago bring dogs with them?

26. What is a dingo?

27. What is whelping?

28. Give five examples of breeds of dogs.

29. Give five examples of working uses of dogs.

30. What is a sniffer dog, and how can it be used?

31. What is a guard dog, and how can it be used?

32. What are hunting dogs used for?

33. How can service dogs be used?

34. How many chromosomes does a dog have?

35. Why is genetic testing done with dogs?

36. Dogs are monoestrous. What does that mean?

37. Provide three examples of rodents and birds that are companion animals. What are some advantages of rodent or avian companion animals?

38. How have rodents been used in biomedical research?

39. What are the restrictions of owning monkeys in the United States?

40. What are the rules for bringing wild birds into the United States?

REFERENCES AND FURTHER READING

American Kennel Club. *Breeds*. Retrieved July 17, 2009, from http://www.akc.org/breeds/index.cfm

Backus, R. C. Cave, N. J., & Keisler, D. H. (2007). Gonadectomy and high dietary fat but not high dietary carbohydrate induce gains in body weight and fat of domestic cats. *The British Journal of Nutrition, 98*, 61–65.

Bartlett, D. J. (1958). Biochemical characteristics of dog semen. *Nature, 182*, 1605–1606.

Campbell, K. L., Campbell, J. R., & Corbin, J. E. (2005). *Companion animals: Their biology, care, health, and management*. Upper Saddle River, NJ: Prentice Hall.

Canine Specialized Search Team. *Forensic evidence and human remains detection dogs*. Retrieved July 15, 2009, from http://www.csst.org

Eden Consulting Group. *K9 officers killed in action*. Retrieved July 15, 2009, from http://www.policek9.com/html/valor.html

howstuffworks. *How police dogs work*. Retrieved July 15, 2009, from http://people.howstuffworks.com/police-dog.htm

The Humane Society of the United States. *Dogfighting fact sheet*. Retrieved July 15, 2009, from http://www.hsus.org/hsus_field/animal_fighting_the_final_round/dogfighting_fact_sheet/

The Humane Society of the United States. *U.S. pet ownership statistics*. Retrieved July 15, 2009, from http://www.hsus.org/pets/issues_affecting_our_pets/pet_overpopulation_and_ownership_statistics/us_pet_ownership_statistics.html

K9 Solutions Center. *Bomb, drug dogs for hire*. Retrieved July 15, 2009, from http://www.dopedog.com

Kanchuk, M. L., Backus, R. C., Calvert, C. C., Morris, J. G., & Rogers, Q. R. (2003). Weight gain in gonadectomized normal and lipoprotein lipase-deficient male domestic cats results from increased food intake and not decreased energy expenditure. *The Journal of Nutrition, 133*, 1866–1874.

McNamara, J. P. (2006). *Principles of companion animal nutrition*. Upper Saddle River, NJ: Prentice Hall.

Mott, M. (2003, September 16). The perils of keeping monkeys as pets. *National Geographic News*. Retrieved from http://news.nationalgeographic.com/news/2003/09/0916_030916_primatepets.html

Nashville Police Department Canine Unit (K-9) Retrieved July 17, 2009 from http://www.police.nashville.org/bureaus/fieldops/canine.asp

National Research Council. Subcommittee on Laboratory Animal Nutrition. (1995). *Nutrient requirements of laboratory animals* (4th ed.). Washington, DC: National Academy Press.

Noakes, D. E., Parkinson, T. J., England, G. C. W., & Arthur, G. H. *(2001)*. *Arthur's veterinary reproduction and obstetrics* (8th ed.). London: Saunders.

Office of Health and Human Services. *Communicable disease control*. Retrieved July 15, 2009, from http://www.mass.gov/dph/cdc/epii/pet_rodents.pdf

Ostrander, E. A., & Wayne, R. K. (2005). The canine genome. *Genome Research, 15*, 1706–1716.

Pickrell, J. (2004, April 8). Oldest known pet cat? 9,500-year-old burial found on Cyprus. *National Geographic News*. Retrieved from http://news.nationalgeographic.com/news/2004/04/0408_040408_oldestpetcat.html

Savolainen, P., Leitner, T., Wilton, A. N., Matisoo-Smith, E., & Lundeberg, J. (2004). A detailed picture of the origin of the Australian dingo, obtained from the study of mitochondrial DNA. *Proceedings of the National Academy of Sciences of the United States of America. 101*, 12387–12390.

Sutter, N. B., Bustamante, C. D., Chase, K., Gray, M. M., Zhao, K., Zhu, L., et al. (2007). A single *IGF1* allele is a major determinant of small size in dogs. *Science, 316*, 112–115.

University of California Davis Veterinary Medicine. *Bomb detection dog association(s)*. Retrieved July 15, 2009, from http://www.vetmed.ucdavis.edu/CCAB/bomb.html#association

Warren, D. M. (2002). *Small animal care and management* (2nd ed.). Clifton Park, NY: Thomson/Delmar Learning.

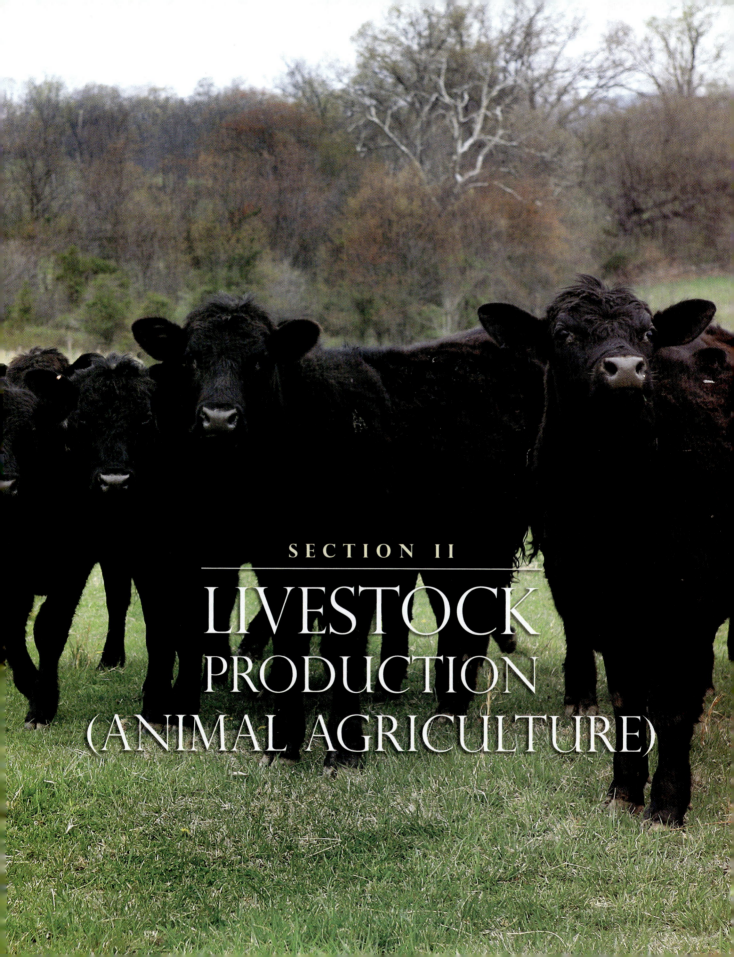

SECTION II
LIVESTOCK
PRODUCTION
(ANIMAL AGRICULTURE)

CHAPTER 7

Cattle

OBJECTIVES

This chapter will consider the following:

- Introduction to cattle
- Cattle domestication and history
- Biology of cattle (genetics, stress, reproduction, digestive system, and nutrition)
- Beef cattle (industry, exports, industry structure, beef production, selection, conformation, and breeds)
- Dairy cattle (history, breeds, dairying in the United States, dairy cattle management, reproduction, mammary gland, milking, biosecurity, infections/ mastitis, behavior, stress, facilities, and manure)

INTRODUCTION

Globally and in the United States, cattle are the most important livestock species in terms of both the value and the weight of their products (meat, milk, and other products). Cattle are ruminants and, therefore, consume fibrous plants that cannot be digested by nonruminants. The plant fiber, including cellulose, is degraded by the bacteria and protozoa that inhabit the rumen. Given the importance of cattle globally and for a long period in human history, it is not surprising that the ox is one of the animals on the Chinese calendar and zodiac. A sign of the zodiac is Taurus.

This chapter will first consider the domestication and classification of cattle. Dairy and beef cattle will then be considered separately. The external anatomy of beef cattle is shown in Figure 7-1.

DOMESTICATION AND CLASSIFICATION

Cattle with 800 different breeds were originally ascribed to two species by Carl Linnaeus (1707–1778): European breeds, known as *Bos taurus*; and Indian, together with Zebu, cattle, known as *Bos indicus*.

There is complete fertility between these two species of cattle; Figure 7-2 provides an example of crossbreed. They are better described as *super breeds*. Evidence from DNA sequence comparisons together with archaeological studies strongly support there being two separate domestications of the now extinct wild ox or aurochs (*Bos primigenius*).

The first domestication occurred around 5800 BC in the Fertile Crescent. The local population of the wild ox (*Bos primigenius primigenius*) was the ancestor of the first domesticated cattle. These then were the ancestors of the European breeds of cattle. A second domestication occurred around 2500 BC in the Indian subcontinent. The ancestral wild ox was the Asia subspecies of the wild ox (*Bos primigenius namadicus*).

> **NOMENCLATURE**
>
> An adult female who has had at least one calf is a *cow*; an adult male is a *bull*. The process of giving birth is *calving*. A group of cattle is a herd. *Ox/oxen* (plural) are cattle used as draft animals pulling plows or carts. *Polled cattle* lack horns. Young cattle are called *calves*. A young female is a *heifer*; a young male castrated before sexual maturity is a *steer*. A *bullock* in the United States is a young bull, and, to confuse matters, is an older steer in England. *Bovine* is an adjective referring to cattle.

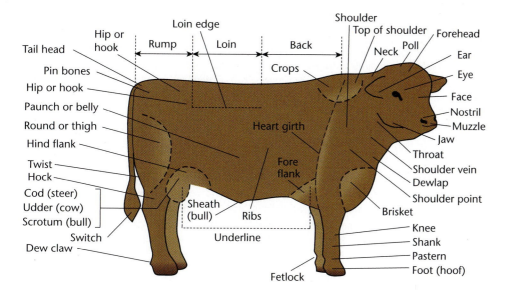

FIGURE 7-1 External anatomy of beef cattle showing nomenclature. Source: Delmar/Cengage Learning.

FIGURE 7-2 Mixed breed bull showing the characteristic hump of the Brahman-like or *Bos indicus* super breed. Courtesy of the USDA.

African Zebu cattle resemble the India cattle and have been considered as *Bos indicus*. However, DNA sequencing data coupled with historical records lead to a different conclusion of their history. The primary genetics are from *Bos taurus*. It is thought that at about AD 670 Arab traders brought *Bos indicus* bulls that were crossed with the native *Bos taurus* cattle. The resulting offspring then did much better in the tropical climate. More crosses occurred, diluting the *Bos indicus* genes but retaining the ability to thrive in the tropical climate. The entire maternal genetic component of Zebu cattle, based on mitochondrial DNA sequencing and comparison, comes from *Bos taurus*.

HISTORY OF CATTLE

- 1493 - Christopher Columbus brought cattle to the Americas (West Indies) in his second voyage to the New World.
- 1624 - The pilgrims brought cattle to New England in 1624.
- 1866–1890 - Cattle drives were made from Texas to Kansas railhead cow towns such as Ellsworth, Dodge City, Newton, Wichita, and Abilene on the Chisholm Trail, Goodnight Trail, and other trails. Cowboys rode ahead, flanking the herds of longhorn cattle. At the end of the U.S. Civil War, cattle could be sold in Texas for $4 per head compared with a price in the North and Midwest of $40. This price differential was the economic basis of the cattle drives.
- 1893 - The great Chicago fire occurred. The fire was said, erroneously, to have been started when a cow kicked over a lamp.
- 1900–1930 - Laws were passed to regulate the milk quality.
- 1930s - Artificial insemination (AI) began in the United States.
- 1949 - The first frozen bull semen was used to successfully inseminate a cow. This led to a rapid expansion in the use of AI, revolutionizing the dairy industry.

BIOLOGY OF CATTLE

Cattle Genetics

Cattle have 60 chromosomes (29 pairs of autosomal chromosomes together with either XX or XY sex chromosomes). There has been tremendous progress mapping the cattle genome. The improvement of cattle genetics, particularly of dairy cows, has resulted in tremendous improvements in production indices such as milk production together with milk composition and overall quality. Sire selection based on its genetics and the heritability of the traits is a primary approach to impact cattle.

Stress in Cattle

There are a number of management techniques, such as ear tagging, branding, dehorning, and tail docking (dairy cattle), that evoke a stress response.

Cattle Reproduction

The primary goals of reproduction differ between dairy and beef production. In dairy production, the primary goal is to have high milk production from lactating cows. In addition, producing a replacement heifer is important. In beef production, the primary goal is to produce numbers of offspring to be grown for meat. In addition, milk production provides the early nutrition for the calves. The cow is the primary focus of reproductive management on most dairy farms, with the bull the second part of the equation.

Reproduction requires the union of a spermatozoa and an egg. The bull produces large numbers of spermatozoa, and these are placed into the vagina of an estrous cow (cow in heat) during mating. The bull identifies an estrous cow by smell. The bull locates an estrous female, then sniffs and licks the vulva area to confirm the heat. He places his chin on her rump to assess her willingness to stand for mounting. If the cow responds, the bull mounts, copulates, and ejaculates within 30 seconds. The bull then dismounts but may service the cow several times in succession.

The female cycle or estrous cycle consists of a growing follicle. At ovulation, the ovum is released, ready to be fertilized. The remnants of the follicle develop into a corpus luteum. This produces progesterone to allow and maintain pregnancy. If pregnancy does not occur, the corpus luteum breaks down, and a new follicle develops. Cattle are polyestrous with cycles of 21 days.

The behavioral signs of heat (or estrus) occur with the regression of the corpus luteum (and decline in progesterone in the blood) and the developing dominant follicle producing estradiol. The behavioral changes include increased vocalization and activity together with attempting to mount or ride other cows. The vulva becomes slightly swollen, moist, and red, with a clear mucus discharge. During standing heat, cows stand to be mounted by other cows and bulls.

The duration of pregnancy is 279–290 days. Normally, a single calf is produced per pregnancy. Cattle have a cotyledonary placenta, with the placenta attaching to 80–120 specialized areas (caruncles) in the uterus. The dam nourishes her fetus through these specialized attachments until the fetus signals its readiness for birth.

The mammary glands, or udder, in the cow consist of four separate glands, called quarters, with the teat hanging down from each quarter. Mammary anatomy and physiology are discussed in Chapter 18. Immediately before the birth of the calf, colostrum synthesis is initiated.

Ruminant Digestive System

There are four stomach compartments: rumen, reticulum, omasum, and abomasum (true stomach). Rumen microorganisms break down cellulose in feeds to volatile fatty acids (VFAs): acetic, propionic, and butyric. The concentration of bacteria and protozoa in the rumen is about 35 billion/mL and 350,000/mL. The VFAs are largely absorbed through the rumen wall, providing 60%–80% of the energy needs. Microbial digestion is why ruminants can be maintained on a roughage-fiber diet (see also Chapter 14).

Cattle Nutrition

Feeds need to meet the energy, protein, and other needs of the animal. Ration formulation involves knowing the requirements, availability, quality, palatability, digestibility, and cost of each feed ingredient. The following nutrients are required by cattle:

- Energy, from forages such as hay, grasses, or grain such as corn
- Nitrogen, either protein or nonprotein nitrogen such as urea that is converted to protein by the rumen microorganisms
- Minerals with calcium and phosphorus being particularly important for dairy cattle
- Vitamins, although many are synthesized by the rumen microorganisms
- Water

These nutrients can be supplied from grazing on a pasture or ranch land, or feeding a high corn/concentrate diet such as with lactating dairy cattle and beef cattle before slaughter at feedlots. Cattle nutrition is covered in more detail in Chapter 14.

 BEEF CATTLE

The U.S. Beef Industry

The beef industry is a very important part of U.S. agriculture, with gross receipts of $35 billion in 2008. There are about 1 million farms and ranches selling cattle and/or calves. Beef is produced as beef steers and heifers or from cull cows. Both beef production and consumption are relatively stable in the United States (see Tables 7-1 and 7-2, and the sidebar).

About 1.4 million people are employed directly and indirectly in beef cattle production and processing. Beef production impacts the economics of virtually all rural counties in the United States. Over 90% of U.S. beef production is consumed domestically, and the remainder (1.9 billion lb) is exported.

TABLE 7-1 U.S. commercial slaughter as 1 million head

YEAR	NO.	BEEF STEERS/ HEIFERS	+ CULL BEEF/ DAIRY COWS
2002	35.7	29.4	5.8
2003	35.5	28.7	6.1
2004	32.7	27	5.2
2005	32.4	27	4.9
2006	33.7	27	5.4
2007	34.3	27.9	5.8
2008	34.4	27.5	6.3

Source: Data from ERS.

TABLE 7-2 U.S. beef production (commercial carcass weight)

YEAR	BILLION lb
2002	27.1
2003	26.2
2004	24.5
2005	24.7
2006	26.1
2007	26.4
2008	26.6

Source: Data from ERS.

Exports of U.S. Beef

There are substantial exports of U.S. beef. Statistics from the USDA on beef exports (commercial carcass weight and value) are as follows:

- 2002: 2.4 billion lb, $2.6 billion
- 2003: 2.5 billion lb, $3.2 billion
- 2004: 460 million lb, $630 million
- 2005: 670 million lb, $1.0 billion
- 2006: 1.1 billion lb, $1.6 billion
- 2007: 1.4 billion lb, $2.2 billion
- 2008: 1.9 billion lb, $3.0 billion

Bovine spongiform encephalitis in North America reduced beef exports in 2004. There are now increases in export and, therefore, consumption of U.S. beef as consumer confidence rebounds, and regulatory agencies allow U.S. beef to be exported to Japan and other countries.

INTERESTING FACTOID

Beef consumption in the United States is stable, as can be readily seen from U.S. Department of Agriculture (USDA) figures on consumption (in billion lb): in 2002, 27.9; 2003, 27.0; 2004, 27.8 ; 2005, 27.8; 2006, 28.0; 2007, 28.1; and 2008, 27.3

Structure of U.S. Beef Production

The structure of the beef cattle industry is summarized in Figure 7-3. The beef industry is primarily divided into three segments: the cow-calf operation, pure-bred breeders, and the feedlot where animals are fed before marketing.

There are also stocker cattle. These are lightweight young cattle raised on forage that can be used either as replacement for cows in cow-calf operations or can go to the feedlot.

The cow-calf operations are found in areas where there is pasture. The feedlots are in areas where feed (corn) is either produced or readily transported to. They may be geographically separated from the cow-calf operations.

For cow-calf operations, there is an average of 40 cows per farm of the 800,000 herds with beef cows in the United States. There is, as might be expected, a mixture of small, medium, and large herds of beef cattle, with the following breakdown by size: 650,000 farms/ranches with 1–49 cows, 100,000 farms/ranches with 50–99 cows, 70,000 farms/ranches with 100–499 cows, and 5,000 farms/ranches with more than 500 cows.

The smallest operations with less than 50 cows represent 80% of herds but produce only 30% of the calves. Inventories of cattle show a cyclic pattern. There are higher prices when supply is restricted and lower prices when supply is high. When inventories are low, producers reinvest.

Feedlots are classified by size: capacity <1,000 (marketing 14% of fed cattle); capacity of 1,000 to 16,000 (marketing 19% of fed cattle); and capacity >16,000 (250 of such feedlots in the United States marketing 67% of fed cattle) (see fig. 7-3).

There is a concentration of beef packing companies, with the three largest processing over 75% of steers and heifers. Live animals are 90% of the processor's costs with the remainder being labor and capital cost of investment. The acceptable range for cattle being marketed is usually between 750 and 1,400 lbs (see Table 7-3).

TABLE 7-3 Slaughter weights (lb) for beef cattle

FRAME	STEERS	HEIFERS
Small	900	750
Medium	1,100	950
Large	1,400	1,250

Producers also receive payment based on the quality of the meat. The federal beef grades are prime (+, o, −), choice (+, o, −), select (+, −), and standard (+, o, −).

Structural Changes in the Beef Industry

The number of beef operations is decreasing, but there are more large producers. The rate of change is much slower than with other agricultural sectors, such as pork production.

Production by State

Cattle are raised for beef in virtually every state of the United States. The top-producing states for beef production are summarized for inventory and for marketed in Tables 7-4 and 7-5, respectively. Texas, Nebraska, and Kansas are consistently the top three beef cattle-producing states.

TABLE 7-4 Ranking of the United States by cattle inventory (January 2009)

RANKING	STATE	MILLION HEAD
1	Texas	13.6
2	Nebraska	6.4
3	Kansas	6.3
4	Oklahoma	5.4
5	California	5.2
6	Missouri	4.2
7	Iowa	3.9
8	South Dakota	3.7
9	Wisconsin	3.3
10	Colorado	2.6

Source: Data from the USDA National Agricultural Statistics Service.

TABLE 7-5 Ranking of the United States by cattle marketed (2008)

RANKING	STATE	BILLION POUNDS
1	Texas	8.3
2	Nebraska	7.7
3	Kansas	6.7
4	Colorado	3.1
5	Iowa	2.8
6	Oklahoma	2.5
7	California	2.4
8	South Dakota	1.9
9	Montana	1.4
10	Idaho	1.3

Source: Data from the USDA National Agricultural Statistics Service.

Beef Cattle Management

Raising beef cattle has the following advantages:

- Beef cattle use roughages (forage) that otherwise would be wasted for feed.
- Labor requirements may be low.
- Capital investment in buildings, but not land, can be small.
- Death losses are usually low.
- Beef are adapted for use in small operations as well as large ones.
- There is high demand for beef.

Disadvantages of raising beef include the following:

- It is a high-risk business.
- Cattle are not efficient converters of concentrated feeds (such as corn) into meat.
- It takes considerable time to develop a cattle herd.
- The capital investment in modern, efficient-feeding operations can be high.

The costs for cattle production include corn, supplements and minerals, hay, pastures and grazing, veterinary/health, machinery/fencing, and labor (see Table 7-6).

Beef Production

There are three main types of beef cattle production systems: cow-calf producers, purebred breeders, and cattle feeders. A producer may specialize in one type of operation or combine several.

Cow-Calf Producers

In the cow-calf system, the producer has a herd of beef cows that are bred each year to calve usually in the spring. The land used is not suitable for growing crops, with most production being in the Western range states and upper Great Plains. The calves are weaned in the fall and sold as feeder calves to cattle feeders. Some calves are fed through the winter and sold the next year as yearlings. Cows and calves are

TABLE 7-6 Costs for beef production

COST ITEM	PERCENTAGE
Corn	16.6
Supplements and minerals	5.8
Hay	29.3
Pastures and grazing	5.3
Veterinary and health	3.6
Machinery and fencing	8.4
Labor	10.5
Miscellaneous, e.g., insurance, interest	20.5
Total	100

Source: Based on Otto and Lawrence (2000).

maintained on roughage with little grain provided. Typical feeds include pasture, hay, silage, straw, and crop residues (see fig. 7-4). Corn silage is widely used as a feed for beef herds. Alfalfa is the most common roughage in the Midwest and West. Coastal Bermuda grass is more common in the southern coastal states. Native range grasses are used to a great extent in the western states. These operations require less labor and investment in equipment and facilities than other types. Beef cows can be wintered outdoors with a minimum of shelter. Cows can be moved to a pasture or lot close to the farmstead just before calving. Open-front calving barns may be necessary in colder parts of the United States. Portable calf shelters can be used for calves on pasture. There is a larger investment in land. The risk is that the price received for the calves follows supply and demand rather than the cost of production.

Purebred Breeders

Purebred breeders keep herds of purebred breeding stock. The calves are reared to provide replacement bulls together with some cows or heifers for cow-calf operations. Purebred breeders have considerable knowledge and skill, and are responsible for genetic improvements in beef breeds. The costs are high because it takes years to develop a high-quality herd.

Cattle Feeders

The objective is to produce high-quality finished cattle for the slaughter market in as short a time as possible. The feeder buys feeder calves or yearlings and finishes them in the feedlot. Some producers feed cattle on pasture before finishing them. Feeder cattle operations require considerable grain, frequently corn, together with some roughage. It usually takes grain to get the quality of finish that is in demand in the marketplace.

The costs for feedlot facilities are more expensive than those required for cow-calf operations. These can be either no confinement facilities or either cold or warm confinement (where cattle are housed in barns in either the winter or summer). Cold confinement barns require less labor than open feedlots and provide some protection from the weather. Manure handling costs are lower if slotted floors are used, and no bedding is needed. Warm confinement barns are closed buildings that are insulated and kept warmer than outside in the winter. They are the most expensive cattle-feeding facility.

FIGURE 7-4 Beef cattle grazing crop residue in a harvested corn field. © 2010 Shutterstock.com.

Selection of Beef Animals

A beef herd is improved by selecting animals that have the desired traits, such as production traits, including rapid growth (average daily gain) and feed efficiency; and carcass quality, including muscling, condition (extent of fat), frame, and meat quality, including tenderness and marbling score.

The producer produces what the consumer demands. Selection is based on performance records of the sire and dam, conformation (likely meat yield), progeny records, and DNA markers for desired traits. Only healthy animals should be brought into the herd after a quarantine period.

Conformation

This is the appearance of the live animal, and includes the skeleton, muscling, and amount of fat. Desirable conformation of beef cattle includes the following:

- Long, trim, deep-sided body
- No excess fat
- No extra hide (for example, around the throat)
- Heavily muscled front legs
- Proper height to the point of the shoulders
- Correct muscling throughout the body
- Maximum development of the round, rump, loin, and rib.

Animals with the best conformations tend to produce maximal amounts of high-value cuts (see fig. 7-5). Ultrasonics uses high-frequency sound waves to measure fat thickness and muscling (loin-eye area) in live animals.

Beef Breeds

Many beef breeds originated in Europe, with selection starting in the late 1700s. New breeds of beef were developed in the 20th century such as the Brahman breed, which is the result of crossing several *Bos indicus* breeds with European-type cattle. There are more than 50 breeds of beef cattle in the United States. The leading breeds in rank order of registrations are Angus, Limousin, Simmental, Hereford, polled Hereford, and Charolais. Each breed has both advantages and disadvantages.

FIGURE 7-5 Location of high- and low-value wholesale cuts of beef. Source: Delmar/Cengage Learning.

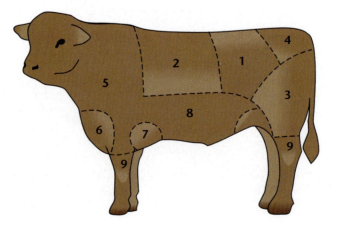

High-Value Wholesale Cuts
1. Loin
2. Rib
3. Round
4. Rump

Low-Value Wholesale Cuts
5. Chuck
6. Brisket
7. Flank
8. Plate or navel
9. Shank

Angus or Black Angus

The official name of the breed is Aberdeen-Angus. It originated in Scotland in Aberdeenshire and Angus. The first Angus cattle were imported into the United States in 1873. Angus cattle are black in color (see fig. 7-6). They have a smooth hair coat and are polled. Angus cattle perform well in the feedlot, producing a high-quality carcass of well-marbled meat. There are also Red Angus herds with the color being recessive.

FIGURE 7-6 *A,* Angus cow. Courtesy of the USDA. Photo by Bill Tarpenning. *B,* Angus bull. Courtesy of the American Angus Association.

Limousin

These originated in west-central France. Limousin cattle entered the United States in 1968, when semen was imported and registered. They have light-yellow hair with lighter circles around the eyes and muzzle (see fig. 7-7). The skin is free of pigmentation. The spread of horns is horizontal, then forward and upward. The Limousin head is small and short, with a broad forehead. The neck is also short. Mature bulls weigh from 2,000–2,400 lb (907–1,088 kg). Mature cows weigh about 1,350 lb (612 kg). Limousin cattle are noted for their carcass leanness and large loin area.

Simmental

This breed originated in the Simmen Valley of Switzerland. It is the most popular breed of cattle in Europe. Simmentals were first brought to the United States from Canada in 1969. They have white to light-straw faces with red to dark-red, spotted bodies (see fig. 7-8). They are a horned breed with medium-size horns. The Simmental is a large-bodied animal and is noted for being docile. Mature bulls weigh from 2,300–2,600 lb (1,043–1,179 kg). Mature cows weigh about 1,450–1,800 lb (658–816 kg). Simmentals have extremely rapid growth, gaining about 3 lb (1.4 kg) per day on roughage. They are thickly muscled and produce a carcass without excess fat. They are adaptable to a wide range of climates.

Hereford

Herefords originated in Herefordshire, England. The early breeders selected for a high yield of beef and efficient production. The first Hereford herd in the United States was established in 1840 in New York. Hereford cattle have white faces, red bodies, and white on the belly, legs, and tail. Herefords are a horned breed (polled Herefords lack horns). They are docile and easily handled. They are well adapted to the western cattle-raising regions of the United States. They have

FIGURE 7-7 Limousin bulls. Courtesy of the North American Limousin Foundation.

FIGURE 7-8 American Simmental bull. Courtesy of the American Simmental Association.

superior foraging ability, vigor, and hardiness. They produce more calves under adverse conditions than do many other breeds. Mature Hereford bulls weigh about 1,840 lb (834 kg) (see fig. 7-9). Mature cows weigh about 1,200 lb (544 kg) (see fig. 7-4). Herefords are popular for their general producing ability.

Charolais

This is a French breed of beef cattle developed around the town of Charolles. The first Charolais cattle were imported in 1936 by the King Ranch in Texas.

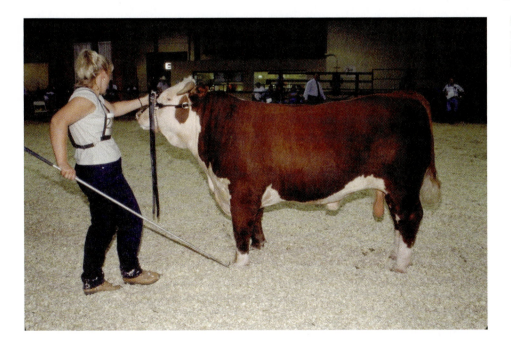

FIGURE 7-9 Hereford bull at the Eastern National Livestock Expo. Courtesy of the USDA. Photo by Bill Tarpenning.

FIGURE 7–10 Charolais bull. Courtesy of the American-International Charolais Association.

Definitions

Artificial insemination (AI) is when spermatozoa are placed into the female by producers or technicians rather than by coitus. The body condition score (BCS) uses a scale of one to five. A score of one indicates an emaciated cow, and a score of five an obese cow. Calving-ease score is a scoring system from easy delivery to extreme dystocia.

Colostrum, which is the first milk produced (first six milkings in cattle), is very high in antibodies and, therefore, a source of passive immunity for the newborn.

Dystocia is abnormal or difficult labor. Dry matter (DM) intake is the feed intake less its water content.

Estrus or *heat* is the time when a female will allow mating.

Eutrophication is when there is tremendous growth of plants (particularly algae) in water due to high plant nutrient levels such as phosphorus. The nutrients frequently come from agricultural practices with livestock waste a significant factor. Decomposition after the death of the plants leads to a lack of oxygen in the water and death of fish.

Fat-corrected milk calculates milk production based on fat concentration of 4%. The standard length of lactation records is 305 days. A cow may be milked longer than 305 days.

The cattle are white to light-straw in color with pink skin. They are a large, heavily muscled breed (see fig. 7-10). Mature bulls weigh 2,000–2,500 lb (907–1,134 kg), mature cows weigh 1,500–1,800 lb (680–816 kg). Most are naturally horned. Charolais cattle have a high feed efficiency and are used in many crossbreeding programs.

 DAIRY CATTLE

History of the Dairy Industry

Mammals, in the wild, produce only enough milk for their offspring. Beginning about 9000 BC, animals, particularly cattle, were domesticated to produce milk for human consumption. Between 1600 and about 1850, dairying was somewhat limited. Farms had one or two cows supplying milk to the family. Milk was largely available in the spring and early summer when pastures were lush. Urban populations had little access to milk because it is perishable. Milk was not subjected to health or quality control regulations. It was often diluted with water and of low quality. Milk surpluses were converted to cheese, preventing spoilage and easing transport.

Dairy production changed markedly beginning in 1850. New techniques in handling, storage, and processing of milk were developed. Between 1900 and 1930, laws were passed to regulate the milk quality. Techniques were developed to improve the safety and nutritive value of milk. Artificial insemination (AI) began in the United States in the 1930s. In 1949, researchers developed the ability to freeze bull semen. This led to a rapid expansion in the use of AI, revolutionizing the dairy industry. Other techniques increased milk production and improved quality control, including production records to accurately predict genetic merit, new milk containers, embryo transfer, and biotechnology.

Breeds of Dairy Cattle

There are many breeds of cattle worldwide. In the United States, commercial dairy producers primarily use six major breeds of dairy cattle: Holstein, Jersey, Brown Swiss, Guernsey, Ayrshire, and Milking Shorthorn. Characteristics of each breed are shown in Table 7-7, with representatives of dairy breeds in Figure 7-11.

Dairying in the United States

Dairy production has distinct advantages, including the following:

1. The dairy industry is stable, with little (1%–2%) variation from year to year.
2. The dairy cow is amazingly efficient, producing 25,000 lb of milk per year with multiple productive years.
3. Dairying provides a steady source of income monthly or biweekly. A grower producing grain, fruit, vegetables, or beef cattle receive income only when products are sold, usually once per year.
4. Dairying provides steady employment for hired labor in contrast to other agricultural work that is highly seasonal. This makes it possible to hire or keep better employees.
5. Dairy cows use a large amount of otherwise unusable forages.
6. By returning the manure to the land, soil fertility is preserved.

The following factors make dairy production a challenging profession to undertake:

1. High capital is required. The amount per cow is double that for a beef cow-calf enterprise.
2. Dairy management requires specialized education and training in business administration, labor management, animal physiology, genetics, and nutrition. This knowledge needs to be melded for a smooth-functioning, efficient production unit.
3. Producers must follow the federal, state, and local regulatory programs.
4. Dairying is labor intensive and provides little free time.

Manure is a mixture of urine, feces, and bedding.

Mastitis is an infection with inflammation of the udder.

Milking parlors are dedicated facilities where cows are milked. Cows are usually milked two times a day. Some are milked three times a day, increasing milk production by 20%.

Registered and grade cows: Registered cattle are pure-bred cattle (a single breed) that are registered along with the sire and dam with a breed association. Grade cattle are nonregistered or cross-bred cattle.

Sire: Male parent.

Superovulation is when cows are induced to develop multiple follicles in the ovary and then to ovulate these.

Transition is the last 2 weeks of pregnancy and the first 2 weeks of lactation.

Volatile fatty acids (acetic, propionic, and butyric acid) are produced by fermentation by microorganisms in the rumen.

TABLE 7-7 Characteristics of different major breeds of dairy cattle

BREED	ORIGIN OF BREED	MATURE SIZE (lb)	COLOR	NUMBER ON DHIA* TEST (ANNUAL RATE OF CHANGE)	MILK PRODUCED/ Y(lb)
Ayrshire	Scotland	1,200	Brown and white	6000 (−10%)	17,400
Brown Swiss	Switzerland	1,400	Solid dark brown or gray	15,000 (−8%)	20,300
Guernsey	Channel Islands (in the English Channel)	1,100	Tan and white	10,000 (−15%)	16,000
Holstein (Holstein-Friesian)	Holland	1,500	Black and white	3.9 million (−1%)	24,500
Jersey	Channel Islands	950	Light gray to dark, fawn to nearly black with dark face	160,000 (+2%)	17,000
Milking Shorthorn	Western Europe	1,350	Red, white, red and white, or roan	<5,000	16,700

DHIA = dairy herd improvement association.

FIGURE 7–11 Representatives of dairy breeds.

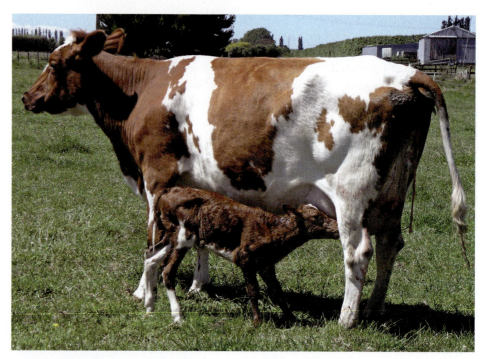

A. Ayrshire cow. Courtesy of the USDA. Photo by Bill Tarpenning.

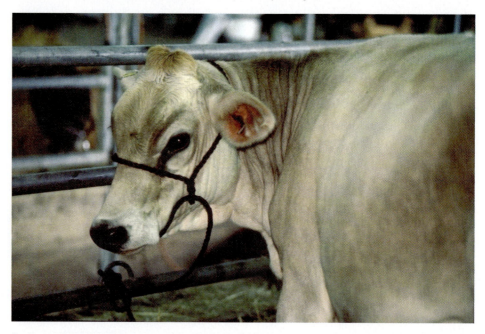

B. Brown Swiss heifer. Courtesy of the USDA. Photo by Bill Tarpenning.

5. Hourly returns to dairy farmers are below returns in other types of farming and for manufacturing.

   Dramatic changes have occurred in the U.S. dairy industry, including the following:

1. Total milk production has increased 30% in the last 40 years.
2. Between 1950 and 2008, the number of dairy farms declined by over 90%, with average herd size increasing by over 10-fold.

C. Guernsey cow. Courtesy of the American Guernsey Association.

D. Holstein cows in the foreground and a Guernsey in the background. Courtesy of the USDA. Photo by Ken Hammond.

3. Between 1960 and 2008, the number of milk cows declined by 50% to about 9 million cows because of increased production per cow (increasing from 3,138 lb in 1920 to over 18,000 lb per cow in 2008). The increase in production per cow is due to improved genetics (made possible by AI and record-keeping systems such as dairy herd improvement [DHI]), nutrition, disease control, and management.

4. The number of farms with more than 200 cows is increasing dramatically.

E. Herd of Jersey cows. Courtesy of the USDA. Photo by Bill Tarpenning.

F. Milking Shorthorn cow. Courtesy of the American Shorthorn Association.

These changes are driven by changes in the economics of dairying. Tight margins with increased labor, land, equipment, and utility costs coupled with relatively static feed costs and milk prices encourage producers to spread expenses over more animal units.

Dairy farmers depend on new technologies to remain competitive. The top milk-producing states are listed in Table 7-8.

TABLE 7-8 Top 10 states for milk production in the United States (2008)

RANK ORDER OF STATES BY MILK PRODUCTION		MILK PRODUCTION (BILLION LB)
1	California	41.2
2	Wisconsin	24.5
3	New York	12.4
4	Idaho	12.3
5	Pennsylvania	10.6
6	Minnesota	8.8
7	Texas	8.4
8	New Mexico	7.9
9	Michigan	7.7
10	Washington	5.7

Source: USDA National Agricultural Statistics Service.

The dairy industry in the western states has experienced tremendous growth over the last 25 years. These have large intensive dairy farms. Many dairy farms in the major dairy areas, especially the lake states, corn belt, and northeast, continue as smaller, more diversified enterprises. Dairying is also driven by the availability of processors.

The USDA estimates that the average dairy cow is worth $1,207, and the average cost to produce 100 lb of milk is $11.84. Feed is the largest cost, accounting for about 55% of the total. Other expenses include bedding, veterinary bills, utilities, breeding fees, DHI associations' expenses, transportation costs, insurance, labor, depreciation on equipment and interest on land, and other costs. Dairying requires the investment of large amounts of capital for land, buildings, silos, machinery, milking equipment, and the animals.

Dairy Management

Record systems are the most important management tool for the producer. They provide a decision-making tool on individual cows, such as how much concentrates to feed each cows, when to breed, when to dry off, and which cows to cull. The National Cooperative DHI program is a voluntary effort to improve milk production and increase dairy profits. State and local DHI associations conduct the program with producers working through Cooperative Extension in cooperation with the USDA. Half of the cows in the United States on production test are part on this program. Records (monthly and cumulative) for individuals and herds include milk production, milk fat and protein, amount and cost of feed, breeding dates, calving dates, and dry dates. Some associations provide somatic cell counts or results of the California Mastitis Test to monitor udder health. Information is provided to producers.

Selecting Herd Sires

A dairy producer has three sources of sires: AI service, purchase of herd sires, or raising herd sires. The producer also decides between proven sires or young sires. The challenge is to select herd sires that will ensure genetic improvement in the herd. Because bulls do not produce milk, there is no direct measure of a sire's individual performance. Evaluation is instead based on the performance of his daughters and from records of all identified relatives in the evaluation. The selection of a sire is extremely important because he will be the parent of many more offspring than an individual cow. A superior sire is responsible for 90% of genetic improvement in a herd. The most reliable source of superior genetics is using AI, and the most frequently used approach is with semen from proven bulls. These are evaluated for a large number of traits such as milk yield, conformation, and calving ease.

Feeding Dairy Cattle

Feeding needs to meet energy and protein requirements. The nutritional requirements of high-producing dairy cows are summarized in Table 7-9. In early lactation cows, the amount of high-moisture feed is reduced as volume of intake is limiting.

TABLE 7-9 Nutrition guidelines for high-producing herds

DM intake	4–5% of body weight
NDF	26–30% of DM
Nonstructural carbohydrates	35–40% of DM
Fat	5–7% of DM
CP	17–19% of DM
Degradable protein	60–65% of CP

Note: CP = crude protein; NDF = neutral detergent fiber.

Dairy Cattle Reproduction

The cow is the primary focus of reproductive management on most dairy farms, with the bull the second part of the equation.

Reproductive Management of the Male

Semen is collected for AI by collection into an artificial vagina, or electro-ejaculation techniques or rectal massage of the ampullae. Bulls are trained to mount dummies for semen collection. Natural mating is considered in the earlier reproduction section for both dairy and beef cattle.

Reproductive Management of the Female

The key to reproductive management is detection of heat or estrus so that insemination occurs at the optimal time, with the greatest probability of an ovum being fertilized and pregnancy resulting. Estrus detection is covered previously in the reproduction section for both dairy and beef cattle.

Assisted Reproductive Technologies

Natural service, the actual mating of a male and a female, is no longer common with dairy cows. Most cows (over 65%) become pregnant by AI. There are also other assisted reproductive technologies, including multiple ovulation and embryo transfer programs, in vitro fertilization, and cloning. In AI, spermatozoa are deposited into the female genitalia (see fig. 7-12). Genetic progress using AI

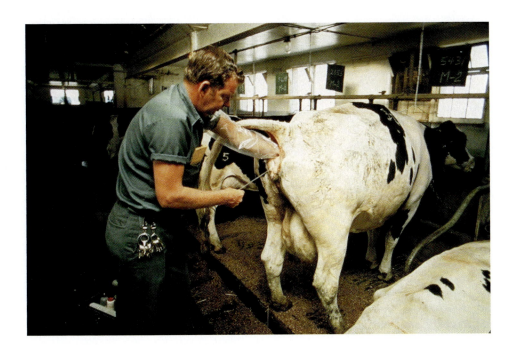

FIGURE 7-12 Artificial inseminator places semen into the female reproductive tract of a cow. Courtesy of the USDA. Photo by George A. Robinson.

sires has been phenomenal. The AI bulls are now advancing cows' genetics at the rate of nearly 250 lb of milk per year.

Embryo Transfer

Embryo transfer is a process in which embryos are transferred from a valuable (genetically superior) donor cow to less valuable recipients. Full-term pregnancies result in offspring with the genetic traits of the donor cow and the bull. Recipients have no genetic influence on the calves they carry. Embryos that are not immediately transferred can be frozen for later transfer. Typically, ethylene glycol or glycerol is used as a cryoprotectant. The embryos can then be thawed and transferred when a suitable recipient is available. Pregnancy rates are not as good after freezing and thawing compared with fresh embryos.

The ability to mature and/or fertilize ova in a culture dish (in vitro) allows even more control over the reproductive processes. Infertile or prepubertal animals can reproduce using such procedures. Immature oocytes are aspirated from the follicles of superovulated animals via guidance by transvaginal ultrasound imaging. These oocytes are cultured for about 24 hours; at maturity, they are fertilized by the addition of capacitated sperm. The fertilized oocytes are cultured further to the early blastocyst stage, when they are either transferred to recipient animals or frozen for later transfer (see fig. 7-13). The cloning of an animal is the production of an exact genetic copy. Identical twins could be called clones because they are derived from a single cell splitting to yield two genetically identical copies. Cloning has been successfully developed with dairy cattle for research purposes. This approach may be used in the industry.

Mammary Gland

The dairy cow produces large amounts of milk. A cow producing 29,000 lb of milk in a year is producing 1,046 lb of milk fat, 1,348 lb of milk sugar, and

FIGURE 7-13
Cryopreservation (ultralow temperature storage) of livestock embryos. Courtesy of Photo Unit, USDA Agricultural Research Service.

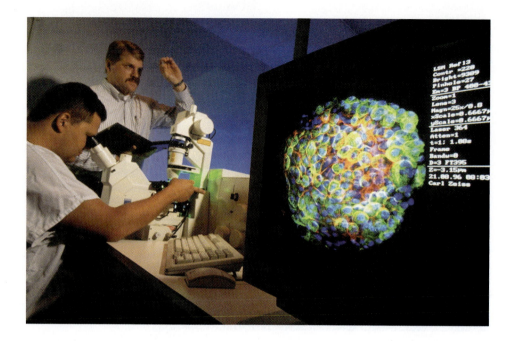

954 lb of milk protein. This is equivalent to the carcass weight gain by five steers in 18 months. Mammary anatomy and physiology are discussed in Chapter 18. Immediately before the birth of the calf, colostrum synthesis is initiated.

Milking Procedures and Processes

Milking is the act of removing milk from the udder. The streak canal must be opened to remove milk from the udder. During machine milking, opening is accomplished by negative pressure. During hand milking, positive pressure is applied to the teat cistern, forcing milk through the streak canal. During suckling, both positive and negative pressures are applied. The milk ejection reflex allows removal of most of the milk in the lumen of the alveoli and smaller ducts. Myoepithelial cells surrounding the alveoli contract in response to oxytocin. Suckling and milking stimulate, causing milk ejection by increasing oxytocin release from the pituitary gland. About 15% of the milk in the udder is not removed when milking is completed.

The bucket and the pipeline systems are two types of milking systems. In the first, the milk is received directly into a vacuumized portable bucket. This may be either of two types: a floor type, or a suspended type. Conventional pipeline systems use a rigid heat-resistant glass or stainless steel sanitary pipe to carry vacuum from the milk receiver to the individual milking units and the milk from the units to the receiver. Pipeline milkers are used in stanchion and tie-stall barns or milking parlors. The mechanical milking systems are separated into three major parts: vacuum supply, pulsation, and milking unit.

Milking is the most important single job on a dairy farm. Cows are milked at regular times, preferably by the same milker. Cows like to be milked, if it is done properly. The release of milk is a delicate process, requiring close cooperation between the milker and the cow. A milking program consists of the following coordinated steps:

1. Preparation. The equipment should be assembled, checked, and sanitized.
2. Stimulating milk letdown and cleaning the udder. The cow is stimulated by cleaning the cow's teats with warm water (49–54° Celsius [C] or 120–130°F) or a predip, then massaging and drying.
3. Checking for mastitis. Two or three streams of milk are removed from each quarter into a strip cup and examined for visible evidence of mastitis. This process also stimulates milk letdown. Full milk letdown requires 10 seconds of teat stimulation.
4. Attaching the teat cup. About 45–60 seconds after washing the udder, the teat cups are attached and milking begins. Most cows milk out in 3–6 minutes, depending on the amount of milk and the individual cow. Both incomplete milking and overmilking should be avoided. Automatic takeoffs control the timing of removal by electronically monitoring the rate of milk flow and automatically shutting off vacuum to the milking unit when milk flow drops below a preset flow rate. The unit is then retracted away from the cow after vacuum levels at the teat end diminish, minimizing stress on the teat.
5. After milking. The teats are dipped in a fresh disinfectant solution to prevent the invasion of bacteria into the udder and to avoid attracting flies. After milking the last cow, all milking equipment is thoroughly cleaned and sanitized.

Milk Quality

Milk quality determines consumer satisfaction. A bad experience with poor-quality milk can result in the loss of a customer for life. Quality milk is produced when the producer pays special attention to the following:

1. Herd health. The herd should be free from diseases that might be spread to people through the milk. Bacteria in milk cows must be eliminated. Mastitis is the most important herd health problem.
2. Clean animals. It is important to have clean flanks and udders to prevent dirt from getting into the milk.
3. Clean equipment. Milking equipment is kept clean and free from bacteria.
4. Cool and store milk. Milk is cooled rapidly from 101–40°F and held at that temperature. Bacteria grow exponentially (1-2-4-8-16-32) by dividing. The doubling time varies with bacterium species and can be as little as 20 minutes. Each bacterium has an optimal temperature for maximum growth (often around 37°C or 98°F) and a lower temperature at which growth does not occur.

Regulatory Programs

These include federal, state, and local programs. Producers are issued permits allowing them to ship grade A milk. The permit is revoked if either the bacteria count or somatic cell count of raw milk exceeds current standards. By law, all fluid milk sold for human consumption must be pasteurized. At dairy-processing plants, milk is pasteurized to kill disease-causing organisms. Antibiotic residues are a major concern. Processors must dump any milk found to have antibiotic residues from tankers or storage silos. The financial consequences to producers can be severe. Producers are careful in adhering to the required withholding periods for antibiotics. On-farm tests are available to verify that milk is antibiotic free before leaving the farm. The Food and Drug Administration inspects dairy products and processing plants for contamination and adulteration.

Dairy Farm Biosecurity and Animal Health

Animal diseases cost producers over 15% of cash receipts. Health programs focus on sanitation, biosecurity, and vaccination. Vaccination programs are essential to herd health. Vaccines are costly and do not confer absolute immunity. Effective plans for reducing pathogen load and exposure can dramatically affect herd health. Biosecurity reduces the risk of introducing new pathogens to a dairy herd and the spread of pathogens within a herd. A good biosecurity program decreases culling rates, subclinical infections (lower milk yield and reproductive problems), and veterinary costs.

A biosecurity program reduces potential contamination. Sources of pathogens are visitors' cloths, boots or vehicles; shared or borrowed equipment between farms; purchased animals; or even a shared fence line with another cattle herd. All visitors should wear clean boots (or plastic boot covers) and pass through a decontaminating footbath before gaining access to cattle. Cattle should be quarantined for 30 days after arrival. This requires isolation from shared feeders, waterers, or other equipment. People can spread pathogens between production units on a farm or from quarantined animals to the herd. Because calves are susceptible to pathogens, their exposure should be minimized. Calf chores should be completed before working with other cattle on a farm to prevent transfer of pathogens from cows to calves. Feeding colostrum is critical to optimize immunity for the first few weeks of life. However, colostrum and milk are potential vectors for infection (e.g., bovine lymphoma virus and Johne's disease). Diseased animals should be isolated to minimize the risk of disease transmission, and persistently infected animals should be culled.

Infections of the Mammary Gland

Mastitis is infectious inflammation of the udder that interferes with the normal flow of milk and/or its quality. It costs the industry more than $2 billion per year ($225 per afflicted cow) with decreased milk production and quality. The average mastitis infection rate is 35%. On average, every cow has two new infections per lactation. In acute mastitis, the udder is hot, very hard, and tender. The animal has an increase in temperature, dull eyes, and a rough coat together with a refusal to eat. Milk production is greatly reduced. The milk may be lumpy or watery. Death often occurs with untreated, acute mastitis. In chronic mastitis, there may be thick or lumpy milk. Mastitis may be either infectious or noninfectious. Noninfectious mastitis is the result of injury or bruising. Infectious mastitis results from the invasion of bacteria into the gland. Infectious mastitis may be subclinical (hidden) or overt clinical

---

DAIRY CATTLE HEALTH GOALS

Producers must address all the primary goals for dairy cattle health:

1. Minimize the use of antibiotics.
2. Minimize pathogen load to the animal. Sanitation is very important to maintaining animal health.
3. Maximize the immune function of the animal using a vaccination program specific for the pathogens associated with the herd and with the geographic region.
4. Minimize the potential for disease transmission between animals and by people.
5. Obtain veterinary assistance to get early diagnosis and treatment of diseases.
6. Monitor animal health by production records, body temperature, heart rate, etc.

infection. In subclinical cases, bacteria are present in the gland, but both the udder and milk seem normal. There are between 15 and 40 cases of subclinical mastitis for every case of clinical mastitis. The losses from subclinical mastitis account for two thirds of the losses from mastitis.

Dairy Cow Behavior

Knowledge of cattle behavior ("cow sense") is necessary for the successful management of dairy cows. Dairy cattle behavior has received less attention than the quantity and quality of milk produced. Recently, there has been renewed interest in behavior, especially as a factor in obtaining maximum production. With the increasing confinement of cows, some abnormal behaviors have emerged, including loss of appetite, pica, stereotyped movements, poor maternal behavior, and excessive aggressiveness.

Social behavior is a behavior caused by or affecting another animal, usually of the same species. Social organization is a stable aggregation of individuals into a group based on the interdependence of the separate animals and on their responses to one another. The social structure of a dairy herd is of practical importance. When several cows are brought together to form a herd, there is a substantial period during which there is much butting and threat posturing to establish a dominance hierarchy. This is disruptive to a dairy herd and results in reduced production. The more aggressive cows are most dominant or have a higher social rank. Older cows generally dominate the younger. Heavier animals dominate lighter ones, and cows with horns tend to be of higher social rank. The higher the social rank, the more likely they are to be near other members of the herd. The dominant individuals crowd the subordinate ones away from the feed bunk, reducing feed intake, growth rates, and production. When moving from the paddock to the milking parlor, dairy cows travel in a consistent order. Mid-dominant cows tend to be in front of the group. The animals at the rear are usually the younger, subordinate heifers.

Communication involves an individual giving a signal to influence the behavior of others.

1. Sound. Cattle have a very acute sense of hearing, so sound is an important method of communication among them. Sounds express hunger (bawling), distress calls like the bellowing of a bull, sexual behavior, and mother–young interrelations to evoke care behavior.
2. Odor. Cattle can smell at a greater distance than people, even up to 6 miles away. In cattle, females in estrus secrete a substance (pheromones) that attracts males. Bulls locate cows that are in heat by their sense of smell.
3. Visual displays/sight. The wide-set eyes of cattle enable them to have a large panoramic field of vision. When several strange cows are brought together, there is threat posturing as well as butting to establish a dominance hierarchy. In addition, bulls strike a hostile stance before fighting.

Welfare Issues in the Dairy Industry

Dairying has not received sustained criticism by animal rights activists. The welfare of cows leads to improved production and, therefore, income. Management

MASTITIS REDUCTION

Mastitis can be reduced by management. A milking program should embrace the following:

1. Injuries to the udder and teats predispose to mastitis. One preventive is dry bedding.
2. Clean udders and teats are essential to the production of clean milk of high quality.
3. Proper milking procedure.
4. Post-milking teat dip.
5. Keep milking machines in good operating condition at all times.

systems that reduce labor or housing costs may result in animal problems. However, by reducing behavioral and environmental stresses, it is possible to ensure that lower labor and housing costs are not offset by losses in productivity. Raising young calves for veal is a scrutinized practice. It involves several emotional "hot button" issues, namely, removing young animals from their mothers, raising young animals in confinement to produce white veal (with low iron intake), and slaughtering these young calves for their meat.

Separating the calf and dam at birth may seem inhumane. Terminating the bond between mother and offspring soon after birth is not acceptable to some people. The act of separation is both a management issue and a welfare issue. Cows that are separated from their calves at birth are more easily integrated into the milking herd. They let down their milk more readily than those associated with their calves. Most calves are raised in individual housing (see fig. 7-14), limiting their opportunities for socialization with other calves while limiting their exposure to pathogens.

Cows are increasingly spending extended time in confinement on concrete. Housing systems are being improved to meet the behavioral needs of the cow and the economic needs of the producer. Concrete floors can adversely affect hoof health. The ideal size and design of stalls are still being researched. Another issue is the handling of downer animals (cows that cannot walk) because more of these downers originate from the dairy industry. Downer cows should not be used for meat. Instead, they should be euthanized humanely on the farm rather than transporting them to a slaughter facility. The lost income is of little consequence when compared with the damage done to the public image of the dairy industry.

Tail docking is another management practice that has come under close scrutiny. Tails are amputated for the comfort of milkers because it prevents them being hit by manure-laden cows' tails. The practice of removing tails does present issues with fly control for the cows. With suitable sanitation and facility design, the procedure is unnecessary.

FIGURE 7-14 Holstein calf at organic dairy farm. Courtesy of the USDA. Photo by Bill Tarpenning.

Animal Stress

Among the factors that stress cattle are changes in feeding, water, space, housing, or herd mates; the number of animals housed together; transportation; presence of strangers; fatigue; infection; management; weaning; temperature; and abrupt weather changes. Cows are very sensitive to any changes in their environment or routine, or even changes in the people who are milkers. Problems with an animal's surroundings, including confinement, can lead to abnormal or unusual behavior. Homosexual behavior is common among all species where adult mammals of one sex are confined together. Pica (consumption of dirt, hair, bones, or feces) may develop perhaps because of boredom, nutritional inadequacies, or physiologic stress. Milk cows may kick because they are in pain, are frightened, or have been mistreated.

Dairy Cattle Facilities

Facilities that provide optimal environments are critical to milk production and for cattle to meet their genetic potential for production. Buildings are becoming progressively more important with the shift to high-density production. Modern dairy cattle buildings and equipment are designed for the comfort, health, and productivity of cows. In addition, buildings and equipment should minimize maintenance needs, labor requirements, bedding costs, and facilitate manure handling. Together, these considerations lead to profitability.

The most critical considerations for optimizing cow performance are ventilation; quality and accessibility of feed and water; lighting; and cow comfort, including both stall configuration and bedding quality.

To minimize labor costs, the facility should be designed for cow movement and minimize the time needed for daily procedures such as feeding, bedding, and milking.

The primary reason for livestock buildings is to modify the animal's environment. Barns and other shelters may have insulation, ventilation, and air conditioning, and, therefore, approach an ideal environment. Environmental control is becoming more common for dairy facilities. In hot climates, the use of shades, fans, sprinklers, sprayers, foggers, mechanical ventilation systems, and windbreaks is increasing. The cost per head is higher for environmentally controlled facilities. Environmental control can be justified by the economics with the cows producing more milk on less feed.

Housing Systems at Different Life Stages

Calf housing must be clean, dry, and well ventilated at all times. Calves have strict requirements for ventilation because they are susceptible to respiratory problems when ventilation is not adequate. Strict isolation is needed to prevent pathogen transfer from sick to healthy calves. Some preweaned calves are housed in indoor facilities, others in outdoor facilities, but these expose feeders to severe weather. Calves should be housed separately (in individual pens or huts) from birth until at least 1 week after weaning. After weaning, they may be raised in groups of six to eight calves. Each group should feature a maximum age difference of 2 months and a weight difference of 100 lb. Facilities should keep heifers dry at all times. Replacement heifers do not need a closed barn. Housing should protect them from drafts, rain, snow, and winds. Artificial shade is required in hot climates if natural shade is not available.

The two systems for cows are stall barns and loose housing. Similar milk production can be achieved with either system. The stall barn allows the cows to be observed more easily. Loose housing systems require less labor, bedding, and construction costs. Adequate ventilation is critical for cow productivity with the number of air changes set to ensure moisture removal. The removal of large quantities of water, especially in the winter when barns are closed, can be challenging. Sidewalls are typically open, with curtains to control airflow. Building free-stall facilities on an east–west orientation reduces radiant energy load when compared with barns built with a north–south orientation. Two- or four-row barns have improved ventilation and increased bunk space per cow compared with three- or six-row barns.

For larger herds, cows are grouped to permit faster movement of each group through the milking parlor. This reduces stress, increases feed intake, and improves profitability. Group sizes depend on parlor size and parlor throughput. Each group is a multiple of the parlor size and can be milked in less than 1 hour.

Milking Facilities

There are two system of milking: milking parlor and stall milking. Free stalls and loose housing arrangements lend themselves to milking parlors. Cows housed in stanchion or tie-stall barns are milked at the stall.

The holding area confines cows before milking. It should be paved, easy to clean daily, funneled to the parlor, and well ventilated. The milking parlor (see fig. 7-15) improves labor efficiency, working conditions, and sanitation surrounding the milking operation. When milking is done by hired labor, parlor size should maximize

FIGURE 7-15 Holstein cows in milking parlor. Courtesy of the USDA. Photo by Bill Tarpenning.

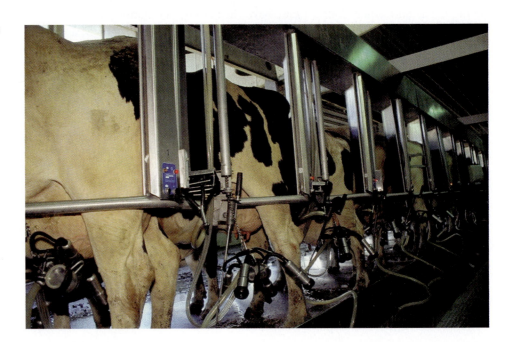

usage (to more than 20 hours per day) to give a maximum return on the investment. In farms with limited labor availability, such as those that rely strictly on family labor, parlors are designed to minimize time spent milking. Parlor size depends on the number of cows to be milked and the number of daily milkings. Each milking can take up to 10 hours for twice a day milking and 6.5 hours for three times a day milking. This leaves time for special-needs cows and parlor cleaning.

Assessing Cow Comfort

Assessing this is a primary management area because comfort affects feed intake, growth rates, milk production, herd health, and reproductive efficiency. Factors critical to maintaining cow comfort include adequacy of stalls, choice of bedding, feed and water accessibility, quality of walking surfaces, adequacy of ventilation, heat abatement, adequacy of lighting, and control of stray voltage. Indirect assessment of cow comfort can be made from ventilation air changes, stall dimensions, and bedding quality. Although these measurements are valuable, our observations should not be limited to measurements. The most accurate indicators are the cows themselves. Comfortable cows are eating, drinking, milking or waiting to be milked, moving to a specific destination, exhibiting estrus, or lying down and ruminating. In ideal conditions, cows lie down for more than 14 hours daily. Herds with a high percentage of cows that are standing are herds with a cow comfort issue.

On pasture, cattle lie into a slope and lay uphill. Cattle typically lie on the stomach (sternal recumbency), or they may tilt to one side, with the forelimbs folded under the body, and one hind limb extends forward, while the other sticks out. When rising, the cow must lunge forward, using the knees as a pivot point. This motion transfers much of its weight off the rear legs. It allows the animal to raise its hindquarters then rise to a full standing position.

Stall Design

The ideal stall design integrates an understanding of cow behavior, cost considerations, ventilation, and labor requirements for maintenance. Critical considerations in stall design include the provision of a spacious yet defined resting space, adequate lunge room so the cow can easily rise, and an adequate amount of bedding. Reductions in lying time by as little as 2 hours affect performance. The choice of bedding is important. Sand is often best. Other choices include straw, shavings, rice hulls, and shredded newspaper. Bedding provides adequate cushioning and reduction in friction, reduces microbial growth, and removes moisture from the cow. In facilities with well-designed and well-maintained stalls, more than 80% of cows that are not eating or drinking are lying down. This reduces hock lesions and joint swelling, cows are cleaner, the incidence of mastitis is lower, detection of estrus is easier, and milk production is higher.

Manure Management Practices

Manure management is an important part of dairy production. The collection, transport, storage, and use of manure must meet sanitary and pollution control regulations. Nitrogen and phosphorus are the plant nutrients of primary concern to meet land application regulations. These are directed to the quality of surface water and groundwater. Phosphorus is the surface water contaminant of highest concern. Phosphorus stimulates the excessive growth of algae and eutrophication of surface waters. Rain or snowfalls onto manure can cause severe pollution in streams or lakes. Runoff should be kept from reaching waterways. Many states also regulate air quality.

Gases released from manure include carbon dioxide, ammonia, and methane together with gases with an unpleasant odor such as hydrogen sulfide. Odors are an issue only if they affect the quality of life of the people exposed to them. Proper facility design, especially the location of manure-handling and storage facilities, can minimize problems associated with nuisance odors. The direction of prevailing winds should be considered when selecting a site for manure storage. Windbreaks can be used to diffuse and deflect the airborne emissions away from populated areas. Mechanical aeration in lagoon systems can reduce some odors. Injecting manure directly into soils rather than surface application helps minimize odor emissions and problems to groundwater. Fly populations can be dramatically affected by the manure-handling and storage practices.

When stored inside a building, gases from liquid wastes create a potential health hazard. Animals and people can be killed (asphyxiated) because the odorless gases, methane and carbon dioxide, displace oxygen. No one should enter a storage tank unless the space over the wastes is first ventilated with a fan and another person is standing by to give assistance. Methane emissions also pose a threat to the environment because greenhouse gases contribute to global warming.

 ## MANAGEMENT OF DAIRY CATTLE

Managing Preweaned Calves

The first day of life is the critical period; it can affect the lifetime profitability of an animal. Optimal management improves feed efficiency, rate of weight gain, and health throughout the rearing period, and allows optimal milk production during the first lactation. Calf management during the first day of life focuses on reducing the stress from the birth process and maximizing passive immunity.

About 80% of calf deaths are directly related to calving difficulty. There is a scoring system for calving ease used. Scores range from one, indicating no assistance was provided, and a score of five, indicating extreme difficulty or dystocia. High calving-ease scores are associated with calf mortality. With a calving-ease score of five, about 50% of calves are stillborn, and many of those surviving die within 48 hours. All calvings should be monitored. The position of the calf should be determined to ensure the correct presentation, front feet and head first. When this is not the case, it should be corrected early. It is easier to reposition early than after several hours of uterine contractions. With uncorrected posterior presentation (hind feet first), the delivery is almost certainly going to be difficult. There is considerable risk of the calf suffocating due to the premature rupture of the umbilical cord or by strangulation. If the position of the calf is normal, assistance should be provided when the cow appears to be in distress.

Following birth, calves should be removed immediately from their dams to a thermoneutral environment (between 15°C and 24°C or 60°F and 75°F). After breathing has started, calves should be dried, the umbilical cord should be dipped in antiseptic/bacteriocidal solution (e.g., 7% iodine or chlorhexidine), and high-quality colostrum should be provided. Because the bovine placenta is not permeable to maternal antibodies, passive immunity comes from antibodies in the colostrum crossing the intestinal wall of the newborn calf. Failure of this results in mortality rates in excess of 50% and long-term impairment of productivity of survivors. The first six milkings from fresh cows are considered colostrum for marketing and cannot be sold. However, the most important colostrum for the newborn calf is the first milking.

Managing Replacement Heifers

Cows in the United States are generally milked for less than 4 years before they are culled. For a 100-cow dairy, 25 first-calf heifers must be available to replace these culled cows each year. Not all heifers become cows. About 20% of calves die before reaching maturity, and others are culled. To maintain the status quo in a milking herd, the dairy producer replaces one third of the cows each year. Cows are culled if they are less profitable, or have health problems or reproductive failure.

Replacement heifers enter the milking herd at a young age when there are no negative effects on health or lactational performance. This requires fast growth. Well-grown heifers are bred at 14 months. Early breeding of heifers is more profitable because it shortens the time from birth to lactation, lowering the cost of feeding and managing a nonproducing heifer, and lifetime production is greater. Early calving permits faster genetic progress. Heifers should be raised to maximize profit. The total cost of raising replacement heifers to 24 months is about $1,200.00. Managers can decrease rearing costs by $200.

Managing Dry Cows

During the dry period when lactation is not occurring, there are important changes to the biology of the cow, including regression of old alveolar (milk-producing) cells and then regeneration of new alveolar cells, development of the fetal calf, and the reconditioning of the digestive tract after the rigors of fermenting large amounts of the high-concentrate feeds used during lactation. Nutrition is very important in late lactation and during the dry period. Both at the drying-off time and calving, the BCS should be about 3.5. Lactation is terminated (or cows are "dried off") 30–45 days before the expected calving date, otherwise the next lactation is reduced by about 25%. Roughage intake is important during the dry period to stimulate muscle tone of the rumen. During the last few weeks, the rumen is prepared for lactation. DM intake is reduced by about 30% immediately before calving.

Managing Metabolic Disorders in Transition Cows

Management during this critical period can determine profitability for the entire lactation. Transition cows have several major hormonal and metabolic changes suppressing immune function and increasing susceptibility to metabolic disorders.

Maximizing intake of a properly balanced ration in the early lactation period is critical to minimize these problems.

Managing Lactating Cows

Optimizing milk production entails maximizing milk production in early lactation, but not necessarily during late lactation. In early lactations, rations should maximize DM and nutrient intake. Rations are designed to maximize DM and nutrient intakes. The high-producing early lactation cow cannot fully meet the energy needs of milk production through feed intake. There is a negative energy balance with mobilization of body reserves and a loss of the BCS. Dairy cattle should lose no more than one point in their BCS during the early lactation period. Ideally, cows should calve at a BCS of about 3.5–3.75. The key factors affecting the rate of BCS loss are milk production and intake of nutrient-dense ration. Fresh, clean water should be readily available. Cows consume about 4 lb of water for every pound of milk produced. In hot, dry environments, this amount increases dramatically.

In mid lactation, cows need to reach a positive energy balance. At this time, bovine somatotropin (marketed under the name Posilac) can be used to increase milk production and its efficiency. The feeding program shifts to preparing the cow for the next lactation together with milk production in the current lactation. The effectiveness of the program can be judged by milk production and the increase of the BCS. When the cows reach a positive energy balance, they are ready for reproduction.

In late lactation, cows are fed to maintain milk production (persistency of lactation) without gaining excessive weight. Cows are dried off at a BCS of 3.5.

REVIEW QUESTIONS

1. What are the following: a cow, bull, polled cattle, steer, and heifer?
2. From what species were cattle domesticated, and where did this occur?
3. What is the family of cattle?
4. When were cattle introduced to the Americas?
5. What was the Chisholm Trail?
6. How many chromosomes do cattle have?
7. What is estrus?
8. How does a bull identify an estrous cow?
9. What are the behavioral changes of a cow at estrus?
10. How long is gestation in cattle?
11. What are the components of a cow's stomach?
12. What do the microorganisms in the rumen do that is important for the nutrition of cattle?
13. What are the major nutrients cattle require?
14. How many farms and ranches are selling beef cattle in the United States?
15. What is U.S. beef production: 25 million lb, 25 billion lb, or 25 trillion lb?
16. Name the three main types of beef cattle production systems.
17. Briefly describe each system.
18. What is the acceptable weight range for cattle being marketed?
19. What are the top three beef-producing states in the United States?
20. What are the desirable traits when selecting beef cattle?
21. Describe the kind of conformation that is desirable in beef cattle.
22. What are the major breeds of beef cattle?
23. Describe the characteristics of the major breeds.
24. When were the first cattle domesticated to provide milk for human consumption?
25. Why were dairy herds on farms very small between 1600 and 1850?
26. What accounts for the increase in dairy consumption in cities in the 20th century?
27. Give examples of technologies that impacted dairying.
28. What are *calving-ease scores* and *body condition scores* for dairy cattle?
29. What is artificial insemination?
30. What is colostrum?
31. What is dystocia?

32. What is eutrophication?

33. What is fat-corrected milk? What is the standard length of lactation in records?

34. What is manure?

35. What is mastitis?

36. What are milking parlors used for?

37. How many times a day are cows milked, and what is the impact of this on milk production?

38. What is superovulation?

39. What is transition?

40. What are the six major breeds of dairy cattle in the United States? Why is one predominant?

41. Is dairying increasing or decreasing? Are there structural changes in the industry? Comment on the reasons for the changes in the industry.

42. What factors make dairying challenging?

43. Why has dairy production per cow increased from 3,138 lb in 1920 to over 18,000 lb per cow in 2008?

44. What is the dairy herd improvement (DHI) program?

45. How are sires selected?

46. How is semen obtained?

47. Give examples of male and female sexual behavior.

48. What is embryo transfer?

49. What causes milk letdown?

50. What are the major types of milking systems?

51. Why is biosecurity important?

52. What are the major Dairy Cattle Health Goals?

53. What is the impact of mastitis, and how can it be controlled?

54. What are important welfare/stress issues for dairy cattle?

55. What are the most critical considerations in housing to optimize cow performance?

56. What are the management issues for dairy cattle at the different stages of life?

REFERENCES AND FURTHER READING

Beef Production

Brady, C. (2008). *An illustrated guide to animal science terminology*. Clifton Park, NY: Thomson/Delmar Learning.

Field, T. G., & Taylor, R. E. (2007). *Beef production management and decisions* (5th ed.). Upper Saddle River, NJ: Prentice Hall.

Iowa Beef Center. *Economics & business.* Retrieved July 16, 2009, from http://www.iowabeefcenter.org

Otto, D., & Lawrence, J. D. (2000). *Economic impact of the United States beef industry*. Retrieved July 16, 2009, from http://www.beef.org/uDocs/Econ%20Impact%20Beef%20v2.doc

USDA Economic Research Service. *U.S. beef and cattle industry: Background statistics and information*. Retrieved October 22, 2009, from http://www.ers.usda.gov/news/bsecoverage.htm

USDA National Agricultural Statistics Service. *Livestock slaughter 2008 summary*. Retrieved October 22, 2009, from http://usda.mannlib.cornell.edu/usda/current/LiveSlauSu/LiveSlauSu-03-06-2009.pdf

USDA National Agricultural Statistics Service. *Meat animals production, disposition, and income: 2008 summary*. Retrieved October 22, 2009, from http://usda.mannlib.cornell.edu/usda/current/MeatAnimPr/MeatAnimPr-05-29-2009.pdf

Dairy Nutrition

Chase, L. E. (1998). Feeding programs to achieve 13,600 kg of milk. In *Advances in dairy technology* (5th ed.) (pp. 13–20). Western Canadian Dairy Seminar.

National Research Council (2001). *Nutrient requirements of dairy cattle* (7th ed. rev.). Washington, DC: National Academies Press.

Dairy Production

Tyler, H., & Ensminger, M. E. (2006). *Dairy cattle science* (4th ed.). Upper Saddle River, NJ: Prentice Hall.

USDA National Agricultural Statistics Service. *Dairy products 2008 summary*. Retrieved October 22, 2009, from http://usda.mannlib.cornell.edu/usda/nass/DairProdSu//2000s/2009/DairProdSu-05-28-2009.pdf

USDA National Agricultural Statistics Service. *2008 milk production*. Retrieved October 22, 2009, from http://www.nass.usda.gov/Statistics_by_State/Wisconsin/Publications/Dairy/anmkpd.pdf

CHAPTER 8

Pigs

OBJECTIVES

This chapter will consider the following:

- Domestication and classification of pigs
- History of pigs
- Biology of pigs
- U.S. pork production
- Structure of the industry
- Export markets
- Facilities for production
- Genetics and crossbreeding systems
- Production phases
- Animal waste/nutrient management programs
- Contract production
- Future of the industry

INTRODUCTION

Pigs have long been a major source of protein and other nutrients to the people of Eurasia, Africa, and, after the arrival of Europeans, the Americas. This is perhaps best illustrated by the expression "living high on the hog." The global significance of pigs is seen because the pig is one of the animals on the Chinese calendar and zodiac. The external anatomy of pigs is shown in Figure 8-1.

In this chapter, the terms *pig*, *hog*, and *swine* will be used to refer to a domesticated pig. A farmer rearing and breeding pigs will be referred to as a pork producer. The national organizations are the National Pork Producers Council (www.nppc.org) and the National Pork Board (http://www.pork.org). The National Pork Producers Council focuses on legislative and public policy issues, and can lobby the U.S. Congress. The National Pork Board administers the check-off program that funds promotion of pork with, for instance, the "The other white meat" campaign and also research to aid pork producers.

DOMESTICATION AND CLASSIFICATION OF PIGS

Pigs were domesticated from wild Eurasian pigs (*Sus scrufa*). Based on studies of DNA of present breeds of European and Chinese pigs together with populations of wild Eurasian pigs, it seems that domestication occurred independently in two distinct geographic locations—present-day China and the Fertile Crescent—with local wild Eurasian pigs. Domesticated pigs were selected for reduced size and manageability. They were excellent scavengers, living on human refuse and whatever else they could find. Their ability to have large numbers of offspring was another significant advantage.

HISTORY OF PIG PRODUCTION IN THE UNITED STATES

The first pigs were introduced to the New World in 1493 by Christopher Columbus, with the first pigs in what became the continental United States arriving

---

NOMENCLATURE AND TERMINOLOGY

An adult female pig is a *sow*, and an adult male is a *boar*. A young pig is a *baby pig* (*piglet* is used in the United Kingdom) and later a *growing pig*. A young female pig is a *gilt*, and a young castrated male is a *barrow*. The process of giving birth is *farrowing*. *Porcine* is an adjective referring to pigs.

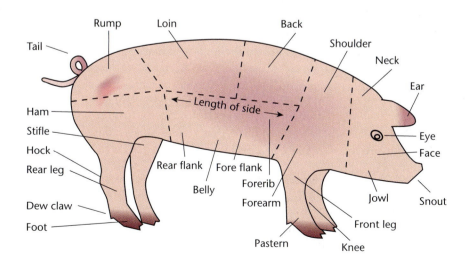

FIGURE 8-1 External anatomy of a pig. Source: Delmar/Cengage Learning.

in Florida with explorer Hernán DeSoto in 1539. Pigs were brought to what is now Hawaii more than 1,200 years ago by the Polynesians. Most farms had pigs in the 1700s and 1800s. With the development of the corn belt, pig production became more focused in the Midwest beginning in the mid-1800s. Escaped or abandoned domesticated pigs have become established as wild populations in the United States. There are many of these feral pigs in Southern states, with it estimated that there are 1.5 million in Texas alone. These feral pigs are hunted.

 ## BIOLOGY OF PIGS

Pig Genetics

Pigs have 38 chromosomes (18 pairs of autosomal chromosomes together with either XX or XY sex chromosomes). There has been tremendous progress mapping the pig genome with its 2 billion base pairs. In November 2009, an international team of researchers with Professor Larry Schook from the University of Illinois as project leader announced the draft sequence of the pig genome.

Pig Reproduction

Male Reproduction

Boars produce a very large volume of semen (100–500 mL, or up to a pint per ejaculation), which contains huge numbers of spermatozoa (10–100 billion per ejaculation). These are placed into the cervix of an estrous sow or gilt during mating. The boar identifies an estrous sow by smell. Collection of semen for artificial insemination (AI) is widespread. Semen can be collected using an artificial vagina or a gloved hand. The semen is evaluated for spermatozoa number, motility (percent moving), and morphology (appearance of the spermatozoa). In addition, semen with more than 25% abnormal spermatozoa or containing pathogens may be rejected.

Female Reproduction

Pigs are polyestrous with an estrous cycle of 21 days (range of 17–23 days). The duration of pregnancy or gestational length is 114 days (range of 112–115 days). A litter of about nine baby pigs is produced per pregnancy, with two litters possible per year. Pigs have a diffuse epitheliochorial placenta.

Mammary Glands

There are about 16 (between 12 and 20 pairs) mammary glands in the sow arranged in pairs along the underside (or ventral surface), from front (anterior or thoracic area), to middle (abdominal), to rear (or inguinal). Mammary anatomy and physiology are discussed in Chapter 18.

Digestive System and Nutrition

Pigs are omnivorous with a simple stomach. Feeds need to meet the carbohydrate, protein, fats, minerals, and vitamins of the animal. Ration formulation involves knowing the requirements, availability, quality, palatability, digestibility, and cost of each feed ingredient. In addition to feed, a critically important nutrient is water.

U.S. PORK INDUSTRY

The U.S. pork industry continues to experience rapid change and phenomenal growth as it strives to meet worldwide consumer demand for what has become the world's most popular meat product. The key to continued growth in the U.S. pork industry is the American pork producer, who works to become more efficient through modern production practices, state-of-the-art innovations, and scientific advancements in technology. U.S. pork producers have a strong commitment to the high standards of animal care and well-being, quality programs in food safety, and following good environmental stewardship practices. Advancements in management practices allow producers to develop safe and wholesome products with consistent flavor and superior eating quality. As the pork industry changes in scope and complexity, a major challenge to pork producers is to adapt and to continue to be profitable. Today's producers quickly implement new technologies and combine them with a hands-on approach to enhance production efficiencies. The U.S. pork industry remains an integral part of the U.S. economy, providing hundreds of thousands of jobs each year, and adding to the nation's wealth as it produces safe, high-quality, and abundant pork products.

STRUCTURE OF THE PORK INDUSTRY

The structure of the U.S. pork industry has changed over the past 30–40 years. Structural changes provide both challenges and opportunities to those in the industry.

Number of Producers

The number of farms in the United States raising hogs has shown a significant decline over the past 40 years (see fig. 8-2). Hogs were raised on a total of 4.9 million U.S. farms in 1920, and 3 million farms in 1950 (*Pork Industry Handbook*). In 1965, hogs were raised on over 1 million farms, and in 2005, hogs were raised on only 67,330 U.S. farms (*Pork Industry Handbook*). While he number of U.S. hog operations has declined, the average inventory per farm has continued to increase (see fig. 8-3). Economies of scale have driven farms to get bigger, and whereas the rate of decline in the number of farms has decreased slightly, there is little indication that the trend in reduction in the number of farms will end soon.

Size of Producers

The chart in Figure 8-4 outlines the percentage of operations in each of six different size categories and the percentage of the U.S. inventory that is included in each. Note that 76% of U.S. farms in the smaller farm category (less than 500 head) make up only 5.5% of the U.S. inventory, whereas the farms that market over 5,000 head per year (3.1% of operations) make up over half the U.S. inventory.

Where Pigs Are Raised

Figure 8-5 lists the percentage of market hog inventory in the United States from 1970 to 2005. Iowa has maintained its position as the leading state in pig production,

Definitions

Hogs are pigs in production systems.

Outdoor production systems and other alternative systems include hogs on pasture, hoop structures, and other systems. These require low capital investment.

Pork producers are farmers raising hogs (pigs).

Total-confinement, intensive production systems for hogs involve hogs raised indoors in controlled environment facilities. These require high capital investment.

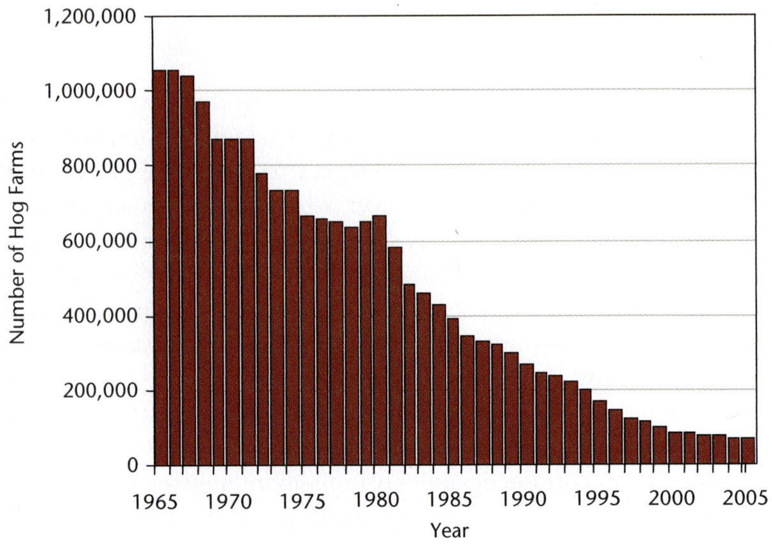

FIGURE 8-2 Changes in the number of hog farms in the United States over 40 years. Source: U.S. Department of Agriculture National Agricultural Statistics Service.

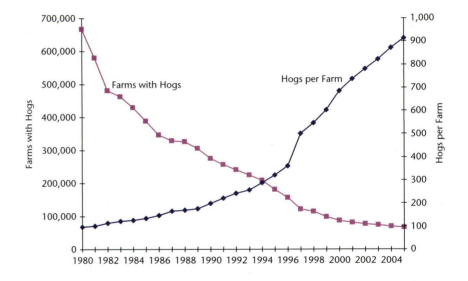

FIGURE 8-3 As the number of farms with pigs has declined over time, the number of pigs per farm has gone up. Source: U.S. Department of Agriculture.

producing 25–30% of market hogs in the United States each year. North Carolina had a rapid increase in pig production in the 1990s and has been the second-leading hog state over the past decade.

Table 8-1 lists the market hog inventory and number of sows farrowing in the 10 leading hog states in 2008. Iowa has a significant lead in the number of market hogs produced, followed by North Carolina and Minnesota. Iowa also is the leading state in the number of sows farrowing.

Trends in Production

U.S. pork production has increased at the rate of approximately 1.5% per year over the past 75 years (see fig. 8-6), and annual production per sow in total pounds of pork produced has increased at the rate of 3.5% per year since 1960

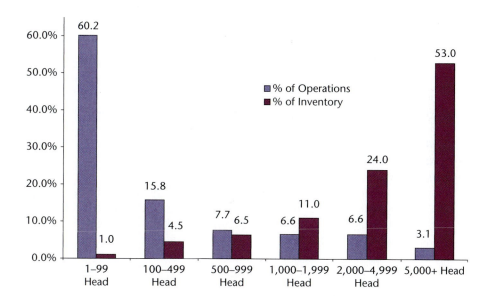

FIGURE 8-4 Size of hog operations in the United States can be seen from the percentage of hog operations and inventory by size in 2006. Source: U.S. Department of Agriculture.

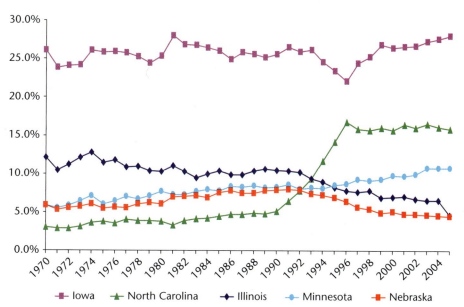

FIGURE 8-5 Changes in pork production in selected states in the United States. Source: U.S. Department of Agriculture.

(see figs. 8-7 and 8-8). The trend in annual production per sow is due to increases in the number of pigs weaned per litter, litters per sow per year, and market weights.

 ## EXPORT MARKETS

The ability to expand exports has been one of the pork industry's greatest success stories, and export growth occurred alongside increases in production and productivity. There have been increases in annual pork exports from 1990–2006 for 16 consecutive years. More than any other single factor, this trend provides strong evidence of the global competitiveness of the U.S. pork industry (see fig. 8-9).

ISSUES *for discussion*

1. What are the reasons that the number of farms raising hogs has declined in the United States, whereas production has increased with farms that market more than 5,000 head per year producing over half the hogs?
2. What is the impact of this on the competitiveness of U.S. pork production globally?
3. What is the impact on U.S. exports?
4. What is the impact of this on rural communities?
5. What are the alternatives?
6. What are the consequences of having fewer but larger producers of hogs?
7. What are the advantages and disadvantages of contract production? Include in your discussion the various affected groups: pork producers, communities, banks, consumers, and packers.
8. What are the consequences of the trends in pork production?

TABLE 8–1 Ranking of states in 2008 by pig production (in billion lb) and inventory of breeding sows (in thousands)

RANK	STATE	MARKET INVENTORY	RANK	STATE	SOWS FARROWED
1	IA	9.4	1	IA	1,070
2	NC	4.2	2	NC	980
3	MN	4.1	3	MN	580
4	MO	1.8	4	IL	490
5	IN	1.8	5	OK	400
6	IL	1.7	6	MO	370
7	NE	1.4	7	NE	370
8	OK	1.3	8	IN	280
9	KS	1.0	9	OH	170
10	OH	1.0	10	KS	165

Source: Data from the U.S. Department of Agriculture.

FIGURE 8–6 Changes in pork production by the United States. Source: U.S. Department of Agriculture.

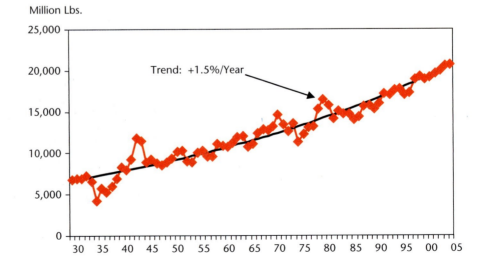

Pounds

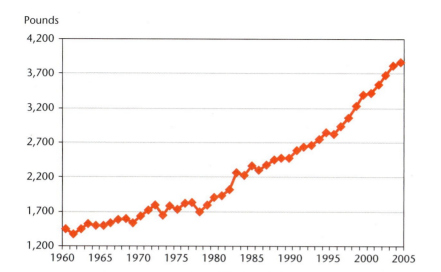

FIGURE 8–7 Changes in sow productivity according to annual U.S. production per sow. Source: U.S. Department of Agriculture.

FIGURE 8–8 Baby pigs suckling from sow. Per sow productivity has increased due to the number of offspring produced together with the higher growth rate. Courtesy of the U.S. Department of Agriculture. Photo by Ken Hammond.

Million Lbs.

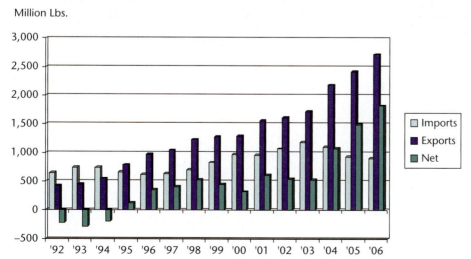

FIGURE 8–9 Changes in U.S. exports and imports of pork. Source: U.S. Department of Agriculture National Agricultural Statistics Service.

FACILITIES FOR PRODUCTION

Pigs in the United States are produced in a variety of production systems and under many different conditions. Facilities and production systems can be classified into two main categories: total-confinement, intensive production systems; and low-investment, outdoor production, and other alternative systems.

There are systems intermediate between these two; however, the majority of pigs in the United States are in either two main categories of production systems. Advances in technology and production have made both total confinement and outdoor systems viable alternatives for today's producer.

The majority of pigs produced in the United States are produced in total-confinement (see fig. 8-10), intensively managed systems because they offer several significant advantages. Totally enclosed confinement facilities maintain a constant, controlled environment with proper ventilation, resulting in more efficient production. These systems maximize the use of the large capital investment that is required and allow the use of specialized labor, often resulting in lower labor requirements per pig produced. Intensively managed systems also offer improved marketing opportunities because of the large, consistent supply of pigs.

Confinement systems do have several potential disadvantages. A large capital outlay is required to start up a system, and once the system of buildings is constructed, there is little short-term flexibility to change the size of the system if market situations warrant such a change. Confinement buildings generally have high energy requirements and a need for alternate/backup energy sources in case of power failure. A high degree of mechanization is involved that requires periodic maintenance and replacement of equipment. A final potential disadvantage of confinement systems is the requirement that sufficient land must be available to dispose of the large amounts of manure that are produced by the system.

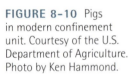

FIGURE 8-10 Pigs in modern confinement unit. Courtesy of the U.S. Department of Agriculture. Photo by Ken Hammond.

Outdoor production and other alternative systems (see figs. 8-11 through 8-13) are favored by many small to midsize producers. These alternative systems include open-ended hoop structures with deep straw, hogs outdoors on pasture, and other systems. Use of these types of systems is increasing in number but still accounts for only a small percentage of pigs produced annually in the United States. Producers use outdoor systems because of their lower capital investment in buildings and equipment, and the fact that the system is flexible and relatively easy to modify due to lower fixed costs. There is less dependence on mechanical equipment, and the disagreeable aspects of environmentally regulated buildings (odor and manure handling) are largely avoided.

Outdoor systems also have several potential disadvantages or shortcomings. Pigs are often exposed to the elements, and these systems offer less control over the pig's environment, which may adversely affect its performance and efficiency. Exposure to variable weather conditions produces less desirable working conditions for the caretakers and often results in higher labor requirements per pig produced. Land for pasture lots is required, and a source of bedding is also needed. Because of the reductions in efficiency, outdoor systems generally result in lower profit opportunities for producers.

A type of finishing building for outdoor production that has increased in popularity is the hoop building (see figs. 8-12 and 8-13). It appeals to producers because of its lower cost and versatility. In these buildings, pigs are housed in one large group (150–200) in a deep bedded system. Pigs are allowed free access to roam and have at least some protection from weather conditions because of the hoop covering over them. Bedding is added as needed during the finishing phase, and manure is handled as a solid material after all pigs are marketed.

FIGURE 8–11 Hogs in semi-outdoor facility. Courtesy of the U.S. Department of Agriculture. Photo by Gene Alexander.

FIGURE 8-12 Exterior view of alternative rearing situation for raising hogs: hoop barns. Courtesy of Iowa State University.

FIGURE 8-13 Hogs being raised on straw in a hoop structure. Courtesy of Iowa State University.

 ## GENETICS AND CROSSBREEDING SYSTEMS

The primary system of swine breeding in the United States is the terminal cross-breeding system. In this system, terminal sire lines are selected for growth, efficiency, muscle content, and meat quality traits. They are mated to specialized maternal lines that are primarily selected for reproductive performance with some emphasis on growth, muscle content, and meat quality. All animals produced from the matings of these specialized lines are considered terminal market hogs, with the goal to produce efficiently large numbers of animals that are fast growing, uniform in size and characteristics, and with high lean content together with good meat quality. The diagram in Figure 8-14 is an example of a terminal crossbreeding system in which all offspring of the parent terminal sire and the parent female are terminal market pigs.

The U.S. swine seed stock industry (performing breeding and genetics) can be divided into two main categories: breeding stock companies, and independent seed stock producers.

Breeding stock companies produce and market specialized terminal and maternal boars and gilts that are produced from hybrid or crossbred lines that have been selected over a number of generations for traits of economic importance. These companies maintain nucleus pure lines that are crossed to produce the parent animals that are sold to commercial producers. The origin or basis of these hybrid lines is the pure breeds of swine that were either imported from Europe or developed in the United States, predominantly between 1800 and 1880. Independent seed stock producers maintain lines of purebred animals of the major pure breeds and have also developed crossbred lines, particularly for production of maternal females.

There are eight major recognized breeds of pigs in the United States that are registered in their respective breed associations. Examples are illustrated in Figure 8-15: white breeds, which include Chester White, Landrace, and Yorkshire; and dark breeds, which include Berkshire, Duroc, Hampshire, Poland China, and spot.

The two maternal lines of major importance are the Yorkshire and Landrace breeds. Both of the breeds are white in color and are known for superiority in maternal and reproductive traits, making them the basis for the maternal side

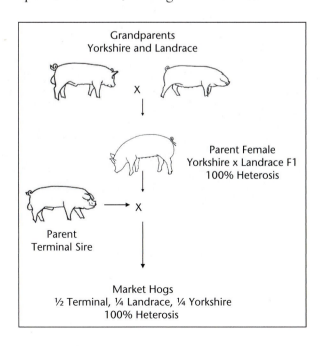

FIGURE 8-14 Example of the terminal crossbreeding system.

FIGURE 8–15 Various breeds of pigs.

A. Chester White pig. Courtesy of Mapes Livestock Photos.

B. Landrace pig. Courtesy of Mapes Livestock Photos.

of nearly all commercial crossbreeding systems. The Chester White is another white maternal breed that is used to a lesser extent in some crossbreeding systems. Duroc is the most popular terminal sire breed in the United States and throughout the world. Purebred Durocs are red in color, and are known for their

C. Yorkshire pig. Courtesy of Mapes Livestock Photos.

D. Berkshire pig. Courtesy of Mapes Livestock Photos.

superior growth, efficiency, durability, and excellent meat quality. The Hampshire is another terminal breed that is black in color with a white band or "belt" around its shoulder. It is known for its leanness and muscling, and is used to a lesser extent in crossbreeding systems.

E. Duroc pig. Courtesy of the National Swine Registry.

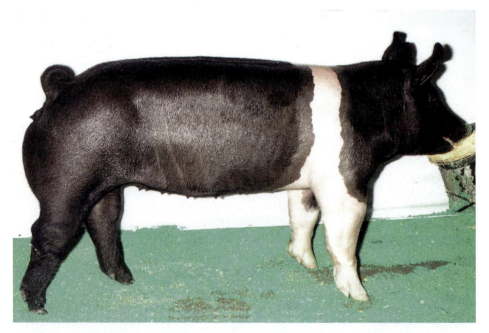

F. Hampshire pig. Courtesy of Mapes Livestock Photos.

 ## PRODUCTION PHASES

Figure 8-16 outlines the three main production phases that are currently used by most producers in the United States: breeding/gestation/farrowing, nursery, and finishing. The breeding production phase is made up of all producing females and

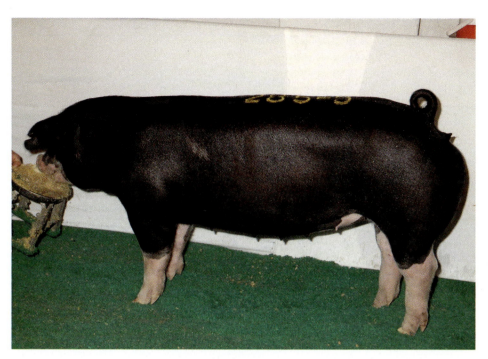

G. Poland China pig. Courtesy of Mapes Livestock Photos.

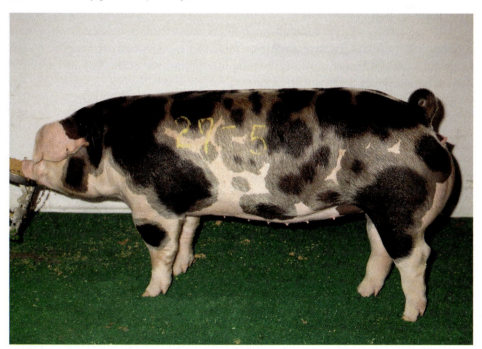

H. Spotted swine. Courtesy of Mapes Livestock Photos.

goes from the time of entry into the herd until litters are weaned. The product of this phase is the weaned pig, and most pigs are weaned at 18–21 days of age. The nursery phase is from the time of weaning until the pig reaches 50–60 lb in weight at approximately 8–10 weeks of age. The finishing phase begins at the end of the nursery phase and goes until pigs are marketed at 260–270 lb. In modern

FIGURE 8-16 The three main production phases that are currently used by most producers in the United States: breeding/gestation/farrowing, nursery, and finishing. Source: Delmar/Cengage Learning.

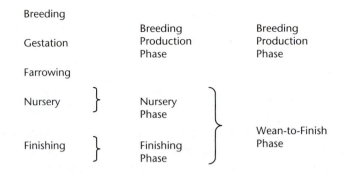

systems, all three of these phases may be located on one site or at three separate sites. The main advantages of a three-site system include improved health and the ability to practice all-in/all-out management, along with the opportunity to specialize in the labor and management required at each site.

An additional option that has become popular with U.S. pork producers is wean-to-finish production in which the nursery and finishing phases are combined into a single stage of production, resulting in two-site production. In this system, the product of the breeding production phase, the 18- to 21-day-old weaned pig, is moved directly into the finishing facility. Wean-to-finish production requires additional management to succeed but does have the following significant advantages over traditional nursery finishing production:

- Reduced transportation costs through fewer moves and less labor to move pigs
- Reduced labor for power washing and disinfecting between groups
- Reduced stress of moving and commingling of pigs, resulting in improved average daily gain and feed efficiency
- Increased flexibility in facilities because finishing buildings are easier to modify than nurseries
- Reduced downtime between groups

Potential disadvantages of wean-to-finish include the requirement for a higher level of management, increased facility costs to accommodate newly weaned pigs compared with traditional feeder pigs, less efficient use of space in finishing barns from 10–50 lb, and higher utility costs because of the additional heat required for newly weaned pigs. The choice to move to wean-to-finish production is often farm dependent, but production advantages and labor savings will frequently more than offset additional utility costs.

Breeding Production Phase

Females in the breeding and gestation phase are housed in a variety of facility types ranging from environmentally controlled, total confinement buildings to outside lots and hoop structures. In the former, over the past 30–40 years, females have been housed in individual stalls. Recent concerns regarding animal welfare have led to a movement toward group housing during gestation. Females housed in outside lots and hoop structures are kept in groups with access to bedding.

Intensive production systems require strong management in the breeding phase. Over 80% of females in the United States are mated via artificial insemination (AI), with the remaining females individually hand mated or bred in groups with boars housed together with the females. Semen used in AI is either collected from boars housed on the farm, or obtained from regional boar studs that offer either same-day delivery of semen or next-day delivery via United Parcel Service of America (UPS) or Federal Express (FedEx).

AI has increased in popularity because of multiple factors. Improved breeding facilities and management skills have made AI possible for all producers. Long-life semen extenders and low-cost, rapid delivery methods have made it possible to obtain and store semen for 5–7 days before it is used. Modern methods of genetic evaluation have made it easier to identify superior genetics, and AI has made it possible to maximize the use of better quality boars on more females. Production systems that require large groups of pigs with minimum age spread are possible because larger groups of females can be bred in a short period time than what would be possible with hand mating. The use of AI lowers disease risk and improves the health of swine herds by eliminating the need for continuous introduction of new animals into the herd.

Most farrowing in modern production systems takes in place in stalls in environmentally controlled facilities. Weaning age varies from farm to farm, but most litters are weaned at 18–21 days of age. Litters have been weaned as early as 15 days of age, but weaning ages closer to 21 days improve conception rates for females for subsequent litters, especially for parity one females (females with their first pregnancy).

Nursery Production Phase

Segregated early weaning is the process of weaning pigs at less than 21 days of age and moving them to a separate site. The goal of early weaning is to control chronic swine diseases by isolating the young pig from its dam at an early age. This procedure improves health levels by stopping the vertical transfer of disease organisms from the older dam to the young piglet. Pigs weaned at less than 21 days of age have high levels of antibodies from colostrum, which gives them protection from infectious agents. The process of segregated early weaning requires physical separation of buildings to have complete isolation from older, potentially disease-carrying swine and to minimize aerosol transmission of disease. It also requires control of human traffic between buildings, and minimization of contact with rodents, birds, and wildlife that may carry disease. The best-case scenario for segregated early weaning is to maintain no more than a 7-day range in age for pigs weaned into a nursery site, building, or room. The weaning age is affected by factors such as management level, facility type, and health status of the herd.

Finishing Phase

The primary focus in the finishing phase is on feed efficiency because the cash cost of feed accounts for 50–60+% of the total cost of production. Choices of facility type, nutritional programs, and management all have a significant impact on the ultimate efficiency of an operation. Feed efficiency is influenced by non-nutritional factors, including feed wastage, stress due to temperature and crowding, health status of the pigs, and adequacy of feed preparation (particle size and form). Nutritional factors, including composition of the diet and adequacy of the diet for the genetic type of the pig and the production environment used, also have a major impact on feed efficiency.

The conventional finishing stage includes a range of 100–140 days on feed. Pigs enter the finisher weighing 50–60 lb at 50–60 days of age and are marketed at 150–210 days of age, at a range of 250–300 lb in weight. The wean-to-finish stage is a longer period of time in which pigs enter at 10–20 lb (15–21 days of age) and are marketed at 250–300 lb (150–210 days of age). Regardless of the type of finishing stage used, all-in/all-out management is critical. This involves following strict sanitation and biosecurity rules, and removal of all pigs from the facility before new pigs are introduced. All-in/all-out management results in a 6–10% increase in average daily gain, reduction of 6–10 days to market, and 5–7% improvement in feed efficiency.

Feeding and nutritional programs in the finishing stage are designed using the concepts of lean growth modeling and phase feeding. Lean growth modeling is using a mathematical model to attempt to accurately quantify the daily nutrient requirements of the pig for maximum protein accretion based on its genetic potential, nutrient intake, and the environmental conditions under which it is raised. Phase feeding then is the use of multiple diets that are adjusted frequently during the finishing phase to closely match the pig's requirements and minimize overfeeding of essential nutrients. Phase feeding results in optimizing the pig's genetic potential for lean growth and production efficiency.

Split-sex feeding may be used in the finishing stage because barrows and gilts differ in feed intake, growth rate, and rate of lean deposition. Barrows generally grow faster and have a higher rate of daily feed intake, whereas gilts generally have improved feed efficiency and higher lean meat percentage. Feeding gilts diets with higher energy and protein levels during the finishing stage more closely meets their genetic capacity for lean growth and results in improvements in feed efficiency.

ANIMAL WASTE/NUTRIENT MANAGEMENT PROGRAMS

Management of the manure produced by swine operations is a critical issue in today's swine industry. Pork production is increasing, and larger facilities have resulted in a greater concentration of manure in one place. This has resulted in larger quantities of manure to handle and more potential for environmental problems due to air and groundwater pollution. Larger production units also offer more opportunities for utilization and management of nutrients produced by swine operations. Manure can now frequently be treated as an asset for crop production because it returns valuable nutrients back to the soil and reduces the need for commercial fertilizer (see fig. 8-17).

Capabilities and requirements of a waste management system must be evaluated before implementation. Each system must include timely collection of manure, adequate waste treatment and/or storage, maximum utilization/management of

nutrients, and efficient land application of nutrients. The choice of the type of system used will depend on the phase of production on the farm, production and management practices, water/energy use, climate, land availability, and value of the waste for the cropping operation.

The use of deep pits under slotted floors in total confinement facilities is one of the most popular waste-handling systems in the United States. Manure is stored in the pit in liquid form and is applied to the land in the spring or fall. These systems offer easy collection and storage of manure, high retention of nutrients in the manure, and relatively low management requirement and operating cost. These systems have a high initial investment in the building and the potential for objectionable odors due to close concentration of manure. There is also a labor requirement for application of the manure and the potential accumulation of sludge in the pit. Some production units may use a shallow pit under slotted floors with the manure stored in a storage tank outside the building. These buildings have a reduced initial cost for the building and less potential for odor inside the building but require an extra outside storage facility for the manure.

Flush systems with manure stored in large lagoons are used, particularly, in the southeastern United States (e.g., North Carolina) (see fig. 8-18). These systems offer greater labor flexibility for land application of nutrients because manure is less concentrated and can be stored for longer periods of time. Manure from lagoons may be applied using an irrigation spray system. Disadvantages of lagoons include the potential for odor from the lagoon, loss of fertilizer value of the manure, and possible problems because of leaching of the manure into groundwater. Manure from hoop buildings or open concrete lots is handled as a solid because it also includes the bedding that is used in these facilities. It is applied with a manure spreader and is less concentrated, so it has reduced nutrient value as fertilizer.

FIGURE 8-18 Pork producer checking the operation of his lagoon pump in North Carolina. Courtesy of the U.S. Department of Agriculture. Photo by Ken Hammond.

Definition

Contract production is a relationship between growers (swine producers) and owners (contractors or integrators). Contractors provide feeder pigs, feed, transportation, and veterinary care, and growers provide the facilities, utilities, and labor. Contractors retain ownership of the pigs and compensate growers based on a value per pig produced.

 CONTRACT PRODUCTION

Contract production in the swine industry is specified by relationships between growers (swine producers) and owners (contractors or integrators). Contractors provide certain inputs (feeder pigs, feed, transportation, veterinary care), and growers provide the facilities, utilities, and labor. Contractors retain ownership of the pigs, bear most of risk involved, and compensate growers on a fee-for-service agreement based on a value per pig produced. Contracting has allowed individual producers to increase the size of their operation while specializing in one specific phase of production.

Production contracts differ from marketing contracts in that they are between a grower and a contractor, whereas marketing contracts are between a contractor/owner and a packer. Marketing contracts specify the number of pigs to be delivered to the packer, the location and time of delivery, and the price based on a predetermined formula or criteria. Production contracts refer to the actual production (growing) of the pig and are generally independent of marketing contracts.

Early attempts to introduce contracting occurred in the 1950s, at the same time that contract production of broilers became common in the United States. These early attempts were unsuccessful, and feed companies and packers were also generally unsuccessful in introducing the idea in the Midwest in the 1960s and 1970s. Contract production became widespread in the Southeast and has continued to be a common business arrangement in that part of the country. The farm crisis of the early 1980s helped to initiate the movement of contract production to the Midwest by feed companies, feed dealers, and investors. Independent producers with adequate financing have also contributed to the increase in contracting over the past 2 decades.

TABLE 8-2 Percentage of U.S. hogs raised under contract

FIRM SIZE (THOUSAND HEAD MARKET)	FARROWED				FINISHED			
	1997	2000	2003	2006	1997	2000	2003	2006
1–50	1%	2%	7%	1%	8%	3%	5%	7%
50–500	4%	7%	5%	4%	7%	10%	11%	14%
500+	11%	13%	17%	15%	16%	21%	25%	25%
Total	17%	22%	23%	20%	30%	34%	41%	46%

Source: University of Missouri, Illinois State University, PORK Magazine, Pig Improvement Company, and the National Pork Board.

TABLE 8-3 Percentage of U.S. hogs raised by firms that are contractors

FIRM SIZE (THOUSAND HEAD MARKET)	FARROWED BY CONTRACTORS				FINISHED BY CONTRACTORS			
	1997	2000	2003	2006	1997	2000	2003	2006
1–50	10%	5%	15%	3%	14%	9%	13%	12%
50–500	8%	8%	13%	14%	9%	13%	12%	14%
500+	22%	26%	40%	41%	22%	33%	39%	40%
Total	40%	39%	68%	55%	45%	55%	64%	66%

Source: University of Missouri, Illinois State University, PORK Magazine, Pig Improvement Company, and the National Pork Board.

Tables 8-2 and 8-3 summarize the level of contracting in the pork industry. In 2006, 20% of the pigs farrowed in the United States were farrowed under a contract arrangement, and nearly half (46%) of pigs were finished under some type of production contract. In addition, 55% of the pigs in the United States were farrowed by firms that are contractors, and 66% of pigs were finished by contractors. As the size of the firm increases, the percentage of pigs that are contracted also increases dramatically.

 ## FUTURE OF THE INDUSTRY

The trend toward larger production units will likely continue as technology, modern facilities, and economies of scale have made efficient production of a large number of pigs at one location feasible. The problems related to the disposal of large amounts of manure, environmental control, and proximity to the human population will limit growth of the industry in some parts of the country. Recent trends have indicated that a greater percentage of pigs in the United States are produced by the largest farms every year. This trend will likely be maintained as the majority of swine production in the United States will be owned by producers marketing over 50,000 pigs per year.

Definition

Economies of scale exist when the average cost to produce one unit decreases as production increases. This is due to the ability to spread costs over more units produced. Such costs include capital costs of facilities, labor, management, and marketing. There is also a scope for bulk buying at lower cost, specialized managers, and the ability to negotiate lower interest rates.

REVIEW QUESTIONS

1. What are the major organizations for pork producers?

2. What is a sow, boar, gilt, and barrow?

3. Where were pigs domesticated?

4. What are the genus and species names for the pig?

5. When were the first pigs brought to the Americas?

6. What is the source of wild pigs in the United States?

7. How many chromosomes do pigs have?

8. Do boars ejaculate small volumes of semen into the vagina of a sow?

9. Assuming boar semen is used for artificial insemination, how is it evaluated?

10. What are the lengths of the estrous cycle and pregnancy?

11. How many mammary glands does a sow have?

12. Do pigs have rumens or simple stomachs?

13. What have been the major changes in the U.S. pork industry in the past 30–40 years?

14. What are hogs, pork producers, total confinement, outdoor production, and alternative production systems?

15. Has the number of farms in the United States raising hogs declined? If so, by how much?

16. What is farrowing?

17. What are the top five states in the United States for number of hogs produced (based on inventory) and number of sows farrowed?

18. The amount of pork produced per sow is increasing at 3.5% per year. What are the reasons for the change in number of baby pigs produced per farrowing and the increased growth rate of the pigs?

19. Have U.S. exports of pork increased? If yes, why?

20. What are the production systems for hogs?

21. What are the advantages of total confinement systems?

22. What are the advantages of outdoor production systems?

23. What are the eight major recognized breeds of pigs in the United States?

24. What are the phases of pig production?

25. What are the advantages of three-site production systems?

26. What is the wean-to-finish production system?

27. What is segregated early weaning?

28. Why is artificial insemination used in hog production?

29. What is the market weight for pigs?

30. What systems are used for manure management?

31. What is contract production?

32. How prevalent is contract production in the United States?

REFERENCES AND FURTHER READING

Brady, C. (2008). *An illustrated guide to animal science terminology*. Clifton Park, NY: Thomson/Delmar Learning.

Holden, P. J., & Ensminger, M. E. (2006). *Swine science* (7th ed.). Upper Saddle River, NJ: Prentice Hall.

McGlone, J., & Pond, W. (2003). *Pig production: Biological principles and applications*. Clifton Park, NY: Thomson/Delmar Learning.

Pig–Pork Information

Purdue Extension and U.S. Pork Center of Excellence. (2008). *2008 pork industry handbook*. West Lafayette, IN: Purdue University.

Alternative Production Systems

Iowa State University of Science and Technology. *Hoop structures for livestock*. Retrieved July 17, 2009, from http://www.abe.iastate.edu/hoop_structures

Minnesota Extension (University of Minnesota, St.Paul). *Hogs your way: Choosing a production system in the Upper Midwest*. Retrieved July 26, 2009, from http://www.misa.umn.edu/vd/publications/HogsYourWay_2009.pdf

CHAPTER 9

Sheep and Goats: Production and Utilization

OBJECTIVES

This chapter will consider the following:

- Introduction to global sheep and goat production
- Sheep (introduction, classification, domestication, breeds, judging sheep, meat, predators of sheep, wool, milk, sheep-goat hybrids, and transgenic sheep)
- Goats (introduction, classification, domestication, breeds, milk, goat hair, meat, and other products)

INTRODUCTION TO GLOBAL SHEEP AND GOAT PRODUCTION

Sheep and goats are frequently produced together in much of the world, particularly in developing counties. Both are used for meat, milk, fiber (wool and hair), and leather. The United Nation's Food and Agriculture Organization combines the meat production data for sheep and goats because the meat is processed and sold together. Global sheep and goat meat production in 2006 was 13.8 million metric tons (compared to 61 million metric tons of beef). Data for different countries are shown in Table 9-1.

TABLE 9-1 Ranking of sheep and goat meat production by country in 2006

COUNTRY	PRODUCTION IN MILLION METRIC TONS
China	4.70
India	0.76
Australia	0.64
Pakistan	0.56
Iran	0.50
New Zealand	0.54
United Kingdom	0.33
Sudan	0.33
Turkey	0.32
Spain	0.24

Source: FAO.

SHEEP

Introduction

Sheep have long been one of the major species of agricultural livestock. Sheep are, for instance, one of the animal species on the Chinese calendar and zodiac, and appear as Aries (the ram) in the Western zodiac. They still are very important globally, but their significance has declined in the United States. In 2006, there were 6.2 million head of sheep (and over 60,000 sheep operations) in the United States. This compares with between 40 and 50 million sheep in the United States between 1900 and 1906. Sheep meat (lamb and mutton) represents less than 5% of all meat produced in the United States.

About 3 million lambs are produced each year in the United States, with an average weight of 134 lb at market, together with 1.5 million lb of sheep milk and 38 million lb of wool. Sheep are raised for meat, wool, and milk, and are important to the people raising them as a primary source of income, or to supplement other income from agricultural enterprises or from outside employment. Sheep production in the United States is unique among all sheep-producing countries, because the U.S. market emphasis is on meat, rather than wool. The external anatomy of sheep is shown in Figure 9-1.

FIGURE 9-1 External anatomy of sheep. Source: Delmar/Cengage Learning.

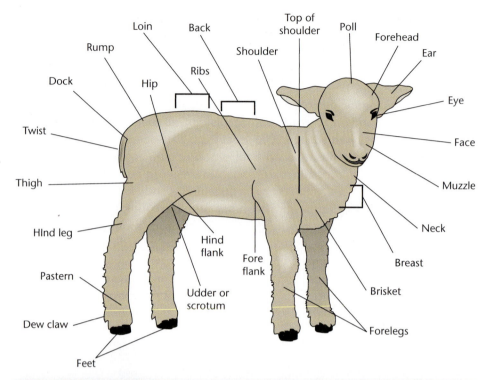

FIGURE 9-2 Sheep on rangeland in Nevada. Courtesy of the USDA.

Today, sheep production in the United States is predominantly found in the Western and mountain states (see fig 9-2), as can be seen from the statistics from the U.S. Department of Agriculture (USDA) for 2006 in Table 9-2.

We sometimes think of sheep as a minor species. There are 1.02 billion sheep in the world. Although this represents less than 5% of meat production globally, sheep are very important in developing countries. Sheep provide protein for human nutrition together with a mechanism of accumulating resources, capital, and wealth, and thus of escaping poverty. Australia is a major wool producer.

TABLE 9-2 Top 10 sheep-producing states based on number of sheep and lambs in 2009

RANKING	STATE	NO. OF SHEEP
1	Texas	1.74 million
2	California	1.32 million
3	Wyoming	0.84 million
4	Colorado	0.82 million
5	South Dakota	0.61 million
6	Utah	0.58 million
7	Montana	0.51 million
8	Oregon	0.44 million
9	Idaho	0.42 million
10	Iowa	0.40 million

Source: USDA.

TABLE 9-3 Top sheep-producing countries with inventory number of sheep in 2005

RANKING	COUNTRY	NO. OF SHEEP
1	China	144 million
2	Australia	98 million
3	India	59 million
4	New Zealand	39 million

Source: FAO.

> **NOMENCLATURE AND TERMINOLOGY**
>
> An adult female sheep is a *ewe*, and an adult male sheep is a *ram*. A castrated male is a *wether*. A young sheep is a *lamb*. A group of sheep is a *flock*. The process of giving birth is *lambing*. *Ovine* is an adjective referring to sheep.

Globally, the top sheep-producing countries are China, Australia, and India (see Table 9-3).

Classification

Sheep are ruminants in the order Artiodactyla (the even-toed ungulates) and family Bovidae (see Classification sidebar). Another member of the genus *Ovis* is the bighorn sheep (*Ovis canadensis*).

Domestication

Sheep were the first livestock species to be domesticated, occurring around 9000 BC. The ancestor of domestic sheep (*O. aries*) was the Asian Mouflon (*Ovis orientatis*). These domesticated in the Fertile Crescent, possibly as hunting decoys or as the result of children adopting "cuddly" orphan lambs. Sheep became important as a source for meat, milk, leather, and wool, and spread to other areas fairly rapidly, reaching the Indian subcontinent by 6500 BC, Europe by 6000 BC, and Britain by 4000 BC. Sheep and goats were an ideal domestic animal early in agriculture because they browsed on vegetation that would provide little or no nutrients to people, cleared land before planting, and are small and relatively easy to manage.

Sheep Reproduction

Sheep are short-day breeders, that is, they breed in the fall such that lambs are born in the spring, and are seasonally polyestrous. The length of gestation is 147 days.

> **CLASSIFICATION**
>
> The classification of sheep is phylum, Chordata; class, Mammalia; order, Artiodactyla; family, Bovidae; genus, *Ovis*; and species, *Ovis aries* (more accurately, *Ovis orientates aries*).

FIGURE 9-3 *A,* Dorset ram. Courtesy of *Sheep Breeder Magazine. B,* Finnsheep. Courtesy of the USDA. Photo by Fred Ward. *C,* Rambouillet ram. Courtesy of *Sheep Breeder Magazine.*

A

B

C

Breeds of Sheep

There are many breeds of sheep (see fig. 9-3), originating predominantly in Europe and Asia. Breeds of sheep include the following:

- Border Leicester
- Cheviot
- Dorset English Leicester
- Finnsheep or Finnish Landrace
- Ile-de-France, the product of crossing the English Leicester and the Rambouillet
- Merino
- Rambouillet Romney
- Scottish blackface
- Texel
- Welsh Mountain

Merinos and Rambouillet are fine-wool breeds. Cheviots, Dorsets, and Finnsheep or Finnish Landrace are medium-wool breeds.

FIGURE 9-4 What to look for in judging sheep. Source: Delmar/Cengage Learning.

Look for width—
1. At the center of leg
2. At the dock
3. Between hind legs
4. Over back and loin

Look for—
5. Long rump
6. Long, bulging stifle
7. Depth of leg
8. Width at chest
9. Heavy bone

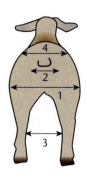
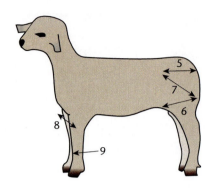

Judging Sheep

Judging sheep is an activity that is useful in itself to know the best livestock. There are collegiate, Future Farmers of America, 4-H, county, and state judging contests. Livestock judging also builds confidence with a sense of accomplishment and improves communication skills.

Items to watch for in judging sheep include the following (see fig. 9-4):
From the back of the animal

1. Width at the center of the leg
2. Width at the dock
3. Width between the hind legs
4. Width over the back and loin

From the side of the animal

1. Long rump
2. Long, budging stifle
3. Depth of leg
4. Width at chest
5. Heavy bone

Sheep and Lamb Meat

Lambs or sheep are marketed directly to a packer or small processor, through terminal markets or by direct sales to the public, including electronic marketing. Sheep and lambs are classified as follows:

- Sheep (>2 years old)
- Yearlings (1–2 years old)
- Lambs, further divided by age as:
  - Hothouse lambs (<3 months old and <60 lb [27.2 kg] weight).
  - Spring lambs (3–7 months old). Although these new-crop lambs are frequently born in the spring, there are production systems for fall lambing.
  - Lambs (7–12 months old; 110–130 lb [49.5–58.9 kg]).

Lamb carcasses are graded according to USDA standards from top to bottom as prime, choice, good, and utility. The wholesale cuts of lamb are shown in Figure 9-5.

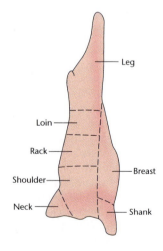

FIGURE 9-5 Wholesale cuts of lamb. Source: Delmar/Cengage Learning.

Leg

Loin

Rack

Breast

Shoulder

Neck

Shank

Predators and Sheep

There are estimates of over 250,000 sheep and lambs being killed every year by predators. Coyotes are responsible for 60% of the losses, domestic dogs 15%, and large cats (such as mountain lions and pumas) 6%.

Methods to protect sheep include guard dogs, guard donkeys, and guard llamas; fencing to exclude predators; and killing or trapping the predators. There is an increasing use of guard dogs together with some use of llamas and good evidence of their effectiveness. Donkeys are also used to protect sheep from predators.

Wool

Wool is the hair or fur of sheep and goats. It is predominantly composed of the protein keratin. The wool or fleece is removed by shearing. The best wool is from merino sheep because the fibers are fine and very crimped (about 100 crimps per inch). Crimping facilitates spinning of the wool into yarn.

Lanolin or wool grease is secreted by the sebaceous glands and provides waterproofing to the wool. Examples of products that may contain lanolin are shave cream, diaper rash ointment, moisturizing lotion, hair remover with aloe and lanolin, lip balm, and cuticle massage cream.

Sheep Milk

Sheep's milk is used extensively worldwide for specific high-value cheeses. For examples, see Chapter 26. The composition of the sheep's milk is highly variable between breeds, and is also affected by season, stage of lactation, and nutrition. Its composition is 5.4% protein, as compared with to 3.3% in cow's milk, and 6.6% fat, as compared with 3.9% in cow's milk.

> **Definitions**
>
> *Wool* is obtained by shearing the sheep. It is spun into yarn and woven into fabric. *Virgin wool* is wool that has not ever been processed into fabric previously. *Shoddy* is wool that has been recycled.

DOLLY THE TRANSGENIC SHEEP

In the late 1980s, transgenic sheep were produced by microinjection of the pronucleus of the ovum with specific DNA constructs. One transgenic sheep, for instance, produced milk with a very high concentration of human α-1 antitrypsin, a protein that may be used to treat people with cystic fibrosis.

The technology of cloning a sheep from somatic cells by nuclear transfer into an ovum was demonstrated with the sheep Dolly, at Roslin, Scotland, in 1996. The next step was to put the transgene into the somatic cells by the process of transfection before the nuclear transfer into enucleated oocytes. The first such transgenic sheep was produced by transfecting with a construct containing the neomycin resistance marker gene (to allow selection of those cells expressing the gene), a human coagulation factor IX gene together with promoters to cause a high level of expression in the mammary gland and then secretion into the milk. There was then successful nuclear transfer leading to the birth of a sheep called Polly that produced human coagulation factor IX, a protein important to people with hemophilia in whom the blood does not clot, and excessive bleeding and ultimately death occurs if the protein is not given to the patient.

Definitions

A *mosaic* is an individual animal made up of more than one genetically distinct population of cells. Mosaics originate from a single zygote.

A *chimera* is an individual animal made up of more than one genetically distinct population of cells. Chimeras originate from more than one zygote.

A *transgenic animal* is one in which the gene of another organism (the transgene) is inserted into its genetic material. The incorporation of the gene into the genome leads to it being transmitted to the next generation via the gametes. Transgenic sheep and goats have been produced that express a transgene for foreign protein that can be used as a drug for therapeutic purposes. To readily obtain large quantities of the protein, the transgene is expressed in the mammary gland and the protein secreted into the milk. To achieve this, a gene construct is made with promoters that have high levels of expression in the mammary gland and restricts expression to that organ. For example, the promoter for a milk protein may be used.

Sheep–Goat Hybrids

It is generally thought not possible to produce a cross or hybrid between a sheep and a goat due to their genetic (sheep with 54 chromosomes and goats with 60 chromosomes) and evolutionary separation. However, there is a documented case of a sheep–goat hybrid with 57 chromosomes resulting from the mating of a male sheep and female goat. Chimeras have been made experimentally with a mix of embryonic sheep and goat cells. These are called *geeps*.

 # GOATS

Introduction to Goats

Goats are relatively uncommon in the United States, with a population of about 2 million (see fig. 9-6). These include milk goats together with Angora goats bred for their wool. The dairy goats produce milk for liquid consumption and for specific cheeses. The trade organization is the American Dairy Goat Association. The external anatomy of goats is shown in Figure 9-7.

Worldwide, there are over half a billion goats, with the highest numbers in Asia, particularly in India and China, and Africa, particularly East Africa. Goats provide milk, meat, and leather, together with cashmere and Angora wool.

Classification

Goats, like sheep, are in the order Artiodactyla and family Bovidae (see "Classification" sidebar on the next page). The mountain goat (*Oreamnos americanus*) is strictly speaking not a goat because it is not in the genus *Capra*.

FIGURE 9-6 Close-up of goat head, full face. Courtesy of the USDA. Photo by Larry Rana.

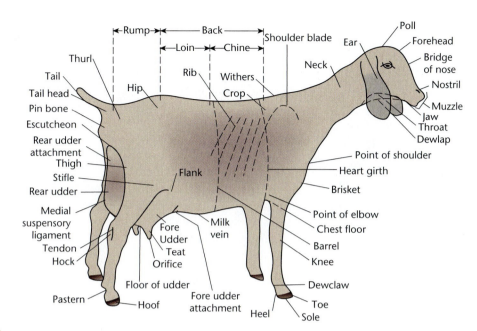

FIGURE 9–7 External anatomy of goats. Courtesy of the American Dairy Goat Association, North Carolina.

Domestication

Goats were domesticated in the Fertile Crescent about 8500 BC. The earliest goat remains found at an archaeological site have been dated to 7900 BC. The ancestor of domestic goats (*Capra hircus hircus*) was the Bezoar goat (*C. hircus*).

Goat Reproduction

Like sheep, goats are short-day breeders, that is, they breed in the fall such that kids are born in the spring, and they are seasonally polyestrous. The length of gestation is 150 days.

Breeds of Goats

There are many breeds of goats, originating predominantly in Europe, Africa, and Asia. Breeds of goats include the following:

- Angora goat
- American Cashmere
- Anglo-Nubian or Nubian (see fig. 9-8)
- Boer (meat goat)
- Border Leicester
- Pygmy goat or Cameroon pygmy goat
- Thuringian
- Toggenburg (see fig. 9-9)

Goat Milk

The composition of goat's milk varies with breed, season, stage of lactation, and nutrition. Average composition has been determined from bulk goat's milk, but this is dependent on the individual goats. The composition of goat milk is shown in Table 9.4.

CLASSIFICATION

The classification of goats is phylum, Chordata; class, Mammalia; order, Artiodactyla; family, Bovidae; genus, *Capra*; and species, *Capra hircus*.

TERMINOLOGY AND NOMENCLATURE

An adult female goat is a *doe*; *nanny goat* is the term used in the United Kingdom. An adult male goat is a *buck*; *billy goat* is the term used in the United Kingdom. A young goat is a *kid*. The process of giving birth is calling *kidding*. *Caprine* is an adjective referring to goats.

FIGURE 9-8 Nubian goat. Courtesy of the American Dairy Goat Association.

FIGURE 9-9 Toggenburg goat. Courtesy of the American Dairy Goat Association.

With the similarity in antigens, 80% of people with a cow's milk allergy also respond to goat's milk.

Goat milk either is used as liquid milk sold through distributors to consumers, who frequently have problems with cow's milk, or is made into cheese (see Chapter 26). Cleanliness and sanitation are essential throughout the process. The sale of goat milk must follow appropriate regulations.

TABLE 9-4 Composition of goat's milk

CONSTITUENT	PERCENTAGE
Water	88.6
Solids	11.4
Fat	3.3 (compared with 3.9% with cow's milk)
Lactose	4.3
Protein	3.2 (compared with 3.3% with cow's milk)
Ash/minerals	0.64
Calcium	0.11
Phosphorous	0.08
Magnesium	0.01
Sodium	0.04
Potassium	0.21

Goat Hair

There are two very valuable products from specific breeds of goats: mohair and cashmere.

Mohair

Mohair is made from the hair of Angora goats. The hair is obtained by shearing and spun into thread. This can then be woven, knitted, or crocheted to produce textiles for clothing (e.g., a mohair sweater), throws, tapestries, and blankets. Mohair garments are light but provide good insulation. The mohair gives the clothing a "silky" feel and look. Angora goats are named from the part of Turkey where they originated.

Cashmere

Cashmere is the wool or hair of the cashmere goats. These were originally found in Kashmir, in the northwest of the Indian subcontinent. What makes the wool of the cashmere goat so attractive for making clothing is the fine fibers (<18.5 μm in diameter). Cashmere is used for luxury clothing, including sweaters, socks, gloves, scarves, hats, pants, shawls, and overcoats.

Other Goat-Related Products

Goat Meat and Cheese

There is increasing demand for goat meat in the United States. Goat meat from 48- to 60-lb meat goats (6–9 months old) is called *chevon*. *Caprito* is meat from young goats. Goat leather is also a significant product.

There are a variety of goat cheeses. These are collectively called *chèvre* or goat cheese. Typically goat cheese is soft and can be spread easily.

Pharmaceutical Proteins

Transgenic goats have been produced. One has been reported to produce the enzyme butyrylcholinesterase. This is a protein that protects the human body against the nerve gas sarin.

REVIEW QUESTIONS

1. How many sheep are there in the United States?

2. Has the number of sheep declined since 1900?

3. What are the major uses of sheep in the world?

4. How does the United States differ with this?

5. How many lambs are produced each year in the United States?

6. What states in the United States have the highest number of sheep?

7. What countries have the highest number of sheep?

8. What is sheep milk used for?

9. What is the name for an adult female sheep?

10. What is the name for an adult male sheep?

11. What is the name for a castrated adult male sheep?

12. What is the species name for a sheep?

13. What was the species that was domesticated to produce the sheep?

14. Where and when were sheep domesticated?

15. Why were sheep domesticated?

16. Give the names of five breeds of sheep.

17. What are the age classifications of sheep for meat production?

18. Why is wool from merino sheep considered very high quality?

19. What is lanolin?

20. How does sheep milk differ from that of cows?

21. What are the following: a geep, a mosaic, and a chimera?

22. Sheep are short-day breeders. When do they breed?

23. How many goats are there in the United States?

24. How many goats are there in the world?

25. What countries have the highest number of goats?

26. Why are goats so important in developing countries?

27. What is goat milk used for?

28. What is the name for an adult female goat?

29. What is the name for an adult male goat?

30. What is the species name for a goat?

31. What was the species that was domesticated to produce the goat?

32. Where and when were goats domesticated?

33. Why were goats domesticated?
34. Give the names of five breeds of goats.
35. How does goat milk differ from that of cows?
36. What are the goat products mohair and cashmere?
37. What is a transgenic animal?
38. How are they produced?
39. What was the significance of Dolly and Polly?
40. Why are transgenic sheep and goats being produced?

REFERENCES AND FURTHER READING

Brady, C. (2008). *An illustrated guide to animal science terminology*. Clifton Park, NY: Thomson/Delmar Learning.

Composition of Sheep and Goat Milk

Agriculture Canada. Retrieved October 19, 2009, from http://sci.agr.ca/crda/pubs/goat2000-chevre200_e.htm

Bencini, R., & Pulina, G. (1997). The quality of sheep milk: A review. *Wool Technology and Sheep Breeding, 45*(3), Article 5.

U.S. Department of Agriculture, Agricultural Research Service. (2008). *USDA national nutrient database for standard reference, release 21*. Retrieved July 25, 2009, from http://www.ars.usda.gov/ba/bhnrc/ndl.

Breeds of Sheep

Oklahoma State University Board of Regents. *Breeds of livestock. Sheep: (Ovis aries)*. Retrieved July 17, 2009, from http://www.ansi.okstate.edu/breeds/sheep/

Breeds of Goats

Oklahoma State University Board of Regents. *Breeds of livestock. Goats: (Capra hircus)*. Retrieved July 17, 2009, from http://www.ansi.okstate.edu/breeds/goats/

Predators and Sheep

Andelt, W.F. (2004). *Livestock guard dogs, llamas and donkeys*. Colorado State Extension Service Livestock management series no. 1.218. Retrieved July 25, 2009, from http://www.ext.colostate.edu/pubs/livestk/01218.pdf

U.S Department of Agriculture. (2004). *Highlights of the sheep and lamb predator death loss in the United States*. Retrieved July 25, 2009, from www.aphis.usda.gov/.../sheep/sheep_pred_deathloss_2004_highlights.pdf

Transgenic Sheep

Schnieke, A. E., Kind, A. J., Ritchie, W. A., Mycock, K., Scott, A. R., Ritchie, M., et al. (1997). Human factor IX transgenic sheep produced by transfer of nuclei from transfected fetal fibroblasts. *Science, 278*, 2130–2133.

CHAPTER 10

Alternative Mammalian Livestock: Buffalo, Llamas and Alpacas, Rabbits, and Deer

OBJECTIVES

This chapter will consider the following alternate mammalian livestock species:

- Buffalo
- Llamas and alpacas
- Rabbits
- Deer

BISON ("BUFFALO")

It is a misnomer to call the American bison the American "buffalo." The American bison is only distantly related to the water buffalo, that is, the domesticated water buffalo that produces the milk for mozzarella cheese, or the Cape buffalo. In this chapter, the word *bison* will be used.

Bison today are in reserves or on ranches (see fig. 10-1). Consumption of bison meat is increasing. The National Bison Association (http://www.bisoncentral.com) promotes the production and marketing of bison in the United States.

The external anatomy of bison is shown in Figure 10-2.

History of Bison

The American bison has a fascinating history. It is thought that the American bison originated in Eurasia and crossed the Bering Strait land bridge about 10,000 years ago at the end of the last Ice Age. By the 16th century, herds of bison (70 million animals) roamed the Great Plains of North America. These were the basis of the economies of Native Americans on the plains. The bison were hunted aggressively in the 19th century because of commercial hunting, as government policy to remove Native Americans from lands to be settled by European Americans, and to prevent bison-related damage to the railroads. By 1880, the population of American bison was down to only about 1,000 animals, and the species risked extinction. Through the efforts of conservationists and ranchers, the species was saved. Today, it is estimated that there are 380,000 bison in the United States and Canada, with over 90% privately owned.

FIGURE 10–1 American bison (buffalo) on rangeland in Montana. Courtesy of the U.S. Fish & Wildlife Service. Photo by Jesse Achtenberg.

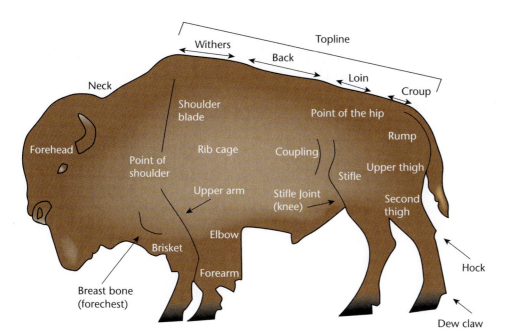

FIGURE 10–2 External anatomy of a bison. Courtesy of The National Bison Association.

Bison Production

Bison production is increasing in the United States. In 2006, 42,687 animals were harvested in federally inspected commercial facilities, which was a 21% increase on the previous year. During the same year, bison meat sales totaled approximately $239 million at the retail and restaurant level.

The American bison is well suited to the Great Plains. As a ruminant, it can prosper on pastures of the native grasses. It can withstand the extremes of temperature better than domestic cattle.

Bison Meat

Bison meat is low in fat and high in protein. A 100 g portion (3.5 oz) of cooked bison meat contains only 2.4% fat (and, therefore, much less calories than either chicken or beef), and has 28% protein together with iron and vitamins.

Bison Reproduction

The American bison is a seasonally polyestrous species, with a breeding season that starts between the middle of July and early August such that the calves are born in the spring. The gestation period is 266 days. American bison are sexually mature by 24 months old. The semen of breeding bulls is often tested to ensure fertility.

Bison Growth

Bison bull calves are not castrated. The bulls are sold for meat before they reach breeding age. There is a sexual dimorphism in the size of bison (see sidebar). Bison grow at a steady rate and can be finished in feedlots with grain diets.

SIZE AND GROWTH OF BISON

An adult male bison is 1,200 lb, and an adult female 950 lb. At 8 months, a male weighs 475 lb. A 12-month male weighs 550 lb, and a 24-month male 1,050 lb.

Bison Health

Bison can be vaccinated in a manner similar to that with cattle, for example, to brucellosis. Drugs approved for cattle are considered safe for bison.

Beefalo (Cattleloo): Hybridization between American Bison and Domestic Cattle

Despite being in separate genera, there are fertile crosses between domestic cattle and the American bison. This is more understandable when it is realized that the common ancestor for both existed 1–1.5 million years ago and that they have the same number of chromosomes (30 pairs). Genome studies support initial hybridization favoring matings between male bison and female cattle. Full-blood beefalo are such hybrids with three eighths of their genetic material from bison and five eighths from cattle. Beefalo may also have a three sixteenth bison background.

The American Beefalo Association (http://www.americanbeefalo.org) supports the production and marketing of beefalo. It claims greater hardiness for beefalo together with improved meat. A 100-g portion of cooked beefalo meat contains 6.3% fat according to the U.S. Department of Agriculture nutrient database (http://www.nal.usda.gov/fnic/foodcomp/search/). An alternate name for beefalo is *cattleloo,* that is, cattle with a "buffalo" appearance.

WATER BUFFALO

The water buffalo (see fig. 10-3) is only distantly related to the bison (the so-called American buffalo). It is the major milk-producing species in India. This animal is also used as a draft animal. In addition, water buffalo are important in Europe,

FIGURE 10-3 *A* and *B,* Water buffalo are a major milk-producing species globally. Their milk is used to make mozzarella cheese. *A* is reproduced by permission from Bimbel. © 2010 by Shutterstock .com. *B* is reproduced by permission from Muellek. © 2010 by Shutterstock.com.

A

B

particularly Italy, where the milk is used for mozzarella cheese. The milk from water buffalo has a different composition from that of cow's milk, as can be readily seen in Table 10-1.

TABLE 10-1 Comparison of the composition of cow's milk with that of water buffalo milk in g/100 mL

CONSTITUENT	COW'S MILK	WATER BUFFALO MILK
Protein	3.2	4.5
Lactose (carbohydrate)	4.8	4.9
Total fat	3.9	8.0
Saturated fat	2.4	4.2
Monounsaturated fat	1.1	1.7
Polyunsaturated fat	0.1	0.2

Source: http://www.northwalesbuffalo.co.uk/milk_analysis.htm.

LLAMAS AND ALPACAS

Like camels, llamas (see fig. 10-4) and alpacas (fig. 10-5) are members of the Camelidae family (see Classification sidebar). Llamas are considerably bigger than alpacas. Alpacas are traditionally produced for their fiber. There are about 70,000 llamas and over 5,000 alpacas in the United States and Canada.

Traditionally, llamas have been used as pack animals and alpacas for their fiber. Alpacas have hollow fleece fibers (not true wool) that can be used to weave blankets and other textile products (seee fig. 10-6). The fiber comes in multiple natural colors, including black, white, brown, red, and gray. Recently, llamas have been used as guard animals protecting sheep from dogs and coyotes.

CLASSIFICATION

The classification of water buffalo is phylum, Chordata; class, Mammalia; order, Artiodactyla; family, Bovidae; genus, *Bubalus*; and species, *Bubalus bubalis*.

NOMENCLATURE AND TERMINOLOGY

Kush is when a llama lays down. This is also the position for mating. An adult female llama is sometimes called a *girl*, and an adult male is called a *stud*. A castrated llama is called a *gelding*. A young llama is a *cria*.

CLASSIFICATION

The classification of llamas is class, Mammalia; order, Artiodactyla; family, Camelidae; genus, *Lama*; and species, *Lama glama*.

The classification of alpacas is class, Mammalia; order, Artiodactyla; family, Camelidae; genus, *Vicugna*; and species, *Vicugna paco*.

History

The camelid ancestors originated in the Americas. Some migrated to Eurasia and were the ancestors of the true camels, that is, the dromedary and Bactrian camels. Llamas were found in both North and South America but disappeared from North America between 10,000 and 11,500 years ago because of the arrival of people and/or climate change at the end of the last Ice Age. Llamas and alpacas

FIGURE 10-6 Alpaca wool is used to weave blankets and other textile products. Reproduced by permission from Sootra. © 2010 by Shutterstock.com.

are domesticated South American or New World camelids. They were domesticated about 6,500 years ago from wild alpacas and llamas in the high Andes of present-day Bolivia, Ecuador, northern Chile, and Peru. At the time of the Spanish conquest, llamas were pack animals used by the Incas for transportation of ore from mines. Llamas and alpacas are now only known in their domesticated state.

Until recently, alpacas were viewed in the same genus as llamas, and their species name was *Lama paco*, but it is now recognized that alpacas are more closely related to another camelid, the vicuña.

Hybrids

Fertile hybrids are produced by breeding llamas and alpacas.

Alpaca and Llama Reproduction

There are a number of unusual features to reproduction in llamas and alpacas. They are seasonal breeders in native Andes, but not in temperate conditions with adequate feed. Mating occurs in a lying position. Males are dribble ejaculators, producing about 3 mL of semen release over the 30-minute period of ejaculation.

Both llamas and alpacas are induced ovulators. Some 20% of females do not become pregnant after mating.

The gestational length is about 343 days. The placenta is diffuse and epitheliochorial. Surprisingly, at least in alpacas, there is a seasonal difference in the length of pregnancy being longer with pregnancies starting in the spring (349 days) than in the fall (336 days). Newborns are called *crias* (see fig. 10-7).

INTERESTING FACTOID

Induced ovulators are animals that ovulate after being bred. The biology of this is that during intromission, the penis physically or mechanically stimulates the female genitalia, resulting in nervous impulses passing to the brain. There is then a release of gonadotropin-releasing hormone from neurosecretory terminals in the median eminence, which is a part of the hypothalamus in the brain. Subsequently, the anterior pituitary releases luteinizing hormone, and this induces ovulation. There is evidence also of a factor in the semen that stimulates luteinizing hormone release and ovulation.

Reproductive technologies, such as artificial insemination, embryo transfer, superovulation, pregnancy detection measuring the hormone, progesterone, and diagnosis of pregnancy/fetal development by ultrasound, have not been used extensively but are applicable to llamas and alpacas.

Nutrition

Llamas and alpacas have a digestive system that functions in a manner similar to ruminants, but the forestomach has three compartments, compared with four in ruminants. They chew their cud. Ingesta stay in the stomach for a longer time than that in the ruminant, and there is a higher rate of digestion. There is a lower nitrogen/protein requirement because llamas retain more nitrogen.

Llamas will eat both forage and browse; alpacas are less likely to eat browse. The feeding program is based on roughage, with concentrates added as a supplement to the diet. The total daily dry matter intake should be in the range of 1.8%–2.0% of body weight. The crude fiber content of the diet should be about 25%.

Health

There are relatively few disease problems with llamas and alpacas.

 ## RABBITS

Domestication of Rabbits

Although there are a number of wild species of rabbits and hares, the rabbit was domesticated only about 1,400 years ago from a species of European wild rabbits native to Spain and North Africa.

Use of Rabbits

Rabbits are raised for a variety of uses, including meat, fur, or wool; as laboratory animals; and as companion animals. The American Veterinary Medical Association U.S. pet ownership survey estimated that there were 6.2 million pet rabbits in the United States in 2007.

Rabbit meat is a white meat with low fat and low cholesterol, similar to chicken meat. Rabbit meat production is relatively uncommon in the United States, with only about 370 metric tons produced in 2004. However, production of rabbits for meat consumption in other countries is more common. Global consumption of rabbit meat is about 1 million metric tons per year, with 500,000 metric tons of rabbit meat produced in China, 87,000 metric tons in France, and 70,000 metric tons in Egypt, for instance, in 2005. The skin or pelt of rabbits after drying and curing is used for fur, ranging from the less expensive fur jackets, to slippers and glove linings. Worldwide, it is estimated that about 1 billion rabbit pelts are used. There is significant specialization with rabbits either for meat or the pelt. Few pelts are obtained from slaughterhouses. Examples of countries producing rabbit pelts are France and China. Interestingly, the rabbit is one of the 12 species in the Chinese zodiac and calendar with some years being the "year of the rabbit" (e.g., 1999 and 2011). Specific types of wool are produced from rabbit hair, specifically from Angora rabbits, and also from the breed the Jersey woolie. The rabbits are either sheared or plucked gently to harvest the wool. After spinning, the wool is used to make such items as Angora wool sweaters. These sweaters use wool exclusively from Angora rabbits and are very soft to the touch.

Breeds

The American Rabbit Breeders Association (ARBA) (http://www.arba.net/index.htm) recognizes more than 50 breeds and varieties of rabbits with differences in size (large, medium, small, and dwarf), ears (some breeds have large ears or lops), color, and color pattern. Their mature body weights are 14–16 lb (6.3–7.2 kg) for large breeds, 9–12 lb (4–5.4 kg) for medium-sized breeds, and 3–4 lb (1.4–1.8 kg) for small breeds. Breeds for meat production are large breeds, including the New Zealand (white) (see fig. 10-8) and the Californian. An example of the breed used for the pelt is the Rex. There are a number of strains of the wool-producing Angora rabbit (see fig. 10-9), with the Chinese strains of Angora rabbit, including the German strain in China, accounting for 95% of wool production. The chinchilla rabbit has a luxuriant fur but is not to be confused with the chinchilla, which is a South American rodent used to produce fur coats. Rabbit enthusiasts develop new breeds and varieties every year. When accepted by the ARBA there is a standard of perfection for each breed or variety.

Rabbit Nutrition

Although rabbits are herbivorous, they are simple stomached but with a large cecum, that is, a hind gut where fermentation of ingesta occurs. Rabbits can digest plant fiber, but they are less efficient than ruminants and horses. Moreover, fiber in the diet is necessary for good health. To more fully utilize plant fiber, rabbits have the strategy

---

NOMENCLATURE AND TERMINOLOGY

An adult female rabbit is a *doe*, and an adult male is a *buck*. A young rabbit is called a *kit*. Young hares are referred to as *leverets*, a term that is sometimes also used for the young of rabbits. Parturition or giving birth is *kindling*. Housing for a single rabbit is a *hutch*, but for groups of rabbits is a *rabbitry*.

CLASSIFICATION

The classification of rabbits is class, Mammalia; order, Lagomorpha; family, *Leporidae*; genus, *Oryctolagus*; and species, *O. cuniculus*.

FIGURE 10-8 New Zealand white rabbit. Courtesy of the American Rabbit Breeders Association.

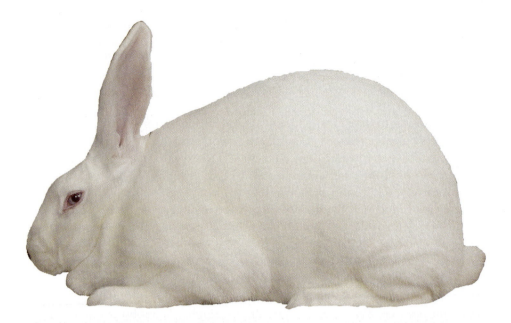

FIGURE 10-9 (English) Angora rabbit. Courtesy of the American Rabbit Breeders Association.

of *coprophagy*, or eating feces. In the case of the rabbit, it produces two types of feces: soft, which are consumed; and hard. The advantage of this strategy is to provide more time for digestion/fermentation and an opportunity to digest the microorganisms that ferment the plant fiber. This might be viewed as analogous to the rumen.

Rabbit feed includes a cereal such as oats, barley, or corn, and a protein source such as soybean meal supplemented with lysine and methionine. Rabbit feed is best provided in a pellet form. Examples of rabbit feed are 16.0% crude protein, 1.5% crude fat, 21–24% crude fiber, 0.6–1.1% calcium, 0.4 % phosphorus, and 0.5–1.0% salt (iodized).

Trace minerals and vitamins with vitamin A at 2,000 International Units/lb (4,400 International Units/kg) are also included.

Riboflavin, pantothenic acid, biotin, and folic acid, together with vitamins B_{12} and C, are synthesized by rabbits themselves and/or the microflora (bacteria and protozoa) in the cecum and, therefore, do not need to be part of a trace mineral and vitamin mix. For pregnant or lactating does, the diet is formulated with higher (18%) protein. This is important particularly with an accelerated breeding program.

Pet rabbits may also be fed alfalfa hay and/or leafy vegetables together with root vegetables such as carrots.

Rabbit Reproduction

Rabbits are reflex or induced ovulators, that is, ovulating 10 hours after coitus. Fertilization occurs shortly thereafter. Rabbits are also season breeders; breeding begins in the early spring (February) and continues until October. The conception rate is higher in the spring than the fall. The length of gestation is 30 days, with litter sizes of five to nine. Parturition occurs more frequently at night. Newborn rabbits or kits are altricial, being blind for about 10 days. Milk is produced from diffuse mammary glands. These are thin and under the skin of much of the ventral (under) surface. The hormone prolactin induces milk production. Weaning starts at about 8 weeks of age. Although a doe is still lactating, she can be rebred 3–6 weeks after giving birth provided that she is receiving optimal nutrition.

The doe can exhibit pseudopregnancy (false pregnancy) because of ovulation being induced by infertile mating or by doe-on-doe sexual contact. It lasts for 17 days. Pregnancy can be confirmed by palpating the doe 12–14 days after mating.

Rabbit Genetics

Rabbits have 44 chromosomes (22 pairs). Efforts on the genetics of rabbits have focused on color and breeds, together with growth rate and other production traits. The latter is gaining more interest recently for commercial rabbit raising, particularly in developing countries.

Care of Rabbits

It is advisable to handle commercial meat or wool rabbits as little as possible to prevent stress. For pet rabbits, there is habituation and regular handling, and stroking occurs regularly. Rabbits should be lifted by the loose skin in the scruff of the neck, between the shoulders, and then the weight supported. They should not be picked up by the legs or ears.

Rabbits are particularly sensitive to high temperature. Controlled environmental conditions should include good ventilation and sanitation. Adequate water supplies are also critically important.

Rabbit Health

Critical to rabbit care are sanitation and good nutrition. One of the common diseases of rabbits is coccidiosis, which can be remedied using coccidiostats.

INTERESTING FACTOID

There are six species of jackrabbit in North America. All are hares and are species in the genus *Lepus*. This is a different species and genus from the domestic rabbit (*Oryctolagus cuniculus*). Several figures in U.S. popular lore are modeled on the jackrabbit, including Bugs Bunny (the cartoon figure making his first movie debut in 1938), Brer or Br'er Rabbit (from the Uncle Remus stories), and the Easter Bunny. Hares have been long associated with fertility. It is not surprising that an expression is "mad as a March hare" because this is the time when mating behavior starts in earnest.

DEER

Deer Farming

Commercial deer farming is relatively new, with annual income globally at more than $100 million. The major deer-producing countries are New Zealand, Ireland, Great Britain, and Germany. It is estimated that 50,000 deer are being farmed in the United States. New Zealand is the major deer-farming country, with about 5,000 farms and 1.7 million deer in 2004. Annual production is 27,000 metric tons of venison. Of New Zealand's deer herd, 85% are red deer (*Cervus elaphus*). In addition, there are elk and a small number of fallow deer.

The two main deer species being farmed in the United States are the fallow and red deer. The North American Deer Farming Association (http://www.nadefa.org) provides information to deer farmers.

Deer Species and Classification

Deer, like cattle, sheep, and llamas, are classified in the order Artiodactyla. There are five species of deer that are farmed:

- Red deer (*C. elaphus*). This is one of the largest species of deer in the world and comes from Europe.
- Elk (also known as Wapiti or Elk-Wapiti) (*Cervus canadensis*), originating in East Asia/North America (see fig. 10-10).
- Fallow deer (*Dama dama* or *Cervus dama*) is native to Europe but has been introduced to North America.

FIGURE 10-10 Deer farming is growing in importance, with farmed elk, for instance, released for hunting or used for meat (venison). Courtesy of the U.S. Fish & Wildlife Service. Photo by Karen Laubenstein.

FIGURE 10-11 Deer farming is growing in importance, with farmed white-tailed deer used for meat (venison). An example of white-tailed deer in an outdoor but experimental setting. Courtesy of Susan Becker, Department of Animal Science, Rutgers University.

- White-tailed deer or white tail (*Odocoileus virginianus*) is medium-sized and found throughout the contigious United States (see fig. 10-11).
- Reindeer (*Rangifer tarandus*) is farmed predominantly outside the United States.

In New Zealand, there are also hybrids between the elk and red deer.

Products from Deer

There are two major products from deer:

1. Venison (deer meat), which is the major product. A growing product is organic venison. Venison has a low-fat content, as can be seen in Table 10-2.
2. "Velvet antler," which is a by-product. The entire growing antler is harvested and used in traditional Asian medicines.

TABLE 10-2 Comparison of the composition of venison and ground beef in a 3½ oz portion (or 100 g portion)

	CALORIES	CHOLES-TEROL (MG)	FAT (MG)	PROTEIN (MG)
Venison loin	164	73	6	26
Ground beef	250	77	14	29

Source: U.S. Department of Agriculture National Nutrient Database.

Restrictions on Deer Farming

In some states in the United States, people who wish to farm deer must get a permit from the state's fish and game agency. Deer are susceptible to chronic wasting disease, which is a prion-caused disease. The movement of deer and interaction with native wild deer need to be limited.

REVIEW QUESTIONS

1. What is the difference between a buffalo and a bison?

2. Is buffalo meat expanding, and, if so, why?

3. What is the origin of the American bison?

4. What are the nutritional advantages of bison and beefalo meat?

5. When is the breeding season for bison?

6. Why aren't buffalo castrated?

7. What is the origin of llamas and alpacas?

8. When were they domesticated?

9. What are the traditional uses of llamas and alpacas?

10. What is a new use for llamas?

11. When is the breeding season for llamas and alpacas?

12. Name three unusual features of llama/alpaca reproduction.

13. Which are the major rabbit-producing countries?

14. List the uses of rabbits in the world.

15. What are the major uses of each of the following breeds of rabbits: (a) Angora, (b) Rex, and (c) New Zealand?

16. What are the mature weights of (a) small breeds, (b) medium breeds, and (c) large breeds?

17. Why is coprophagy important to rabbits?

18. Can rabbits use more forage in their diet than other simple-stomached animals, such as pigs and poultry?

19. What vitamins are produced by the rabbit and/or by its cecal microflora?

20. What is the usual source of protein in rabbit feeds?

21. What is the difference in diet for a pregnant or lactating doe?

22. What is an induced ovulation?

23. When do rabbits breed?

24. What is the gestation period for rabbits?

25. List several good management practices to prevent the transmission of disease in rabbits.

26. What are the major species of deer farmed?

27. What are the products of deer production?

28. What are the major deer-producing countries?

29. Is deer farming restricted?

30. What is chronic wasting disease?

REFERENCES AND FURTHER READING

Adams, G. P., Ratto, M. H., Huanca, W., & Singh, J. (2005). Ovulation-inducing factor in the seminal plasma of alpacas and llamas. *Biology of Reproduction, 73*, 452–457.

Davis, G. H., Dodds, K. G., Moore, G. H., & Bruce, G. D. (1997). Seasonal effects on gestation length and birth weight in alpacas. *Animal Reproduction Science, 46*, 297–303.

Driskell, J. A., Yuan, X., Giraud, D. W., Hadley, M., & Marchello, M. J. (1997). Concentrations of selected vitamins and selenium in bison cuts. *Journal of Animal Science, 75*, 2950–2954.

Deer Farming

Pennsylvania State Extension. *Elk production.* Retrieved July 25, 2009, from http://agalternatives.aers.psu.edu/Publications/elk.pdf

Pennsylvania State Extension. *Fallow deer production.* Retrieved July 25, 2009, from http://agalternatives.aers.psu.edu/Publications/fallowdeer.pdf

Food and Agriculture Organization

FAO AGA Livestock Atlas Series 1. *Global livestock geography: New perspectives on global resources.* Retrieved July 25, 2009, from http://ergodd.zoo.ox.ac.uk/livatl2/index.htm

Food and Agriculture Organization. *Chapter 8: Production of rabbit skins and hair for textiles.* Retrieved July 17, 2009, from http://www.fao.org/docrep/t1690E/t1690e0a.htm

Food and Agriculture Organization. *The rabbit—Husbandry, health and production.* Retrieved July 17, 2009, from http://www.fao.org/docrep/t1690E/t1690e00.htm

Musani, S. K., Halbert, N. D., Redden, D. T., Allison, D. B., & Derr, J. N. (2006). Marker genotypes and population admixture and their association with body weight, height and relative body mass in United States federal bison herds. *Genetics, 174*, 775–783.

Sumar, J. B. (1999). Reproduction in female South American domestic camelids. *Journal of Reproduction and Fertility. Supplement, 54*, 169–178.

Vervaecke, H., & Schwarzenberger, F. (2006). Endocrine and behavioral observations during transition of non-breeding into breeding season in female American bison (Bison bison). *Theriogenology, 66*, 1107–1114.

CHAPTER 11

Chickens and Other Poultry

OBJECTIVES

This chapter will consider the following:

- Introduction to poultry
- Domestication of poultry
- Classification of poultry
- Poultry biology (reproduction, genetics, plumage)
- Global poultry production
- Poultry production in the United States, covering chickens for meat, chickens for eggs, and turkeys
- Other poultry: Ducks, geese, ratites (ostriches and emus), game birds, specialty poultry, and transgenic chickens

INTRODUCTION TO POULTRY

Chickens have been part of the rural scene for thousands of years. One of the animals of the Chinese calendar and Chinese zodiac is the rooster. Poultry are a major source of high-quality protein throughout the world.

Poultry have a distinct terminology (see the Nomenclature and Terminology sidebar). Being birds, the biology of poultry species exhibits unique features. This includes the absence of teeth, presence of feathers, skeletal and muscular requirements for flight, presence of internal testes in the male, and adaptations for egg laying in the female. The external anatomy of a male chicken is shown in Figure 11-1. The digestive and reproductive systems are also covered in Chapters 13 and 16, respectively.

Poultry production in the United States, Europe, and increasingly also in Asia is vertically coordinated. This involves contractual relationships between the genetics companies, breeders, growers or egg producers, and integrators or processors (meat or egg). The most common models are either production contracting or ownership of all stages of production (complete vertical integration).

Production contracting has the following characteristics:

- The integrator owns some or all of the production inputs, such as feed mills and hatcheries for baby chicks.
- The integrator provides veterinary services.
- The integrator processes and markets the meat or eggs.

FIGURE 11-1 External anatomy of a male chicken. Source: Delmar/Cengage Learning.

- The grower raises the birds in the grower's own facilities.
- The grower provides labor and a poultry house.
- The integrator plays a significant role in the grower's decision making.

The system of production contracting provides the processor with chickens and turkeys of the desired weight and quality. The grower receives a guaranteed market at a known price per pound (or kilogram) and premiums/deductions based on performance/quality metrics.

The poultry industry is segmented into the following:

- Genetics companies/primary breeders
- Egg producers
- Meat chicken ("broiler") producers
- Turkey producers
- Other poultry, including ducks, geese, ratites (ostriches and emus), together with game birds
- Specialty poultry (such as free-range and organic)

 # DOMESTICATION OF POULTRY

Table 11-1 summarizes the whens and wheres of domestication of different poultry species.

TABLE 11-1 The where's and when's of poultry domestication

Chickens	Northeast area of present-day China along the Yellow River	5000 BC
Turkeys	Meso-America	200–2500 BC
Ducks	East Asia – present-day China	3000 BC
	Fertile Crescent	3000 BC
Muscovys	South America	1000 BC
Geese		
Anser cynoides	East Asia – present-day China	3000 BC
Anser anser	Fertile Crescent	3000 BC
Guinea fowl	Sub-Saharan Africa	>1000 BC
Pigeons	Fertile Crescent	3000 BC
Ostriches	South Africa	1857
Emus	Australia	Within the last 100 years

Chickens

Based on archaeology, chickens were domesticated from jungle fowl (*Gallus gallus*) along the Yellow River in what is today Northeast China around 5500 BC (see Table 11-1). Based on DNA sequencing comparisons between modern breeds of chickens and various populations of jungle fowl, it is thought that the ancestor of the domestic chicken is the red jungle fowl (*Gallus gallus gallus*), a subspecies found in Southeast Asia (today's Thailand, southern Vietnam, and Cambodia). How, then, did the red jungle fowl get to Northeast China? It is speculated that captive jungle fowl were traded, probably multiple times.

Domesticated chickens were then traded, arriving in present-day Iran about 3800 BC, the Indian subcontinent 2500 BC, the Middle East about 2000 BC, present-day Greece 2200 BC, Western Europe 1000 BC, England about 175 BC, and Japan about 150 BC.

There is evidence that chickens were first brought to the Americas by Polynesian traders before the arrival of the Spanish and Portuguese.

Turkeys

Turkeys were domesticated about 4,500 years ago in Mesoamerica. They were taken to Europe by the conquistadors, and there (in the United Kingdom) the process of genetic improvement was begun. It was continued in the United States with crosses to wild turkeys and then scientific genetic improvement.

CLASSIFICATION OF POULTRY

Chickens are classified as *G. gallus*; this is that same species as the jungle fowl. The chicken has been previously classified as *Gallus domesticus*, as, therefore, of a different species than the jungle fowl, but with recent genomic research, it is clear that chickens and jungle fowl are the same species and completely interfertile. Other poultry within the order Galliformes are turkeys together with game birds such as pheasants, guinea fowl, and quail. Ducks and geese are from a separate order (Anseriformes).

REPRODUCTION

Poultry become reproductively mature when exposed to day lengths, that is, lengths of, say, 14 hours of light per day and 10 hours of darkness per night. This is known as photostimulation. This is the same as wild birds becoming reproductively active in the spring. It is important that broiler breeders, which are the parent stock for the production of meat chickens, and breeder turkeys are not overweight when photostimulated. This is achieved by restricting feed/energy availability. When hens are lit or photostimulated, the calcium concentration of the feed is greatly increased to meet the anticipated needs of egg production and to allow for calcium to be stored in medullary bone, which are the centers of long bones. The storage of calcium into the medullary bones occurs under stimulation by two ovarian hormones: estrogens, such as estradiol; and androgens, such as testosterone.

Artificial Insemination

Because of the size difference between the male and female and the potential for the male to damage the female during mating in today's very large turkeys and broiler breeders, artificial insemination is used widely in the industry.

Sexing Poultry or Gender Identification

Because poultry do not have external genitalia, the sex of newly hatched chicks or poults is not readily apparent. It is possible to identify the sex by inspection of the vent, but this is labor intensive and requires considerable skill. An alternative

approach used by the industry is one in which there are differences in the colors of feathers because of sex-linked genes associated with color.

Secondary Sex Characteristics

An adult rooster is easily distinguished from a hen by its secondary sex characteristics, namely the comb, wattle, and spurs (see fig. 11-2).

A sexually mature tom turkey has a characteristic blue appearance to the wattle around the head together with a snood hanging down from the top of the head. These attributes of male poultry are induced by androgens such as testosterone produced by the testes.

Forced or Induced Molting

Forced or induced molting is a management technique to increase the number and size of eggs. Forced molting causes egg production to cease. Afterward, both egg production and egg size rebound to markedly higher levels than before the forced molt. Molting is induced by reducing feed availability and the hours of light so that the ovaries regress, and stop producing estrogens, progesterone, and androgens. This, in turn, leads to the oviduct regressing to the size in immature animals together with loss of feathers. In the United States, about 3% of all laying hens are forced molted every month.

Caponization

Caponization is the removal of the testes or castration of a male chicken. This is analogous to castration in barrows, steers, cats, and dogs. Because it involves opening the abdomen, it is much more akin to spaying female dogs and cats, and there are concomitant animal health and welfare issues. Caponization used to be done extensively because of the improved meat quality in capons. Today, it is very rare because of the tremendous improvement in poultry growth rates and excellent meat quality together with such issues as high labor costs and animal welfare.

GENETICS

Poultry genetics contributed much to early genetic research because of the ease of breeding chickens and for the simple traits with single gene inheritance such as type of comb or color. The chicken genome is well characterized, has been sequenced, and has 1.2 billion base pairs. Chickens and turkeys have 78 (39 pairs) and 82 (41 pairs) chromosomes, respectively. The chromosomes of poultry (see fig. 11-3) are macro-chromosomes (eight pairs), sex chromosomes (ZZ in males and ZW in females), and micro-chromosomes (30 pairs in chickens and 32 pairs in turkeys).

Much of the variation seen in poultry comes from genetic differences. Strong programs of genetic selection have led to the greatly increased growth rate of poultry (discussed in more detail in Chapter 15), number of eggs produced, and feed efficiency.

Selection occurs in grandparent stock with the parent stock serving as a multiplication stage, and to cross such, there is hybrid vigor.

BREEDS OF POULTRY

There are many breeds of poultry, including white leghorn, which is ancestor stock for laying strains of chickens that produce white eggs (see fig. 11-4); and the adult Plymouth Rock chicken, which is one of the ancestors for today's meat chickens.

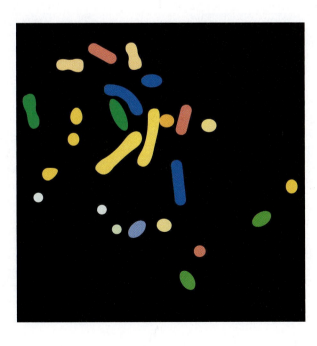

FIGURE 11-3 Chromosomes of chicken "painted" to show different chromosomes.

FIGURE 11-4 Adult white leghorn rooster. Courtesy of the U.S. Department of Agriculture. Photo by Stephen Ausmus.

Commercial chickens for either eggs or meat are lines selected by commercial breeding companies. Chicks are supplied from hatcheries to either growers (meat chickens) or to be raised as pullets (ultimately for egg production) (see fig. 11-5).

FIGURE 11-5 Baby chicks. Reproduced by permission from Ariusz Nawrocki. © 2010 by Shutterstock.com.

PLUMAGE COLOR

The color of skin, hair/fur, and feathers is predominantly because of the different amounts of two different pigments: eumelanin and pheomelanin. The synthesis of both these melanins begins with the oxidation of the amino acid tyrosine catalyzed by the enzyme tyrosinase. Changes in the tyrosinase gene have been linked to white plumage in chickens.

GLOBAL POULTRY PRODUCTION

Production of poultry meat and eggs is increasing globally. Table 11-2 shows the increase in the production of poultry and eggs in the world.

Poultry are produced worldwide by four systems:

1. Commercial large-scale poultry production systems (in developed countries, countries in transition, and developing countries)
2. Traditional village scavenging poultry (largely in developing countries)
3. Semicommercial systems (close to urban areas in developing countries)
4. Alternate systems, including free range and organic (predominantly in developed countries)

TABLE 11–2 World of poultry meat and eggs (in million metric tons) for 2007: Comparison to 1997

COMMODITY	1997	2007
Chicken meat	50.7	75.8
Turkey meat	4.8	5.9
Duck meat	2.4	3.6
Goose and guinea fowl meat	1.7	2.2
Chicken eggs	46.6	59.0
Other eggs	3.6	4.4

Source: FAO.

Traditional Village Scavenging Poultry

In developing countries, traditional poultry are a significant proportion of poultry in the national flock. Village poultry contribute significantly to poverty alleviation and the improvement of food security. They provide a significant source of high-quality protein that is so important for young children. Poultry products can be sold or bartered to provide a source of income, which may be the only sources of cash income. Women and children frequently care for the poultry and receive the benefit. This system of scavenging poultry in the village is similar "backyard" or "farm yard" poultry in Europe and North America of 50 years ago and more. These have markedly decreased but have not disappeared completely.

In 1997, it was estimated that 70% of African poultry (1.5 billion chickens) were of the traditional village scavenging type, and between 15% (Malaysia) and

85% (Myanmar) in Asian countries. Examples of traditional village scavenging poultry include the following:

- Tanzania (East Africa), with 26.6 million scavenging chickens and 1.2 million other poultry, mainly ducks.
- Vietnam, with 86 million chickens and 29 million ducks, with 75% maintained in the traditional village system.
- In Ethiopia, a village chicken provides 12.5 kg of meat per year compared with cattle, with only 5.34 kg.

Are there issues or concerns with traditional village poultry production? Traditional village poultry production may be a "half way house" to being a grower in a vertically integrated system. There are problems, however. Vitamins and other critical nutrients together with vaccines are frequently not available. Traditional village poultry are a potential reservoir for diseases, including zoonotic diseases such as avian influenza.

Poultry Production in the United States

The United States is the world's largest producer and second-largest exporter (after Brazil) of poultry meat. Production of chicken meat has overtaken beef in terms of weight, but beef production represents sales of $35 billion, whereas those of chicken meat are $23 billion. The retail cuts of a chicken are shown in Figure 11-6. Much of the chicken produced in the United States is further processed such that we buy breast (or boned breast meat), thighs, or wings in the grocery store. Poultry meat is considered in this chapter rather than in Chapter 25 for convenience.

Poultry has long been a major feature of agriculture in the United States, but its rate of growth has been impressive. U.S. production of chicken meat in 1970 was 10 billion lb, in 1990 27 billion lb, and in 2008 51 billion lb. U.S. production of eggs in 1996 was 76 billion eggs, and in 2008 90 billion eggs.

Turkey production has shown little change for the last 10 years, being 7.9 billion lb in 2008. (An adult male turkey is illustrated in fig. 11-7.)

The top states for chicken meat production (as of 2008) are the following (data from the USDA National Agricultural Statistics Service):

1. Georgia: 7.5 billion lb
2. Arkansas: 6.4 billion lb
3. Alabama: 5.8 billion lb
4. North Carolina: 5.5 billion lb
5. Mississippi: 4.9 billion lb
6. Texas: 3.5 billion lb
7. The Delmarva states (Delaware, Maryland, and Virginia): 4.4 billion lb

The top states for turkey meat production in 2008 are the following (data from the USDA National Agricultural Statistics Service):

1. Minnesota: 1.3 billion lb
2. North Carolina: 1.2 billion lb
3. Missouri: 0.7 billion lb
4. Arkansas: 0.6 billion lb
5. Virginia: 0.5 billion lb

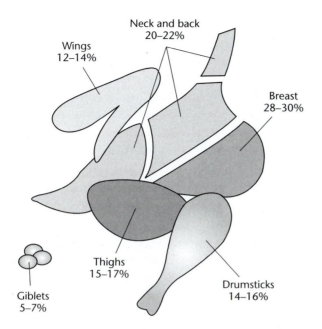

Wings
12–14%

Neck and back
20–22%

Breast
28–30%

Thighs
15–17%

Drumsticks
14–16%

Giblets
5–7%

FIGURE 11-6 *A,* Retail cuts of a chicken. *B,* Roast chicken (This can readily be compared to *A* above.) Reproduced by permission from Tomo Jesenicnik. © 2010 by Shutterstock.com.

And for egg production (2008), the top states are as follows:

1. Iowa: 14.4 billion eggs
2. Ohio: 7.3 billion eggs
3. Indiana: 6.5 billion eggs
4. Pennsylvania: 6.2 billion eggs
5. California: 5.3 billion eggs
6. Texas: 4.9 billion eggs

 PRODUCTION OF MEAT, CHICKENS, AND TURKEYS

Fertilized eggs from either broiler breeders or breeding turkeys are incubated at hatcheries. Vaccination against disease is performed into the shell and/or after hatching. The baby chicks or turkey poults are moved to large houses and placed on litter under brooding condition with elevated temperatures.

Poultry Housing

Chickens are placed in large grow-out facilities with a controlled environment. An example of the type of housing for meat-type chickens with large numbers housed together is shown in Figure 11-8.

Ventilation

Adequate ventilation is especially important for poultry raised on deep litter (see fig. 11-9). Ventilation has multiple beneficial effects, including the following:

- Reducing the humidity of the air around the birds and thereby facilitating evaporation-based cooling
- Allowing drying of the litter, and, consequently, reducing pathogens and aiding the health of the birds' feet

It is very important to eliminate ammonia, carbon dioxide, carbon monoxide, dust, and other noxious elements from the air in poultry houses. A deleterious environment impairs poultry growth and renders the birds susceptible to disease.

FIGURE 11–8 Raising large meat-type chickens on deep litter. Courtesy of William Dozier, Agricultural Research Service, USDA.

FIGURE 11–9 Ventilation fans in a turkey grow-out house. Courtesy of Hess.

COMPOSITION OF A HEN'S EGG

The overall composition of a hen's egg includes the following:

- Albumin or egg white
  - 87% water
  - 11% protein
  - 1% mineral
  - 1% carbohydrate

- Yolk
  - 49% water
  - 16% protein
  - 33% lipid
  - 1% mineral
  - 1% carbohydrate

- Shell
  - 95% mineral calcium carbonate ($CaCO_3$)

The composition of egg white proteins or albumin is as follows:

54% ovalbumin
13% ovotransferrin
11% ovomucoids
8% ovoglobulins
3% lysozyme
<1% avidin

The composition of yolk is as follows:

- Lipid
  - 70% triglyceride
  - 25% phospholipids
  - 5% cholesterol

- Protein (one third of the solids)
  - Granules
    - Phosphoproteins (e.g., phosvitin)
    - Lipoproteins (e.g., lipovitellin)
    - Calcium
  - Soluble proteins (e.g., antibodies)

Litter

Meat chickens and turkeys are raised on deep litter to aid bird health by absorbing water spilled from waters and from excreta. Litter can be wood shavings, sawdust, ground straw, and other fibrous materials. The litter should be 3 to 6 in (approximately 8 cm) deep. There is fermentation of excreta/litter mixture generating some heat.

Litter is now not generally removed after the poultry have been harvested at the end of a run. Fresh litter may be added, particularly in the brooder areas. Litter is ultimately removed after a number of runs when it has reached a critical depth and/or unacceptable moisture level. The used litter can be land applied, where it is an excellent fertilizer providing both nitrogen and phosphorus for plant growth.

Nutrition

Chickens or turkeys are raised with the optimal nutrition for the stages of growth to ensure rapid growth and high-feed efficiency.

Poultry Processing

Birds are collected (or harvested) from the grow-out facility and transported by road to the processing facility, where they are slaughtered, cleaned, and eviscerated. The USDA has a zero-tolerance policy for visible contamination of the carcass with fecal material. To achieve this and reduce food-borne pathogens, processors have a variety of approaches, including inside-outside bird washes and antimicrobial chemicals, such as sodium hypochlorite added to carcass washing water, brushes, and dip tanks.

Production of Eggs

The majority of the eggs produced in the United States are from birds in cages (see fig. 11-10). Each hen produces more than 250 eggs per year. The eggs are collected automatically. Feed is delivered by conveyer belt. Water is available through nipple waterers. Excreta falls from the cages and is later removed.

Eggs are sold as table eggs of various sizes (see Table 11-3) or are broken for the food industry (see fig. 11-11). Small or peewee eggs are rarely available in stores; instead, they are processed. Eggs are components for fast-food breakfasts (e.g., Egg McMuffins), pancakes, cakes, cookies, and many other food products (see fig. 11-12).

Poultry Nutrition

The nutrition of poultry is covered in detail in Chapter 14. The feed for meat-type chickens is formulated for maximal growth rates. Laying hens are placed on an optimal diet for egg production. This is an unusual diet in that it needs to contain so much calcium to allow eggshell formation. The feed for laying hens needs to contain about 2% calcium to accommodate the requirements for eggshell formation.

FIGURE 11-10 Laying hens that produce eggs are frequently housed in battery cages.

TABLE 11-3 Sizes of eggs (USDA definitions)

SIZE OR WEIGHT CLASS	MINIMUM WEIGHT PER DOZEN (OUNCES)
Jumbo	30
Extra-large	27
Large	24
Medium	21
Small	18
Peewee	15

OTHER POULTRY: DUCKS, GEESE, RATITES (OSTRICHES AND EMUS), GAME BIRDS, AND PIGEONS

Ducks and Geese Poultry Production in the United States

Ducks are raised for meat in the United States, with over 100 million lb produced each year. Most ducks go to market at about 7 weeks of age. Duck meat is dark and high in fat. A dressed duck has about 39% fat, 12% protein, and 49% water. Duck is popular in white-tablecloth (fine-dining) restaurants (an example of a duck dish is duck with orange sauce) and for Chinese food.

Peking duck are widely used in duck production because of their rapid growth and white color (see fig. 11-13). When dressed, any remaining white feathers are much less noticeable or offensive. Peking duck is also the name for a Chinese food dish and is served with a special plum sauce.

Production of geese (see fig. 11-14) is smaller than that of ducks in the United States. In contrast, there is major production of both ducks and geese internationally, with about 4 billion lb produced annually in China and significant production of eggs from ducks and geese. Both duck and goose feathers, including the small

FIGURE 11–11 Egg production in the United States is represented here by attractive-looking eggs in a bowl. Courtesy of Iowa Poultry Association/Iowa Egg Council.

under-feathers (i.e., the down), are used to produce winter clothing and bed coverings due to their ability to provide insulation.

Ratite (Ostriches and Emus) Production in the United States

The ratites are large flightless birds such as the ostrich (*Struthio camelus*, originating from Africa), emu (*Dromaius novae-hollandiae*, originating from Australia), and

FIGURE 11–12 *A, B.* Various uses for eggs.

A. An egg salad sandwich. Reproduced by permission from Lorraine Kourafas. © 2010 by Shutterstock.com.

B. Quiche. Reproduced by permission from Laurent Renault. © 2010 by Shutterstock.com.

rhea (*Rhea americana*, originating from South America). The ratites are classified as a super order (Ratatae) in the class Aves (i.e., birds). Ostriches and emus are domesticated, with some rhea also farmed. Figure 11-15 illustrates the ratites with pictures of an ostrich egg, ostrich chick, and adult ostrich. Ratites are raised in ranch-like conditions on pasture or in paddocks.

FIGURE 11-13 Although U.S. production of ducks is low relative to that of chicken or turkeys, globally, ducks are very important for both meat and eggs. A major meat duck is the white Peking duck. Courtesy of Jurgielwicz Duck Farm.

FIGURE 11–14 Toulouse goose. Reproduced by permission from lynnlin. © 2010 by Shutterstock.com.

FIGURE 11–15 *A, B, C.* Production of ostriches is becoming more important.

A. An ostrich egg (*left*) compared to a chicken egg (*right*). Reproduced by permission from jirkaejc. © 2010 by Shutterstock.com.

B. An ostrich chick. Reproduced by permission from Eric Isselee. © 2010 by Shutterstock.com.

C. An adult ostrich. Reproduced by permission from MaxPhoto. © 2010 by Shutterstock.com.

FIGURE 11-16 *A, B, C.*
Products made from ostrich.

A. Boots. Reproduced by permission from Jeff R. Clow. © 2010 by Shutterstock.com.

B. Grilled ostrich steak. Reproduced by permission from erkanupan. © 2010 by Shutterstock.com.

Ostriches are raised principally for their hide, meat, and feathers, and some eggs are also sold (see fig. 11-16.) They go to market at about 190 lb (9 months of age). Ostrich meat is very low in fat, with less than 1% fat, but high in protein (21%). Emus are raised to produce meat, hides (but the tanning quality is inferior to that of ostrich hide), and feathers, but the highest-value product is the fat/oil that is used in the cosmetics industry.

Game Birds and Pigeons

Game birds include pheasants, guinea fowl, partridge, and quail. These are predominantly raised for release and then hunting. In addition, there is some processing

C. Feather duster. Reproduced by permission from Mark Stout Photograph. © 2010 by Shutterstock.com.

for meat. Peacocks are raised as ornamental birds. Pigeons are raised for the sport of racing, showing, and limited meat production in the United States.

 ## SPECIALTY CHICKEN MEAT/EGGS AND TURKEY MEAT

In the United States, free-range chicken (access to the outdoors) and both organic chicken and turkey are available. Moreover, eggs can be sold as organic and from cage-free chickens. Although free-range systems are reputed to be associated with improved animal welfare, there are downsides. For instance, there is a greater incidence of diseases such as coccidiosis (caused by protozoan parasites) and more losses because of predation. Moreover, there are reports of higher concentrations of dioxins and other toxicants in the poultry products because of the consumption of contaminated soil and bio-accumulation in earthworms.

In France and other parts of Western Europe, there are specialty chickens that are claimed to have superior eating qualities. These are the Label Rouge program chickens, requiring slow-growing genetics, access to the outdoors (as in free range), and a low-protein diet.

Research is beginning on these slow-growing chickens in the United States. Early research results suggest that the meat from slow-growing birds is more tender, is higher in protein, has reduced water-holding capacity, and has lower fat content.

The topic of specialty animal products is also discussed in Chapter 27.

 ## TRANSGENIC CHICKENS

Transgenic chickens have been produced by the use of retrovirus vector systems using a germ line placed into host chicken embryos. In the avian embryo, primordial germ cells (PGCs) are initially found in the central area of the zona pellucida and then on the ventral surface of the epiblast. The PGCs move to the hypoblast and then to the germinal crest region, where they enter the circulatory system. They leave the blood vessels at the germinal epithelium and migrate to the area that will become the gonads.

It has been possible to introduce transgenes into embryonic stem cells. If these are administered into chicken embryos (where the PGCs are destroyed chemically or by radiation), these become incorporated into the chicken gonads as PGCs and, once the resulting chicken is sexually mature, into the gametes. The next generation is a transgenic chicken. Helen Sang in Scotland has successfully produced transgenic chickens that produce high concentrations of functional recombinant therapeutic proteins in the egg white. These potential drugs include a humanized mini-antibody for the treatment of malignant melanoma and human interferon-α-1a (hIFNβ1a).

REVIEW QUESTIONS

1. What are the major anatomic features of a chicken?

2. What are the names for adult male, adult female, newly hatched, and castrated chickens?

3. What are the names for adult male, adult female, and newly hatched turkeys?

4. Which are the major states for the production of chicken and turkey meats and eggs?

5. What are the major countries for production of chicken meat and eggs?

6. Where and when were chickens domesticated?

7. What is the system for poultry production in Western countries and increasingly in Asia?

8. How is the poultry industry segmented in Western countries?

9. Why is *Gallus domesticus* not the best term for the chicken species?

10. What are the species names for domestic ducks?

11. Where and when was the turkey, duck, goose, emu, and ostrich domesticated?

12. What are the environmental conditions necessary for poultry to become sexually mature?

13. Are there specific nutritional requirement for sexually mature female poultry?

14. Where is calcium stored in laying hens?

15. Which hormones stimulate the formation of medullary bones?

16. Is artificial insemination performed in any poultry species on a large scale, and, if yes, why?

17. How is the sex of newly hatched chicks determined?

18. What are the secondary sex characteristics of a rooster?

19. What are the secondary sex characteristics of an adult male turkey?

20. What hormones cause the development of secondary sex characteristics in roosters and adult male turkeys?

21. What is forced molting?

22. How are hens induced or forced to molt?

23. Why was poultry genetics important to early research on genetics?

24. How big is the chicken genome?

25. How many chromosomes are there in the chicken?

26. How has genetics affected the poultry industry?

27. Give two examples of breeds of chickens.

28. Are the same genetic lines of chickens used for egg or meat production?

29. What are hatcheries?

30. Is the production of poultry meat and eggs increasing globally?

31. What are the four poultry production systems globally?

32. How do village scavenging poultry contribute significantly to poverty alleviation and the improvement of food security in developing countries?

33. How many chickens are in Africa? What proportion of these is in traditional systems?

34. Which two countries are the top producers of chicken meat?

35. How has production of turkeys changed in the last 10 years?

36. What are the five top-producing states in the United States for chicken meat?

37. What are the three top-producing states in the United States for turkey?

38. What are the six top-producing states in the United States for eggs?

39. When in their lives are chickens first vaccinated?

40. How are baby chicks or turkey poults raised initially?

41. Why is ventilation required for raising poultry raised on deep litter?

42. What is deep litter? How is it used in raising chickens and turkeys?

43. How do processors reduce food-borne pathogens?

44. Are the majority of the eggs produced in the United States from birds in cages?

45. Give examples of foods produced using eggs.

46. What are the components of eggs? What are the chemical constituents of each?

47. At what age do ducks go to market?

48. Is duck meat high in fat?

49. What country produces the highest amounts of duck and goose meat?

50. What is a transgenic chicken, and what are they used for?

51. How are duck and goose feathers used?

52. What are ratites?

53. What are the products from ostriches?

REFERENCES AND FURTHER READING

Farrelly, L. L. (1996). *Transforming poultry production and marketing in developing countries: Lessons learned with implications for sub-Saharan Africa.* International Development Working Papers 63, Department of Agricultural Economics, Michigan State University. Retrieved July 25, 2009, from http://ideas.repec.org/cgi-bin/ref.cgi?handle=RePEc:msu:idpwrk:063&output=0

Kitalyi, A. J. (1997). Village chicken production systems in developing countries: What does the future hold? *World Animal Review, 89*. Retrieved July 25, 2009, from http://www.fao.org/docrep/W6437T/w6437t07.htm

Kyeema Foundation. *Improvements in rural poultry in developing countries*. Retrieved July 25, 2009, from http://www.kyeemafoundation.org/rural_poultry/index.html

Lillico, S. G., Sherman, A., McGrew, M. J., Robertson, C. D., Smith, J., Haslam, C., et al. (2007). Oviduct-specific expression of two therapeutic proteins in transgenic hens. *Proceedings of the National Academy of Sciences of the United States of America, 104*, 1771–1776.

Scanes, C. G., Brant, G., & Ensminger, M. E. (2003). *Poultry science* (4th ed.). Upper Saddle River, NJ: Prentice Hall.

U.S. Department of Agriculture. *Chicken and eggs 2008 summary*. Retrieved October 26, 2009, from http://usda.mannlib.cornell.edu/usda/current/ChickEgg/ChickEgg-02-26-2009.pdf

U.S. Department of Agriculture. *Poultry–production and value 2008 summary*. Retrieved October 26, 2009, from http://www.usda.gov/nass/PUBS/TODAYRPT/plva0509.pdf

U.S. Department of Agriculture National Agricultural Statistics Service. *Broiler pounds produced United States, 1968–2008*. Retrieved July 17, 2009, from http://www.nass.usda.gov/Charts_and_Maps/Poultry/brlprd.asp

CHAPTER 12

Aquaculture

OBJECTIVES

This chapter will consider the following:

- Introduction to aquaculture
- Domestication of aquaculture species
- Importance of aquaculture globally
- Aquaculture in the United States
- Catfish
- Other aquaculture species
- Future of aquaculture

INTRODUCTION TO AQUACULTURE

We get fish, including shellfish, from three sources: commercial landings (also called *catches* or *captures*); aquaculture, including mariculture; and recreational fishing.

What is Aquaculture?

The National Oceanic and Atmospheric Administration, part of the U.S. Department of Commerce, defines *aquaculture* as follows: "The propagation and rearing of aquatic organisms in controlled or selected environments for any commercial, recreational or public purpose." Aquaculture includes the production of freshwater, brackish water, and seawater species, the latter being called *marine aquaculture* or *mariculture*. Aquaculture species are categorized into two groups: finfish (vertebrates) and shellfish. Shellfish are then divided under two subcategories: crustaceans, such as crabs, shrimp, lobster, and crawfish (see fig. 12-1); and mollusks, including clams, mussels, oysters, and scallops (see fig. 12-2).

DOMESTICATION OF AQUACULTURE SPECIES

Carp and the Nile tilapia were domesticated in the Fertile Crescent over 2,000 years ago, and aquaculture systems were developed to raise these fish. Carp were also domesticated in what is now China. In contrast, most species produced today

Definitions

Aquaculture is defined as "the farming of aquatic organisms in inland and coastal areas, involving intervention in the rearing process to enhance production and the individual or corporate ownership of the stock being cultivated."

Mariculture is defined as "the cultivation, management and harvesting of marine organisms in their natural habitat or in specially constructed rearing units, e.g. ponds, cages, pens, enclosures or tanks."

Good aquaculture practices are defined as "those practices of the aquaculture sector that are necessary to produce quality food products conforming to food laws and regulations."

Source: United Nations Food and Agriculture Organization.

FIGURE 12–1 Shrimp are major aquaculture species. Courtesy of the U.S. Department of Agriculture. Photo by Ken Hammond.

FIGURE 12-2 Littleneck clams are produced by aquaculture. Courtesy of the U.S. Department of Agriculture. Photo by Ken Hammond.

in freshwater aquaculture and virtually all of those in marine aquaculture were domesticated during your or your parents' lifetime (see Tables 12-1 and 12-2). It is clear that most of the technology and approaches are also of a recent vintage.

 IMPORTANCE OF AQUACULTURE GLOBALLY

We are moving from the hunter-gatherer stage of development to fish farming/ranching with aquaculture. This reflects the growing consumer demand and the

TABLE 12-1 Date of domestication of selected freshwater animals for aquaculture

SPECIES	YEAR
Finfish or vertebrate fish	
Carp	800 BC
Tilapia (Nile)	2000 BC
Rainbow trout	AD 1875
Channel catfish	AD 1890
Striped bass	1985
Crustaceans	
Crayfish species	1950–1985

Source: Based on data summarized by Duarte et al. (2007).

TABLE 12-2 Date of domestication of selected marine animals for mariculture

SPECIES	YEAR (AD)
Mollusks	
Abalone	1970
Oyster	1900
Mussels	1950–1980
Crustaceans	
Shrimp and prawns	1980–1990
Finfish or vertebrate fish	
Salmon (Atlantic)	1965
Salmon (Chinook)	1985
Salmon (Coho)	1960
Tuna (yellowtail)	1999

Source: Based on data summarized by Duarte et al. (2007).

is the Pacific salmon species, e.g., *Oncorhychus kisutch* (Coho salmon), *Oncorhychus tshawytscha* (Chinook salmon) (fig. 12-3), and *Oncorhychus mikiss* (rainbow trout).

The classification of catfish (finfish) is class, Actinopterygii; order, Siluriformes; and genus, *Ictalurus*, e.g., channel catfish (*Ictalurus punctatus*).

The classification of tilapia (finfish) is class, Actinopterygii; and order, Perciformes. There are multiple species (see fig. 12-4).

problem of overfishing. Marine landing of fish is showing little changes over the last 15 years (see Table 12-3). There is concern that, based on present trends, most of the world's commercial fisheries may collapse in the next 10–30 years.

Many species are now raised by aquaculture, with many domesticated in the 20th century (see the section "Domestication of Aquaculture Species" for discussion). Globally, there are major increases in aquaculture production, with the FAO estimating an annual growth rate of about 10% for all of the last 20 years. During the last 9 years, freshwater/inland aquaculture has grown by 90.4% (see Table 12-3), and marine aquaculture has grown by 98.2% (see Table 12-4).

Over 95% of carp and oysters are produced by aquaculture; over 70% of salmon, trout, and tilapia are produced by aquaculture; and over 40% of shrimp are produced by aquaculture.

FIGURE 12-3 Among the significant species of finfish produced by aquaculture in the United States are salmon. Reproduced by permission from Natalia Bratslavsky. © 2010 by Shutterstock.com.

FIGURE 12-4 Tilapia. Reproduced by permission from LouLouPhotos. © 2010 by Shutterstock.com.

Carp are the number-one group of species raised by aquaculture in the world. There are 26 species of carp that have been domesticated. A few species dominate: Chinese carp (silver, grass, bighead, crucian, black, and mud), common carp, and Indian major carps (rohu [=*Roho labeo*], catla, and mrigal).

China is by far the largest producer of carp, with the following species dominating: silver carp (30%), grass carp (25%), common carp (21%), and bighead carp (15%).

Over 99% of salmonids that are produced by aquaculture worldwide come from only four different species: Atlantic salmon, Coho salmon, Chinook salmon,

TABLE 12-3 The magnitude and growth of freshwater/inland aquaculture in comparison to commercial capture or catching of fish in the world

YEAR	AQUACULTURE PRODUCTION (LIVE WEIGHT) IN MT*	CATCH (MILLION MT)	TOTAL (MILLION MT)	PERCENT AQUACULTURE
1995	13.5	7.3	20.8	64.9
2004	25.7	9.2	34.9	73.6

Source: Data from FAO and National Oceanic and Atmospheric Administration.
*mt = million metric tons.

TABLE 12-4 The magnitude and growth of marine aquaculture in comparison to commercial capture or catching of fish in the world

YEAR	AQUACULTURE PRODUCTION IN MT*	CATCH (MILLION MT)	TOTAL (MILLION MT)	PERCENT AQUACULTURE
1995	10.9	85.0	95.9	11.4
2004	19.7	85.8	105.5	18.7

Source: Data from FAO and National Oceanic and Atmospheric Administration.
*mt = million metric tons.

TABLE 12-5 Top 10 aquaculture countries

RANKING	COUNTRY	PRODUCTION IN 2007 (MILLION METRIC TONS)
1	China	30.6
2	India	2.5
3	Vietnam	1.20
4	Thailand	1.17
5	Indonesia	1.05
6	Bangladesh	0.91
7	Japan	0.78
8	Chile	0.67
9	Norway	0.64
10	United States	0.61

and rainbow trout. Among the leading producers of salmon are Norway, Chile, and the United Kingdom, whereas the leading countries for trout production are France, Chile, Denmark, and Italy. The United States is the major producer of channel catfish (*I. punctatus*), with Vietnam and other Asian countries rapidly increasing production of several catfish species.

The top aquaculture countries are shown in Table 12-5.

Exporters of Seafood

China has both high consumption (in 2007, 57 lb or 26 kg per capita) of fish or "seafood" and is the number-one-ranked country for aquaculture production. It is also the largest exporter of seafood. Chile leads Latin America for aquaculture and is a major exporter of seafood, including sea bass, tilapia, and salmon.

 ## AQUACULTURE IN THE UNITED STATES

In the United States, aquaculture production grosses the industry $1.06 billion. Table 12-6 summarizes aquaculture in the United States. There is an opportunity for the industry to become much larger because the United States is a major importer of shellfish and finfish. For example, in 2005, the United States imported $12.1 billion in shellfish and finfish, and over 40% of the imports are produced by aquaculture. This reflects the growing demand from consumers for fish.

Aquaculture has been growing in the United States. For instance, catfish production has been increasing tremendously with U.S. production (see the "Interesting Factoid" sidebar on this page). More recently, there were large increases in the production of shellfish between 1999 and 2004. Crawfish production was up 64%, clams were up 96%, and shrimp were up 127%.

Infectious Diseases

There is a considerable problem with infectious diseases in aquaculture species. Infectious disease is not restricted to fish in aquaculture. In noncaptive fish, there are viral, bacterial, and protozoa diseases together with parasite infestation. For example, viral hemorrhagic septicemia virus is a lethal fish virus. This is adversely influencing fish populations in the Great Lakes.

INTERESTING FACTOID

U.S. catfish production is increasing. Annual production of catfish in the United States in 1980 was 23,000 metric tons; in 1990, it was 178,000 metric tons; and in 2004, it was 286,000 metric tons.

TABLE 12-6 Aquaculture in the United States

SPECIES	1999 PRODUCTION (IN THOUSAND METRIC TONS)	2004 PRODUCTION (IN THOUSAND METRIC TONS)	VALUE (IN MILLION DOLLARS)
Finfish (vertebrate fish)			
Catfish	270	286	$439
Trout	27.3	24.9	$57.1
Salmon	17.7	15.1	$56.7
Tilapia	8.05	9.07	$40
Bait fish	7.43	6.33	$45.8
Striped bass	4.41	5.22	$31.4
Shellfish			
Crawfish	19.5	31.9	$42.8
Oysters[a]	8.46	11.9	$80.1
Clams[a]	4.85	9.51	$73.3
Shrimp	2.10	4.77	$21.3
Mussels[a]	0.246	0.269	$4.0
Other[b]	11.0	2.47	$174

[a]Production is reported as meat weight.
[b]Includes ornamental fish, alligators, crabs, scallops, and eels.

There are relatively few Food and Drug Administration–approved drugs for aquaculture species. The drugs available include the antibiotics oxytetracycline, sulfadimethoxine/ormetoprim, and sulfamerazine; formalin for external parasites and fungal infections of eggs; and human chorionic gonadotropin (HCG) to aid spawning.

Antibiotics are used in aquaculture to treat disease but not as a growth promoter. Some 65,000 lb of antibiotics are used in the aquaculture industry annually in the United States. Vaccines for diseases impacting aquaculture species have been developed, and some are commercially available. These are largely for inactivated bacterial pathogens, with some for viruses. There are still significant unmet pharmaceutical needs for aquaculture.

Food Safety

Researchers have found higher levels of organochlorine compounds, including polychlorinated biphenyls (PCBs) and dioxins, in farmed versus wild salmon. The FAO and World Health Organization believe that "the risk of consuming contaminated fish must be weighted in view of the beneficial nutritive effects of fish."

 CATFISH

Catfish production is located in the southeastern United States, predominantly in Mississippi, Alabama, Arkansas, and Louisiana, with much of the production in the Mississippi Delta (see fig. 12-5).

FIGURE 12-5 Harvesting catfish produced by aquaculture. Courtesy of Vance Watson, Mississippi Agriculture and Forestry Experiment Station.

Raising Catfish

Catfish are raised in large rectangular ponds (17.5 acres or 7.1 hectares) considerably larger than the size of football fields (see fig. 12-6). Catfish are also intensively raised in raceways with a high flow of water necessary. Of particular importance to raising catfish is the construction of ponds. The ground needs to be flat. The ground and walls of the pond need to be impermeable to water (e.g., clay).

FIGURE 12-6 Catfish ponds. Courtesy of Vance Watson, Mississippi Agriculture and Forestry Experiment Station.

The ambient temperatures should be 25–27°C (77–81°F) for optimum growth. There needs to be a ready source of clean water for new and existing ponds. The water should have low salinity, low concentrations of hydrogen sulfide, ammonia, nitrite, heavy metals, and other toxicants, and hardness of the water should be between 50 and 100 mg/L calcium carbonate equivalents.

In addition, a catfish enterprise requires clean water (as mentioned previously) and a reliable power source (with backup provision) to aerate the ponds. Growers could experience a catastrophic loss when the oxygen level in the water drops if aeration is stopped. Dissolved oxygen concentration for the water must be above 4 mg/L.

A catfish enterprise also requires a local reliable source of high-protein feed containing vitamins and minerals at an economic price. Veterinary care with Food and Drug Administration–approved drugs to combat diseases is necessary. Infectious disease represents a major potential source of loss to producers. In addition, a management system to optimize growth of the fish and minimize mortality and stress, well-trained reliable employees, and a processing plant for the catfish and a market for the fish are also required.

One of primary geographic areas for raising catfish is the Mississippi Delta. The land is remarkably flat, and the soil has a high content of clay, and, therefore, is relatively impermeable to water leakage. Catfish ponds are made with levees (clay "walls"). Also of critical importance is the ready availability of water from aquifers or groundwater. Production of catfish is about 4,600 lb/acre.

Catfish Biology

A catfish female will produce between 3,000 and 50,000 eggs in a single egg mass. The eggs are 3 mm in diameter and contain yolk. These are spawned ("laid") in the spring, when the water temperatures are above 21°C (70°F). The optimal temperature for spawning and incubation is 25–27°C (77–81°F). Under these conditions, embryos hatch after 4–6 days. The newly hatched catfish, or fry, first complete utilization of the remaining yolk, and then consume a provided formulated diet. As the catfish grow, they move from being called *fry* to *fingerlings*, and then *adults.*

Stages of catfish life are, therefore, embryos in eggs, fry at hatching, fingerlings, and then adults.

Catfish Genetics

There are catfish genetic lines available. These are often developed and/or produced by Land Grant Universities or the USDA's Agricultural Research Service.

Hybrid Catfish

In addition to channel catfish, hybrid catfish are now commercially available in the United States. An example is the hybrid of female channel catfish (*I. punctatus*) × male blue catfish (*Ictalurus furcatus*) (*C × B hybrid*). Companies producing them claim that the hybrid catfish have higher growth rates, improved feed efficiencies, and greater resistance to disease.

Triploid Fish

Chinese catfish (*Clarias fuscus*) are raised not only in Asia, but also in Hawaii. In Hawaii, it has been possible to produce triploid Chinese catfish, which contain triplets of chromosomes instead of the usual pairs in which there is a diploid state. These are reported to have improved production performance.

Catfish Genome

Catfish mitochondrial DNA has been fully sequenced. There are now both linkage and physical maps of chromosomal DNA. The USDA has developed a microsatellite genotyping system to identify genetic lines of catfish.

Food Quality

Earthy or musty off-flavors can be found in farm-raised catfish. These chemicals have been characterized and are produced by blue-green algae (Cyanobacteria) in the aquaculture ponds.

Food Safety

Listeria monocytogenes and other food-borne pathogens have been isolated from channel catfish.

Disease

It is estimated that infectious diseases are responsible for $50–70 million in annual losses to catfish producers in the United States. Examples of diseases in the channel catfish are the following:

- Channel catfish virus is a herpes-type virus that infects channel catfish fry and fingerlings.
- Enteric septicemia is caused by the Gram-negative bacteria *Edwardsiella ictaluri* (responsible for 40% of catfish disease).
- Columnaris is caused by the Gram-negative bacteria *Flavobacterium columnare*.

Vaccines have been developed for both the Gram-negative bacterial pathogens (live vaccines) and channel catfish virus. These are administered to eyed egg stage—that is, 24–48 hours before hatching—channel catfish.

 OTHER AQUACULTURE SPECIES IN THE UNITED STATES

There is considerable potential for the growth of the production by aquaculture of other species of fish in North America, including yellow perch (*Perca flavescens*) and walleye (*Stizostedion vitreum*).

Yellow Perch

Yellow perch have been traditionally served at Friday night fish "frys" in restaurants and community centers in Great Lakes shore communities (see fig. 12-7.) In the 1950s and 1960s, there were over 30 million lb (15,000 metric tons) of yellow

FIGURE 12-7 In the northern United States, there is production of yellow perch (*P. flavescens*) by aquaculture. Courtesy of U.S. Fish and Wildlife Service (USF&W). Photo by Eric Engbretson.

perch harvested in the Great Lakes. The market size for yellow perch, which is 115–150 g or about 1 oz of whole fish, is much smaller than for other aquaculture finfish species. Therefore, it has a considerably shorter grow-out period.

Walleye

Walleye were formerly known as *pike perch* or *walleye perch* (see fig. 12-8). This is a species with considerable potential for expansion of aquaculture production. Commercially, 23.8 million lb were harvested from Lake Erie alone in 1956. Today, there is relatively little walleye harvested commercially except for aquaculture production. Walleye have both high acceptability and name recognition by consumers.

Walleye are supplied for sportfishing and the food-fish market. Production of walleye starts with spawning or stripping eggs. A female may produce up to 300,000 eggs. The eggs are incubated for about 42 days. After hatching, the fry receive a formulated feed (starter diet).

Over 35 million Americans fish, and 35% of anglers in the Great Lakes fish for walleye. Walleye are supplied to state agencies, counties, and municipalities for their lakes, to lake associations, and to angler groups. The walleye are supplied as fry, small or large fingerlings, and adults for stocking purposes for sportfishing.

 ## FUTURE OF AQUACULTURE

Commercial companies have lines of a number of aquaculture species with a markedly improved growth rate and other performance traits such as disease resistance. These have been developed using selection and population genetics. An example is the Speed Line shrimp from SyAqua, which is Sygen International's aquaculture division, with a 10–50% improved growth rate. Using candidate genes and quantitative trait loci will accelerate the development of superior genetics. Given the lack of genetic selection in the past, very rapid progress in improving the genetics of commercial aquaculture species is projected. Another area where rapid progress is likely is using interspecies hybrids.

REVIEW QUESTIONS

1. From where do we get finfish and shellfish?

2. What is aquaculture?

3. Give three examples for each finfish and shellfish produced by aquaculture.

4. What is the difference between captive and domesticated fish?

5. Where they exist, give examples of freshwater finfish domesticated over 1,000 years ago, over 100 years ago, and in the last 10 years.

6. Where they exist, give examples of marine finfish domesticated over 1,000 years ago, over 100 years ago, and in the last 10 years.

7. Where they exist, give examples of shellfish domesticated over 1,000 years ago, over 100 years ago, and in the last 10 years.

8. Give examples of mollusks and crustaceans produced by aquaculture.

9. What are the leading aquaculture countries?

10. What is the growth rate for aquaculture globally and in the United States?

11. What is the size of production by aquaculture globally and in the United States?

12. Which is the major state(s) in the United States producing catfish?

13. What are examples of infectious diseases causing problems with aquaculture?

14. Give examples of drugs approved by the Food and Drug Administration for aquaculture species.

15. How can antibiotics be used in aquaculture species?

16. What are some of the food safety issues when eating products of the aquaculture industry?

17. What is required to produce catfish?

18. Is the solubility of oxygen affected by temperature? If so, what is the impact of this in aquaculture?

19. What are the following as stages in the life cycle of catfish: fry and fingerlings?

20. What is a hybrid catfish?

21. What are diseases of catfish?

22. What is the magnitude of losses because of infectious diseases to U.S. catfish producers?

23. Give examples of northern aquaculture species in the United States.

24. What are some of the factors influencing their production?

REFERENCES AND FURTHER READING

Chapman, F. A. (1992, rev. 2006). *Farm-raised channel catfish*. Gainesville: University of Florida IFAS Extension. Retrieved July 18, 2009, from http://edis.ifas.ufl.edu/FA010

Costello, C., Gaines, S. D., & Lynham, J. (2008). Can catch shares prevent fisheries collapse? *Science, 321*, 1678–1681.

Duarte, C. M., Marbá, N., & Holmer, M. (2007). Ecology. Rapid domestication of marine species. *Science. 316*, 382–383.

Food and Agriculture Organization of the United Nations

FAO AGA Livestock Atlas Series 1. *Global livestock geography: New perspectives on global resources*. Retrieved July, 25, 2009, from http://ergodd.zoo.ox.ac.uk/livatl2/index.htm

Food and Agriculture Organization. *Chapter 8: Production of rabbit skins and hair for textiles*. Retrieved July 18, 2009, from http://www.fao.org/docrep/t1690E/t1690e0a.htm

Storey, S. (2005). Challenges with the development and approval of pharmaceuticals for fish. *The AAPS Journal*, 7, E335–E343.

Summerfelt, R. C. For the North Central Regional Aquaculture Center. (2003). *A white paper on the status and needs of walleye aquaculture in the north central region*. Retrieved July 18, 2009, from http://www.ncrac.org/NR/rdonlyres/EBF93AA4-8F47-4BB9-8E4D-0967299A4CB3/0/walleye32900.pdf

BIOLOGY OF DOMESTIC ANIMALS

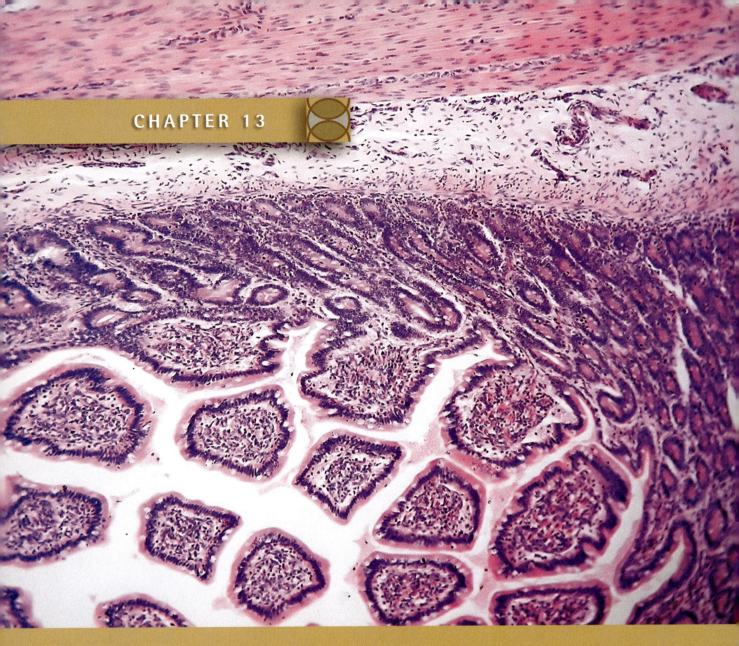

CHAPTER 13

Anatomy and Physiology of Domestic Animals

OBJECTIVES

This chapter will consider the following:

- Cells
- Major organ systems
- The integration of organs/functions by hormones and the nervous system
- Pheromones (signals between members of the same species)

CELLS

The organs of the body are made up of cells. The basic structure of the cell is shown in Figure 13-1. Each cell will normally have the following organelles:

- Nucleus (plural nuclei), which contains the chromosomes. It contains most of the DNA of the cell. This is the site of transcription, that is, production of RNA from the DNA template.
- Cell membrane containing transporters for ions and nutrients, together with receptors for hormones. The phospholipid bilayer allows different ionic environments inside and outside the cell, and an electric potential difference across the cell membrane. These electric potential differences are critically important for nerve conduction, muscle contraction, and hormone release.
- Mitochondria (singular mitochondrion) containing electron transfer proteins that lead to the generation of adenosine 5′-triphosphate (ATP) (i.e., the energy source for the body), some of the metabolic enzymes, and mitochondrial DNA.
- Ribosomes and endoplasmic reticulum are where protein synthesis occurs or translation from the messenger RNA.
- Golgi apparatus is where proteins for export from the cell are packaged into secretory granules.
- Lysosomes are where degradation of waste proteins occurs.
- The cytoplasm contains soluble elements within the cell.

An exception to the rules above is the red blood cell in mammalian species, except camels. Mammalian red blood cells do not contain nuclei or most of the other intracellular organelles. New cells are needed to replace those that die (virtually all cells turn over naturally) or during growth and development. In cell division, it is essential that the daughter cells have the complete set of DNA in the

Definitions

Mitosis is cell division where the daughter cells have the full complement of chromosomes (are diploid). *Meiosis* or *reduction division* is cell division where the daughter cells have half the complement of chromosomes, that is, only one of each pair and are, therefore, haploid. During meiosis there is crossing over or recombination within a pair of chromosomes.

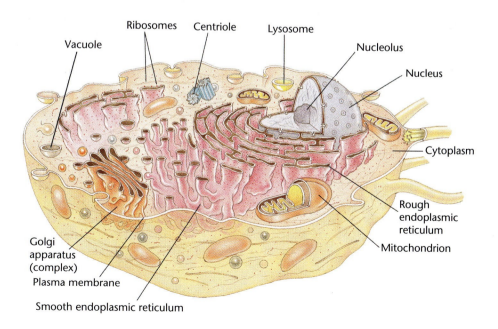

FIGURE 13-1 Basic structure of the cell. Source: Delmar/Cengage Learning.

parent cell. Therefore, there is DNA replication and duplication of the chromosomes by the process of mitosis.

MAJOR ORGAN SYSTEMS

Major organ systems include the gastrointestinal tract (gut), cardiovascular system, respiratory organs, muscle, skeleton, adipose tissue, nervous system and brain, and endocrine organs.

GI Tract (Digestion and Absorption)

It is important to understand the structure and functioning of the GI tract or gut of domestic animals because without it, nutrients will be neither digested nor absorbed. Both the overall size of the gut and its organs varies with whether the animal is herbivorous (large gut) or carnivorous (small gut), with omnivores between the other categories. We shall refer to *feed* as what is eaten by farm animals, *food* as that consumed by people and companion animals, and *ingesta* as what is eaten once it is in the GI track.

The functions of the gut are as follows:

1. Digest the food, first by mechanical grinding of the food and then breaking down complex chemical compounds to simple chemicals by enzymes.
2. Absorb nutrients, including water.
3. Protect against pathogenic organisms.
4. Ferment the feed to provide nutrients.
5. Move ingesta through the GI tract.

The relative importance of fermentation varies with species, with it being essential in ruminants and other plant-eating species. Fermentation also occurs in the hindgut, and this is particularly important in horses and rabbits.

THE FUNCTION AND STRUCTURE OF THE GI TRACT OR GUT

Functions of the GI Tract

The functions of the GI tract are digestion of the food, absorption of the nutrients and water, immunologic protection against pathogenic organisms, fermentation of the feed to provide nutrients, and movement of the ingesta through the gut.

Digestion

Digestion is the chemical breakdown of complex molecules in the food to simple compounds that are absorbed easily. The chemical process is hydrolysis, that is, the addition of water, regardless of whether digesting starch, a branched polymer of glucose and other complex carbohydrates, to glucose, a simple sugar; protein, a linear polymer of different amino acid residues, to amino acids; or triglyceride (fat) to fatty acids and glycerol.

Enzymes catalyze the breakdown of the complex chemicals in digestion. The enzymes that break down starch are called *amylases*; those that break down

Definitions

A *carnivore* is an animal that eats meat. They are typified by very short gastrointestinal (GI) tracts, particularly short hindgut, because meat can be rapidly digested. An *omnivore* is an animal that eats meat and plants. They are typified by medium-length GI tracts. Mixed diets require time for digestion and absorption of the nutrients. They have a hindgut of moderate length to allow some microbial fermentation of ingesta. An *herbivore* is an animal that only eats plants. They are typified by long GI tracts. Plants require considerable time for digestion and absorption of the nutrients. They have adaptations such as a long hindgut and/or rumen to allow microbial fermentation of ingesta, particularly cellulose.

A *ruminant* is a species of animals with a *rumen*, which is a four-chambered organ derived from the stomach. The rumen is the site of microbial fermentation of plant materials, particularly cellulose, to form volatile fatty acids. These are then absorbed by the ruminant.

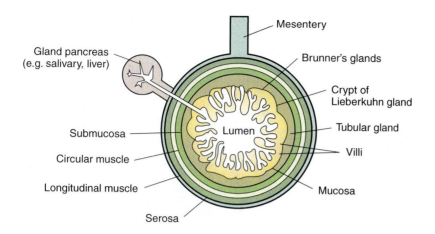

FIGURE 13-2 Section through the small intestine showing the large internal surface areas for absorption of nutrients, the muscular wall that moves the ingesta along the GI tract, and the glandular region producing digestive enzymes.

protein are proteases, and those that break down triglyceride are lipases. This is discussed in more detail in Chapter 14.

The Structure of the GI Tract

The GI tract (or gut) has the same overall structure with a multilayer structure (see fig. 13-2). The layers are the following from inside (lumen) to outside: epithelium, glandular, circular and longitudinal muscle, and external connective tissue layer.

There is the same basic pattern for the organs within the GI tract (see Table 13-2 and fig. 13-3) across species:

1. The mouth (with teeth except in birds). Here, food is chewed, and saliva from the salivary glands is added. The latter lubricates the ingesta by addition of water and mucus, and the presence of the enzyme amylase starts digestion of starch.

2. The esophagus links the mouth with the stomach and, therefore, carries ingesta to the stomach. There is an outgrowth of the esophagus in many birds called the *crop*. It is relatively small (<10% of the gut), with lactic acid being a major product of fermentation of the ingesta stored there.

3. The stomach is the site of the muscular mixing (and grinding in some species) of the ingesta. There is the addition of gastric secretions (or juices) containing hydrochloric acid and pepsinogen. This is converted into the active enzyme pepsin by the acid conditions. Pepsin starts protein digestion. The rumen is a four-chambered structure (see figs. 13-3 and 13-4) formed within the stomach. The avian stomach (see fig. 13-3) has two parts: the gizzard that grinds ingesta and the proventriculus where gastric secretions are added.

4. The small intestine is made up of three parts, from front (anterior) to back (posterior): duodenum, jejunum, and ileum. It is 30% of the gut, and the site of carbohydrates, protein, and fat digestion and absorption of products, together with vitamins and minerals. Digestion occurs at the anterior end, and absorption increases along the small intestine. Digestive enzymes are added to the ingesta. These are produced by the glands in the small intestine and pancreas. The proteases, such as trypsin and chymotrypsin, digest proteins.

FIGURE 13-3 Structure of the GI tract of various domestic animal species showing the different regions of the gut.

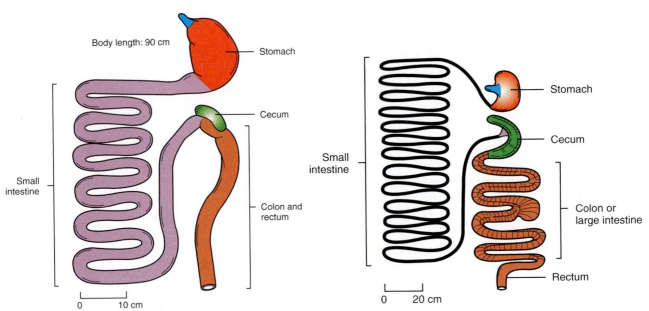

A. Dog (large breed with a height about 30 in and length 35 in). Length of intestine is about 8 1/2 ft or 100 in (2.4 m).

B. Pig. Length of intestine is about 41 1/2 ft or 500 in (12.5 m).

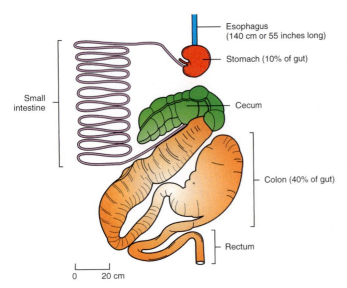

C. Horse. Length of intestine is about 60 ft or 720 in (18 m).

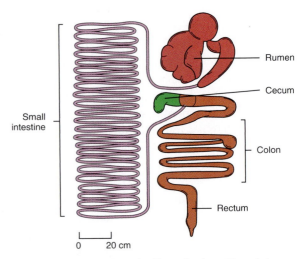

D. Cattle and sheep. Length of intestine in cattle and sheep are, respectively, about 140 ft (43 m) and 65 ft (19.5 m).

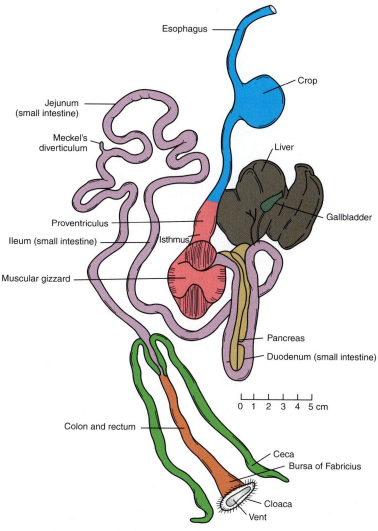

E. Chicken. Length of intestine is about 2 1/2 ft or 32 in (80 cm).

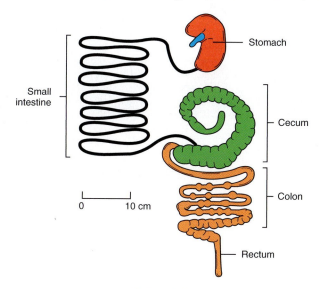

F. Rabbit. Length of intestine is about 15 ft (5 m).

FIGURE 13-4 Structure and function of the rumen. Source: Delmar/Cengage Learning.

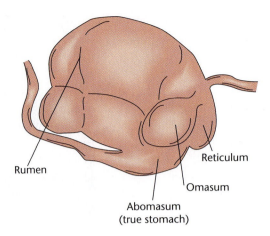

Carbohydrate-digesting enzymes include amylase digesting starch and lactase digesting milk, sugar, and lactose. Lipase digests fat (triglyceride) after it has been emulsified by the addition of bile salts (in the bile produced by the liver).

5. The colon includes the large intestine, ceca, and rectum (see Table 13-1). It is approximately 40% of the gut. Here, there is fermentation of ingesta by microorganisms and absorption of water, together with some minerals and other nutrients. This is particularly important in herbivores without a rumen, such as horses and rabbits. It provides a site for the production of vitamin B_{12} by microorganisms.

Cardiovascular System

The cardiovascular system is made up of three elements:

1. The four-chambered heart. This consists of the right atrium, right ventricle, left atrium, and left ventricle. The right atrium pumps blood to the right ventricle, and this in turn pumps the blood through the lungs to oxygenate the blood. The left atrium pumps blood to the left ventricle, which in turn pumps blood around the body.

TABLE 13-1 The size and capacity of GI tracts in various domestic animal species

SPECIES	STOMACH CAPACITY (L)[a]	SMALL INTESTINE LENGTH IN METERS (% OF GI TRACT LENGTH)	LENGTH OF LARGE INTESTINE, CECUM, AND COLON (M)[b]
Cattle	252	46 (75)	11
Sheep	32.5	26 (80)	7
Horse	18	22 (75)	6.5
Pig	8	18 (78)	4
Dog	4	4 (85)	0.7
Cat	0.3	2 (83)	0.3

[a]For pints, multiple by two.
[b]For feet, multiple by three.

2. The blood vessels, which are arteries taking blood from the heart, veins bringing blood back to the heart, and capillaries bringing nutrients, oxygen, and hormones to cells, removing wastes from cells, and connecting arteries and veins.

3. The blood made up of plasma, which is the watery component containing minerals, nutrients, and proteins such as antibodies and the clotting proteins, and the formed elements, including the red blood cells, which are responsible for transportation of oxygen and carbon dioxide; white blood cells, which are responsible for immune defense (see Chapter 21); and platelets, which are responsible for blood clotting. The composition of blood is illustrated in Figure 13-5.

Plasma, Serum, and Blood Clotting

Plasma is the aqueous constituent of blood containing ions and proteins. Plasma is obtained by centrifugation of blood that has had an anticoagulant such as heparin

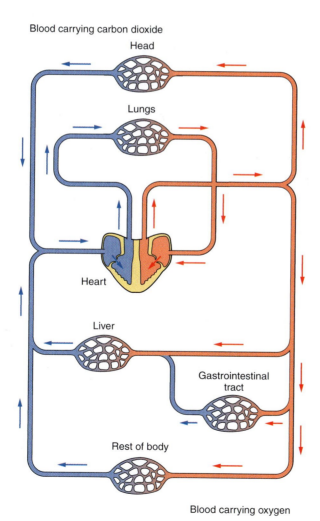

Blood carrying carbon dioxide

Head

Lungs

Heart

Liver

Gastrointestinal tract

Rest of body

Blood carrying oxygen

FIGURE 13-5 Schematic diagram of the circulatory system of domestic animals. Arteries lead from the heart to organs and carry oxygenated blood (red), except from the heart to the lungs. Veins carry blood back to the heart. The blood is deoxygenated, except from the lungs.

FIGURE 13-6 Blood is a mixture of cells and plasma (*A* and *B*). In *C*, It is carried from the heart in arteries. The arteries contain a layer of smooth muscle cells. When the muscles contract, the lumen or opening gets smaller, and blood flow is slowed and blood pressure increases.

The predominant cell type in blood is the red blood cell or erythrocyte. It carries oxygen bound to hemoglobin. The white blood cells or leukocytes are part of the immune system that defends the animal against pathogens and, therefore, disease. The platelets cause the blood to clot, e.g., after an injury.

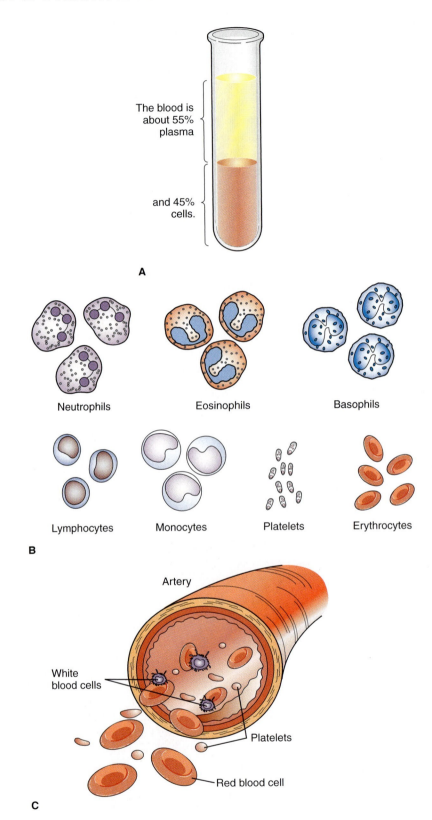

or a calcium chelator added to prevent clotting (see fig. 13-6). Blood outside of a blood vessel rapidly clots. If clotted blood is kept for a period of hours in a refrigerator, the clot retracts, and a fluid becomes separated from the clot. This fluid is serum. The difference between plasma and serum is that serum does not contain the clotting proteins.

Respiratory Organs

There are both similarities and major differences between the respiratory system of mammals and birds. What are the common features?

- There is a trachea bringing air into the lungs and providing an exit for the air.
- There are two lungs where gaseous exchange occurs, that is, hemoglobin in the blood becomes oxygenated, and carbon dioxide is lost; carbon dioxide is transported in the blood as bicarbonate.

In mammals, air is moved in and out of the lungs by contractions of the diaphragm and the intercostal muscles between the ribs. In birds, there is neither a diaphragm nor ribs. Instead, air is moved by the abdominal and other muscles through the lungs into air sacs. These are blind-ending sacs in the abdominal cavity and in bones. Not only are they essential for respiration, but they also reduce the weight of birds, facilitating flight.

Muscle

There are three types of muscle:

1. Skeletal or striated muscle, which are under voluntary control and used for meat
2. Smooth muscle, which are found around the intestines, uterus, and blood vessels, and are under autonomic or involuntary control
3. Cardiac muscle, which contract rhythmically spontaneously but are synchronized to contract at the same time by the cardiac pacemaker or sinoatrial node. The rate and strength of contraction of heart muscles are influenced by the autonomic nervous system.

With the exception of their water content, muscles are predominantly protein. There are three types of protein in muscle: myofibrillar proteins, which are intimately involved in muscle contraction; sarcoplasmic proteins, which are soluble proteins; and stromal proteins, which make up the extracellular matrix that holds the muscle together (see fig. 13-7).

Muscle is composed of cells or fibers, which are sometimes several inches in length. These fibers are arranged in parallel. There are multiple nuclei (up to 100) per cell. The overall structure of muscle is summarized in Figure 13-8:

- Muscle is composed of multinucleated cells or fibers.
- Muscle fibers are composed of myofibrils.
- Myofibrils are composed of a series of sarcomeres along their length.
- Sarcomeres are composed of the contractile elements of muscle: actin and myosin. Myosin molecules slide across the myofilaments made of actin.

FIGURE 13-7 Structure of muscle. Source: Delmar/ Cengage Learning.

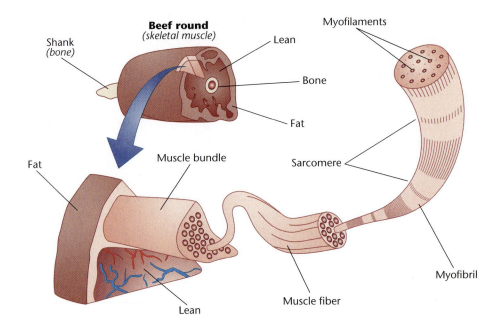

Skeleton

The skeleton has the following functions:

- A framework for the body
- Points of attachment for muscles and, therefore, allowing movement

FIGURE 13-8 Section through the muscle of a turkey showing the fibers surrounded by the extracellular matrix. Courtesy of Sandra Velleman, Ohio State University.

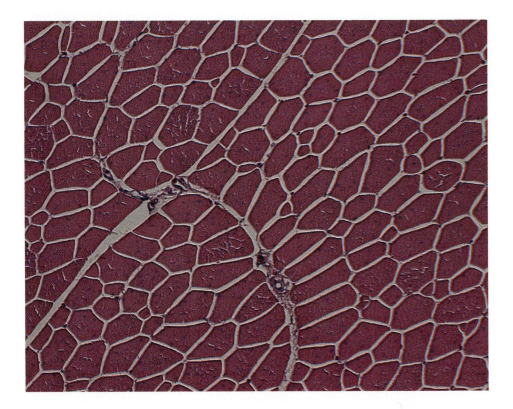

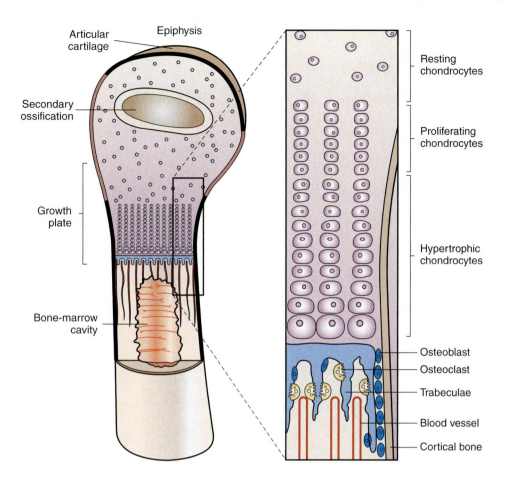

Articular cartilage

Epiphysis

Secondary ossification

Growth plate

Bone-marrow cavity

Resting chondrocytes

Proliferating chondrocytes

Hypertrophic chondrocytes

Osteoblast

Osteoclast

Trabeculae

Blood vessel

Cortical bone

FIGURE 13-9 Development of bone. Growth of the long bones occurs at the growth or epiphyseal plate. This is cartilage and contains chondrocytes. The structure of the bone is shown below the growth plate. The osteoclasts break down bone while the osteoblasts build bone.

- Protection of vital organs, for example, the cranium around the brain, and ribs around the heart and lungs
- Storage of calcium and phosphorus if there are needs for these that are not met from the diet.

There are two principal types of tissue in the skeleton: cartilage (at joints) and bone. The bone contains bone cells (osteoblasts, osteoclasts, and osteocytes) together with a collagen matrix with calcium phosphate deposited. There are two types of bone tissue: compact or cortical bone, and trabecular or spongy bone in the interior of some bones. In the embryo, bones of the skeleton are initially composed of cartilage. Later, bone tissue takes over (see fig. 13-9). Growth of the long bones is at the epiphyseal plates or cartilage plates. At puberty, these plates fuse so that growth does not continue. The anatomy of the skeleton is illustrated in Figure 13-10.

Adipose Tissue

Although people usually think of body fat as a "bad thing," it is essential to the survival of an animal or person. There are two types of adipose tissue: white adipose tissue, which is the common adipose tissue; and brown adipose tissue, which is found at the time of birth and responsible for the production of heat.

FIGURE 13-10 Skeleton of dogs (*A*) and horses (*B*). The skeleton allows for movement with muscles attached to the bones via tendons. In addition, the skeleton protects internal organs with, for instance, the cranium protecting the brain and the ribs protecting the heart and lungs.

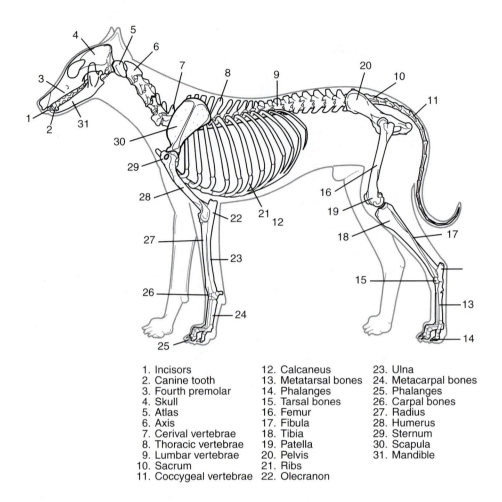

1. Incisors	12. Calcaneus	23. Ulna
2. Canine tooth	13. Metatarsal bones	24. Metacarpal bones
3. Fourth premolar	14. Phalanges	25. Phalanges
4. Skull	15. Tarsal bones	26. Carpal bones
5. Atlas	16. Femur	27. Radius
6. Axis	17. Fibula	28. Humerus
7. Cerival vertebrae	18. Tibia	29. Sternum
8. Thoracic vertebrae	19. Patella	30. Scapula
9. Lumbar vertebrae	20. Pelvis	31. Mandible
10. Sacrum	21. Ribs	
11. Coccygeal vertebrae	22. Olecranon	

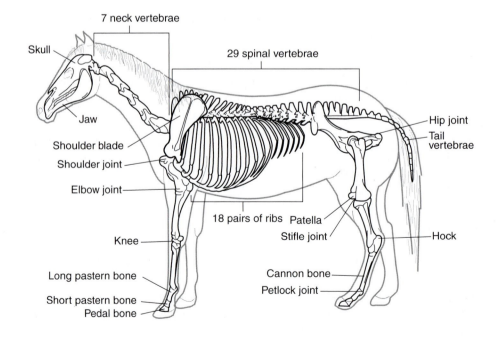

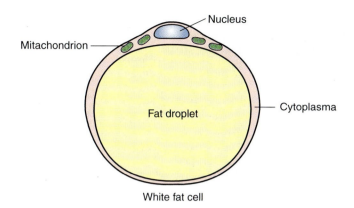

White fat cell

FIGURE 13-11 White adipose cell. Note the very large fat droplet or lipid vacuole containing stored triglyceride.

The function of white adipose cells is to store energy as triglyceride (see fig. 13-11). To store energy, the adipose tissue will either take up fatty acids from the blood or synthesize them from glucose. The fatty acid is then combined with glycerol—in fact, glycerol-3-phosphate that is produced from glucose—to form triglyceride. Three fatty acids plus plusglycerol-3-phosphate equals triglyceride.

At times of inadequate nutrition, the triglyceride is broken down in a manner similar to digestion (see Chapter 14) to form fatty acids that the body can use as an energy source and glycerol that can be converted back into glucose in the liver.

Adipose tissue is found in three anatomic locations:

1. Subcutaneous, which is under the skin
2. Intermuscular and intramuscular, which are responsible for the marbling in meat
3. Abdominal, which is around the intestines in the abdominal cavity and around the kidneys

The largest cellular organelle in an adipose cell is a fat vacuole filled with triglyceride. The major chemical constituent of adipose cells is triglyceride.

Nervous System and Brain

The nervous system is divided into three functionally distinct systems: the central nervous system, including the brain and spinal cord; the peripheral nervous system; and the autonomic nervous system, where there is little conscious control of such functions as reflex responses, breathing, gut function, temperature control, feeding control, and reproduction.

The nervous system is essential for many life processes, including motion or movement (voluntary and reflex), ensuring homeostasis (constant conditions in the body), and playing vital roles in both reproduction and growth.

Endocrine Organs

Hormones are chemical messengers (see "Definition" sidebar) that work closely with the autonomic nervous system to maintain homeostasis and control key

Definitions

Marbling is the adipose tissue (fat) found within muscle. It increases the percentage of fat in the meat but improves its eating qualities, with the steak or chop being more tender, juicy, and flavorful.

A *hormone* is a chemical messenger, is carried in blood, is produced by an endocrine gland or tissue, exerts a specific effect on target cells/tissues/organs, and acts by binding to a specific receptor.

TABLE 13-2 Functions of the regions of the GI tract

LOCATION/ SOURCE	DIGESTIVE JUICE	ENZYME/ SECRETION	ACTION/FUNCTION	COMMENT
Mouth (salivary glands)	Saliva	Salivary amylase	Acts on starch/change to maltose.	Of little importance. None in ruminants.
		Salivary maltase	Acts on maltose/change to glucose.	Saliva adds moisture to feed. Small amount in poultry.
Rumen and reticulum			Microorganisms act on: Protein/nonprotein nitrogen to form essential amino acids. Starch/sucrose/cellulose to form volatile fatty acids (mainly acetic, proprionic, butyric), methane, carbon dioxide, and heat. Fat to form fatty acids and glycerol. Glycerol to form propionic acid.	Synthesize essential amino acids, B complex vitamins, vitamin K.
Omasum			Grinds and squeezes feed/removes some liquid.	Little digestive action in the omasum.
Stomach/abomasum in ruminants/ proventriculus in avians	Gastric juice	Hydrochloric acid Pepsin	Stops action of salivary amylase. Acts on protein/change to proteoses, polypeptides, and peptides.	
Wall of stomach		Rennin Gastric lipase	Acts on milk/curdles the casein. Acts on fat/forms fatty acids and gylcerol.	
Gizzard in avians			Grinds and mixes feed.	Digestive juices continue to act on feed.
Small intestine (pancreas)	Pancreatic juice	Trypsin and Chymotrypsin	Acts on proteins, proteoses, polypeptides, and peptides/produces proteoses, peptones, peptides, and amino acids.	
		Pancreatic amylase Pancreatic lipase	Acts on starch/change to maltose. Acts on fat/forms glycerol, fatty acids, and monoglycerides.	Small amounts in ruminants.
Small intestine (Pancreas) (Liver)	Bile	Carboxypeptidase	Acts on peptides/forms peptides and amino acids. Acts on fats/forms glycerol and soap.	
Intestinal wall	Intestinal juice	Intestinal peptidase (formerly called erepsin)	Acts on remaining proteins, proteoses, peptones, and peptides/produces amino acids.	
		Maltase Sucrase	Acts on maltose/changes to glucose. Acts on sucrose/changes to glucose and fructose.	Small amounts in ruminants. Small amounts in ruminants.
		Lactase	Acts on lactose/changes to glucose, fructose, and galactose.	Large amounts in young mammals.
		Nuclease	Acts to digest DNA and RNA.	
Cecum in horse			Bacterial action digests roughage.	
Large intestine		Cellulase	Acts on cellulose/forms volatile fatty acids. Some digestion continues as material moves from the small intestine to the large intestine.	Mostly in the horse.

TABLE 13-3 Hormones of the hypothalamus and pituitary gland of domestic animals

ORGAN	HORMONE	CHEMISTRY	FUNCTION
Pituitary gland (anterior)			
	ACTH	Polypeptide	Stimulates the adrenal cortex to produce cortisol or corticosterone
	FSH	Glycoprotein	Stimulates follicular growth (females) and spermatogenesis (males)
	GH (also called somatotropin)	Protein	Stimulates growth by increasing liver production of IGF-1, and also milk production in cattle
	LH	Glycoprotein	Stimulates ovulation and formation of corpus luteum in females and production of testosterone in males
	PRL	Protein	Stimulates initiation of lactation
	TSH (also called thyrotropin)	Glycoprotein	Stimulates the thyroid gland to grow and produce thyroxine
Hypothalamus			
	Corticotropin-releasing hormone	Polypeptide	Stimulates release of ACTH
	Gonadotropin-releasing hormone	Peptide	Stimulates release of LH and FSH
	Ghrelin	Polypeptide	Stimulates release of GH
	GH-releasing hormone	Polypeptide	Stimulates release of GH
	PRL-inhibiting factor	Peptide and dopamine	Inhibits release of PRL
	Somatostatin (SRIF)	Peptide	Inhibits release of GH
	Thyrotropin-releasing hormone	Peptide	Stimulates release of TSH
Posterior pituitary gland			
	Oxytocin (mammals)	Peptide	Stimulates uterine contraction and milk letdown
	Antidiuretic hormone	Peptide	Stimulates water retention at the kidneys

Note: ACTH = adrenocorticotropic hormone; FSH = follicle-stimulating hormone; IGF-1 = insulin-like growth factor 1; LH = luteinizing hormone; PRL = prolactin; and TSH = thyroid-stimulating hormone.

functions such as metabolism, growth, reproduction, and lactation. Hormones are produced by endocrine glands or tissues.

The major endocrine tissues are the following:

- Hypothalamus (see Table 13-3 and fig. 13-12)
- Pituitary gland (see Table 13-3 and fig. 13-12)
- Thyroid glands producing thyroxine
- Adrenal glands composed of the medulla (chromaffin cells), producing epinephrine and norepinephrine at times of fight or flight, and the cortex, producing cortisol (corticosterone in birds) to combat stress and aldosterone to increase sodium retention

- Islets of Langerhans in the pancreas (also known as the *endocrine pancreas*)
    - A or α-cells producing glucagon (increases blood glucose)
    - B or β-cells producing insulin (decreases blood glucose)

- Gonads producing estrogens and progesterone (females) and testosterone (males)
- Gut
- Liver producing insulin-like growth factor 1 and converting thyroxine to its active form, triiodothyronine
- Adipose tissue producing a hormone, leptin, that reduces food intake
- Pineal gland producing melatonin that affects seasonal breeding and sleep patterns

INTERESTING FACTOID

Animals have a hypothalamo-pituitary-endocrine organ axis for amplification and more points for control.

FIGURE 13–12 The pituitary gland. The pituitary gland is found under the hypothalamus, which is a part of the brain. The anterior pituitary gland releases hormones controlling growth (somatotropin), reproduction (luteinizing hormone and follicle-stimulating hormone), mammary gland development (prolactin), milk production (somatotropin in cattle, prolactin in many species), coloration (melanotropin), and the response to stress (adrenocorticotropic hormone). The posterior pituitary gland releases hormones produced in the hypothalamus that control kidney functioning (antidiuretic hormone) and both milk let-down (release) and contractions of the uterus in parturition (giving birth) (oxytocin). Release of hormones from the anterior pituitary gland is controlled by releasing hormones produced in the hypothalamus and passing from modified nerve terminals in specialized blood vessels (portal blood vessels).

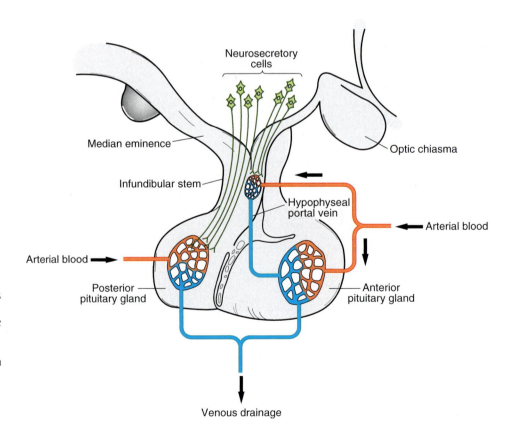

The hypothalamo-pituitary-endocrine organ axis is summarized in Figure 13-13. The hormone from the target endocrine gland exerts a negative feedback on the hypothalamus and/or pituitary gland and thereby suppresses, respectively, the release of hormone. Examples of this system are the following:

1. Hypothalamo-pituitary adrenocorticotropic hormone stimulates the adrenal cortex to produce cortisol release.
2. Hypothalamo-pituitary thyroid-stimulating hormone stimulates thyroid release of thyroxine.
3. Hypothalamo-pituitary growth hormone (GH) increases production of (insulin-like growth factor 1) (growth axis) (see Chapter 17).
4. Hypothalamo-pituitary luteinizing hormone stimulates ovary/corpus luteum (progesterone axis) (see Chapter 16).
5. Hypothalamo-pituitary luteinizing hormone stimulates testis/Leydig cells (testosterone axis) (see Chapter 16).

The advantages of the hypothalamo-pituitary-endocrine organ axis are that a small amount of a hypothalamic-releasing hormone can lead to a very large change, and there are multiple points of control. This is analogous to why we need both a brake and accelerator in a car.

 ## GUT HORMONES

Gut hormones are involved in the control of digestion, causing changes in motility and secretion in the GI tract. In addition, many of the gut hormones may also be involved in appetite control, and in affecting the secretion of the critically important metabolic hormones insulin and glucagon. All the gut hormones are small peptides. Gut hormones include gastrin, cholecystokinin (CCK), ghrelin, secretin, peptide YY, and somatostatin. Other gut hormones include motilin, glucagon-like peptide 1, oxyntomodulin, and glucose-dependent insulinotropic polypeptide.

Gastrin is produced by specific cells (G cells) in the stomach and, to a lesser extent, in the duodenum. Gastrin stimulates the release of gastric acid in the stomach of animals with a simple stomach (nonruminants).

CCK is produced by specific cells—the I cells—in the duodenum. CCK stimulates the release of bile from the gallbladder and enzymes from the pancreas. It was formerly also called *pancreozymin*.

Ghrelin is a peptide hormone produced by the GI tract and also by the hypothalamus. It stimulates appetite and the release of GH. This orexigenic peptide is produced by the stomach between meals.

Secretin is produced by the S cells in the duodenum. It stimulates the release of bicarbonate to neutralize the acid in the ingesta.

Peptide YY is produced by cells in the ileum and colon. It has a number of effects, including reducing gut motility (movement) and inducing satiety.

Somatostatin is produce by the D or δ-cells in the stomach and duodenum, as well as neurosecretory cells in the hypothalamus. It inhibits the release of gastric acid, gastrin, and other gut hormones and GH.

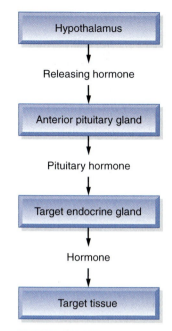

FIGURE 13–13 Schematic representation of how the hypothalamus in the brain controls the anterior pituitary gland and this, in turn, controls target endocrine glands.

Definitions

Anorexigenic means causes anorexia (loss of appetite). *Orexigenic* means causes appetite. *Satiety factor* is a chemical (naturally produced by the body) that when administered before a meal, reduces the amount of food consumed.

Gut hormones also influence the production of insulin. Both glucose-dependent insulinotropic polypeptide and glucagon-like peptide 1 increase insulin release from the β-cells of the islets. Somatostatin inhibits insulin release. Gastrin and CCK promote the development of the islets and, therefore, improve glucose homeostasis.

Gut hormones that inhibit feeding or feed intake are CCK, bombesin, glucagon-like peptide 1, glucagon-like peptide 2, peptide YY, amylin, and somatostatin. The gut hormone that increases food intake or feeding is ghrelin.

 INTEGRATION OF ORGANS/BODY FUNCTIONING

The functioning of the body is controlled by both the nerves (the nervous system, including the central nervous system and brain) and hormones (produced by the endocrine glands). Control of reproduction is discussed in Chapter 16, control of growth in Chapter 17, and control of lactation in Chapter 18.

Control of Feeding

Food intake in many species occurs in bouts or meals. How can an animal or person change the amount it eats?

- Change the number of meals or how quickly an animal or person starts eating after the last meal.
- Change the amount eaten per meal.

Why Do We Control Food Intake?

Why not a simple system of eating until the GI tract is full? It is advantageous to maintain an optimal healthy weight, and different foods have different constituent nutrients and different calorie contents.

How Do We Control Food Intake?

There are two hormones—leptin and insulin—that pass through the circulation to the hypothalamus in the brain and there influence the centers controlling food intake. Leptin is produced by adipose tissue, with the amount released into the blood proportional to the amount of fat. Insulin is produced by the β-cells of the pancreas in response to elevated blood glucose and some gut hormones. A number of gut hormones profoundly influence food intake. These either act via receptors on peripheral nerves, especially the vagus, passing from the gut to the hindbrain or are transported to the brain via the circulatory system.

When an animal eats is controlled by gut–brain signaling. The appetite stimuli are the same orexigenic hormones that are produced by both the gut and brain. The stomach hormone, ghrelin, is a prime orexigenic peptide stimulating appetite.

Satiation or satiety is the feeling of fullness that stops us or animals eating at the end of a meal; that is the control of meal size. There is a satiety center in the brain controlling food intake. Gut hormones from the stomach, small intestine, colon, and pancreas together with brain peptides stimulate the satiety center. These hormones include peptide YY, pancreatic polypeptide, and CCK.

Control of Metabolism

The major hormones controlling metabolism are glucagon, which increases blood glucose by taking energy out of storage, and insulin, which decreases blood glucose by placing energy into storage (see the sidebar "Effects on Metabolism").

Why does an animal have insulin? This hormone stimulates the rapid movement of glucose out of the bloodstream into either temporary storage in the liver as glycogen or longer-term storage such as glycogen in muscles or triglyceride in adipose tissue. This prevents the body being overwhelmed with glucose as would be the case in diabetes (diabetes mellitus) and fosters the efficient storage of energy. In contrast, at the time when there is insufficient energy, there is a release of glucagon, together with epinephrine and norepinephrine. This acts to cause the breakdown of triglyceride and glycogen together with the synthesis of glucose from certain amino acids, lactate, and other chemicals. This occurs in the liver and is called *gluconeogenesis*.

> **EFFECTS ON METABOLISM**
>
> The glucagon effects on metabolism are an increase in circulating glucose, decrease in liver glycogen, decrease in adipose triglyceride, and increase in gluconeogenesis. The insulin effects on metabolism are a decrease in circulating glucose, increase in glycogen, increase in triglyceride, and decrease in gluconeogenesis.

Pheromones

Pheromones are specific chemicals or mixtures. They are produced by one animal and produce an effect with another animal of the same species (or conspecifics). Pheromones may induce effects such as the establishment of a social hierarchy, the induction of estrus, nursing behavior, sexual attraction, marking territory, reducing stress, and allowing coitus.

Pheromones may be volatile, and are found in the air, or are dissolved in sprayed fluids (e.g., saliva) or placed close to the recipient animal by licking. They may act via the olfactory epithelium or through the vomeronasal organ, which is between the nose and mouth in some mammals. A number of examples of pheromones in domestic animals are given below.

Pheromones in Cats

There are at least two pheromones in cats: feline facial pheromone, and territorial marking sprayed pheromone. Cats frequently rub their faces against people or objects such as furniture. As they rub against a surface, they are depositing a pheromone: feline facial pheromone. Synthetic feline facial pheromone (Feliway) is sold commercially, and seems to make cats less likely to spray and calmer in stressful situations, such as when visiting a veterinarian.

Cats spray urine with a specific pheromone to mark territory. Feline (2-amino-7-hydroxy-5,5-dimethyl-4-thiaheptanoic acid) is a precursor for the pheromone and is excreted in cat urine.

Pheromones in Dogs

There are a least two pheromones in dogs: a sexual attractant and arousal (methyl p-hydroxybenzoate) pheromone, and a dog-appeasing pheromone.

Methyl p-hydroxybenzoate was characterized in vaginal secretions of female dogs in heat. If this chemical is put onto the vulvas of females not in heat or spayed females, introduced males placed become sexually aroused and attempt to mount them.

Pheromones in Pigs

There are at least two pheromones in pigs: boar taint steroid or boar pheromone, and maternal pheromone. Both pheromones are detected by the olfactory epithelium.

Boar taint steroid has been chemically defined as 5a-Androst-16-en-3-one. It is found in boar saliva. When delivered as a spray, 5a-Androst-16-en-3-one induces gilts or sows to assume a copulatory standing response. The female responds by the pheromone acting on the olfactory epithelium. Interestingly, the female pig can detect the pheromone at much lower concentrations than the male.

Maternal pheromone is thought to be a mixture of fatty acids that can be synthesized. This pheromone aids recognition of the mother by the newborn pigs. This aids nursing behavior and, therefore, survival of is the neonates.

Pheromones in Sheep and Goats

In sheep and goats, a pheromone effect on reproduction has been well established. Exposure to the male or its odor (its pheromone) induces ovulation in seasonally anovulatory females. This is called the *male effect* and is found in both sheep and goats. The chemical nature of both the sheep and goat pheromones has not yet been determined.

REVIEW QUESTIONS

1. What are the organelles in a cell, and what are their functions?
2. Why is the potential difference across a cell membrane important?
3. What is the difference among a carnivore, herbivore, and omnivore?
4. What is the difference between the gut of a carnivore and herbivore?
5. What are the main functions of the GI tract?
6. What is digestion? How is it performed for protein, carbohydrate, and fat?
7. What are the constituents of blood?
8. What is the difference between plasma and serum?
9. What are the differences in the respiratory system between mammals and birds?
10. What are the two tissue types in the skeleton?
11. What is the significance of the epiphyseal plate?
12. What is the major mineral in bone?
13. What is the major chemical constituent of adipose cells? Why is this important?
14. What is marbling? Why is it important in meat?
15. What is a hormone? Where are hormones produced?
16. What are the major endocrine glands, and what hormones do they produce?
17. What is the hypothalamo-pituitary-endocrine target tissue system, and what are its advantages?
18. What are the three major hormones produced by the gut, and what do they do?
19. How is food/feed intake controlled?
20. How do the hormones produced by the islets of Langerhans (endocrine pancreas) control metabolism in the body?
21. What is gluconeogenesis?
22. What is a pheromone?
23. Give examples of pheromones in pigs, sheep, cats, and dogs.

REFERENCES AND FURTHER READING

Cummings, D. E., & Overduin, J. (2007). Gastrointestinal regulation of food intake. *The Journal of Clinical Investigation, 117*, 13–23.

Espe, D., & Cannon, C. Y. (1940). The length of the intestine of calves and its bearing on the absorption of the nutrients from the chyme. *Journal of Dairy Science, 23*, 1211–1214.

Gerrand, D. E., & Grant, A. L. (2003). *Principles of animal growth and development*. Dubuque, IA: Kendall Hunt.

Murphy, K. G., & Bloom, S. R. (2006). Gut hormones and the regulation of energy homeostasis. *Nature, 444*, 854–885.

Murphy, K. G., Dhillo, W. S., & Bloom, S. R. (2006). Gut peptides in the regulation of food intake and energy homeostasis. *Endocrine Reviews, 27*, 719–727.

Reece, W. O. (2004). *Dukes' physiology of domestic animals* (12th ed.). Ithaca, NY: Comstock.

Scanes, C. G. (2003). *Biology of growth of domestic animals*. Ames, IA: Blackwell.

Sisson, S., Grossman, J. D., & Getty, R. (1975). *Sisson and Grossman's the anatomy of the domestic animals* (5th ed.). Philadelphia: Saunders.

Swenson, M. J. (1979). *Dukes' physiology of domestic animals* (9th ed.). Ithaca, NY: Cornell University Press.

Woods, S. C. (2004). Gastrointestinal satiety signals I. An overview of gastrointestinal signals that influence food intake. *American Journal of Physiology. Gastrointestinal Liver Physiology, 286*, G7–G13.

CHAPTER 14

Animal Nutrition: Feeds and Feeding

CONTENTS

This chapter will consider the following:

- An overview of animal nutrition, including definitions
- The different nutrients, and why they are required
    - Carbohydrates
    - Fats
    - Amino acids and proteins
    - Vitamins
    - Minerals
- Supplements to animal feed and food
- Toxic plants and fungal toxins
- Animal feed, food, and formulation, including grain processing

The mixture of nutrients for dogs and cats is called *dog food* or *cat food*. The mixture of nutrients for horses, livestock (including poultry), and aquaculture species is called *feeds* or *feedstuffs*. A *diet* is all the food/feed provided or consumed by an animal or person. It could also be one of the mixtures of feed components in an experiment. A diet could be a synthetic diet with all the nutrients provided in a chemically pure state or a semisynthetic diet, in which some of the nutrients are provided in a chemically pure state.

Ration is the food given daily to an animal. For livestock, this is can be a synonym for diet. A *nutrient* is a chemical required for life. *Concentrates* are high energy and/or protein supplements provided to ruminants. Definitions of *energy* are as follows:

- *Calorie* (cal) is a unit of heat being the heat required to increase the temperature of 1 g of water from 16.5°C to 17.5°C. A calorie is defined as containing 4.184 international Joules (J).
- *Kilocalorie* (kcal) is 1,000 calories.
- *Megacalorie* (Mcal) is 1 million calories.

 INTRODUCTION

Animals require nutrients for life, with different species having different requirements. All animals require the following broad categories of nutrients: energy, usually in the form of carbohydrates and/or fats; protein; minerals, including macrominerals and trace minerals; vitamins; and water.

SUMMARY *of Nutrients Required by Animals*

1. Energy
   * Carbohydrates from complex polymers (cellulose nondigestible in nonruminants, starch) to disaccharides (e.g., lactose in milk or sucrose in sugar) to simple sugars or monosaccharides (e.g., glucose, galactose, mannose).
   * Triglyercides and other fats.
   * Volatile fatty acids (acetate, propionate, and butyrate) in ruminants.
   * Protein: A carnivore, such as a cat, can use amino acids derived from proteins as its major energy source.
2. Proteins (a polymer of different amino-acid residues)
   * Nonprotein nitrogen (NPN) such as urea can substitute for protein in ruminant feed as the rumen microorganisms synthesize protein from NPN. This protein is ultimately digested by the ruminant animal.
3. Minerals
   * Macrominerals such as calcium, sodium, potassium, and chloride.
   * Microminerals or trace minerals such as copper, zinc, manganese, selenium, and iodine.
4. Vitamins
   * Water-soluble vitamins—vitamins of the B complex (B_1 thiamine, B_2 riboflavin, B_3 niacin, B_5 pantothenic acid, B_7 biotin, folic acid/folate, B_{12}/cobalamins) and ascorbic acid (vitamin C).
   * Fat-soluble vitamins such as vitamin A, D, and E.
5. Water

Animals, including domestic animals, can be categorized into the following three groups based on the type of food they eat:

1. Herbivorous species eat plants, including the fiber (largely not digestible in nonherbivores). Herbivores include the ruminants together with horses, rabbits, and geese. Ruminants are animals that chew cud and have a four-chambered stomach or rumen. Examples of ruminant species are cattle, sheep, goats, and deer.
2. Omnivorous species consume plants (grain) and animals in their feed. Examples include humans, pigs, chickens, and dogs.
3. Carnivorous species fully or almost exclusively consume animals, including mammals, birds, and fish. The domestic cat is a good example.

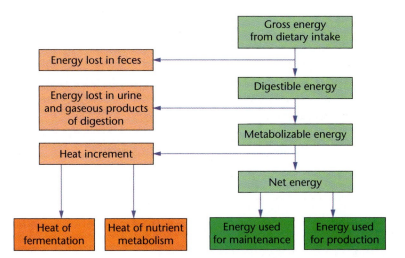

FIGURE 14-1 Utilization of dietary energy by an animal. Source: Delmar/Cengage Learning.

Energy deficiency is the insufficient energy in the feed or food of animals, resulting in the following problems:

- Slower growth
- Delayed puberty/sexual maturation or reduced fertility
- Decreased milk production
- A loss in body weight
- Decreased resistance to infectious diseases and parasites
- Reduced hair growth (in sheep, less wool)

 ## CARBOHYDRATES

Carbohydrates are made of carbon, hydrogen, and oxygen. They have a ratio of carbon, 6; hydrogen, 12; and oxygen, 6. Carbohydrates include starch, simple sugars, hemicellulose, cellulose, pectins, and lignin in plants, together with glucose and glycogen in animals.

Essential to all life is the production of carbohydrates by photosynthesis by plants, which can be expressed as six molecules of carbon dioxide plus six molecules of water plus light energy equals glucose, or $6\,CO_2 + 6\,H_2O \rightarrow C_6H_{12}O_6 + 6\,O_2$.

The most important sugar to animals is glucose, which is a monosaccharide. The simplest carbohydrates are monosaccharides or simple sugars. These are easily absorbed in the digestive system. Hexoses are monosaccharides containing six carbon atoms, whereas pentoses contain five carbon atoms. The most common of the hexoses are glucose (found in blood and other organs of animals), fructose (in fruit), and galactose (part of the milk sugar lactose). The structure of monosaccharides is shown in Figure 14-2.

When two simple sugars are combined, a disaccharide is formed. For instance, lactose is made up of two simple sugars: glucose and galactose. This is illustrated in Figure 14-3.

During digestion, these are broken down to monosaccharides. Polysaccharides contain a number of sugar moieties in a chain or polymer. Examples include glycogen, starch, cellulose, and hemicellulose. Starch is a principal storage form of energy in the seed of plants and, therefore, is a major constituent of grains such as corn. Starch contains a chain of glucose moieties linked by α-linkages, and cellulose

Definitions of energy in feed

Gross energy is the heat released when the feed or a constituent is completely oxidized in a bomb calorimeter under 25–30 atm of oxygen (see fig. 14-1). The gross energy is usually expressed as kilocalories/kilogram or kilocalories/pound of feed.

Digestible energy is the gross energy of the feed consumed less the gross energy of the feces.

Metabolizable energy is the digestible energy less the energy in the urine and in ruminants the gases produced by fermentation (e.g., methane).

Heat increment is that metabolizable energy used for digestion or metabolism. In cold weather, this heat may help keep the animal warm. During hot weather, the heat increment reduces production efficiency because the heat has to be dissipated.

Net energy is the metabolizable energy minus the heat increment. It is energy used either for maintenance, or maintenance plus production or production. The net energy requirement is increased during lactation and pregnancy.

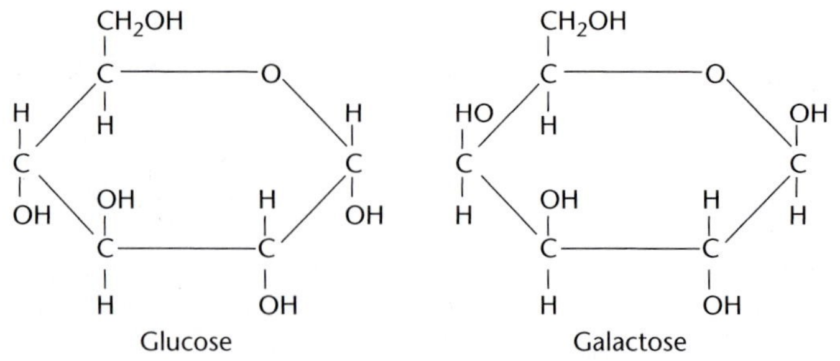

FIGURE 14-2 Structures of three common monosaccharides (hexoses). Source: Delmar/Cengage Learning.

Glucose

Galactose

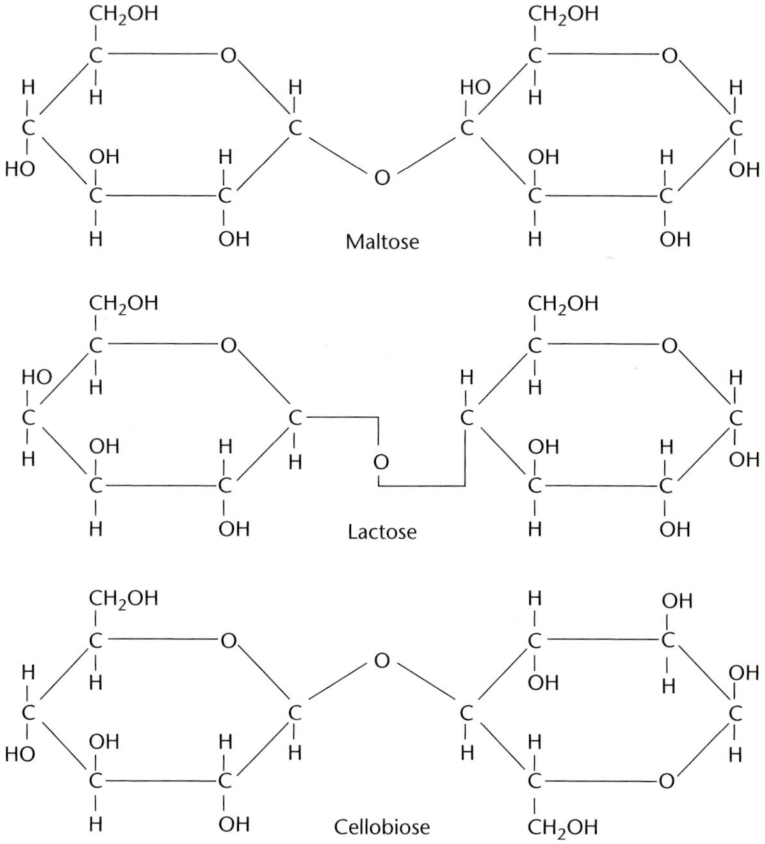

FIGURE 14-3 Structures of three common disaccharides. Source: Delmar/Cengage Learning.

Maltose

Lactose

Cellobiose

is a polymer of glucose moieties with β-linkages. Hemicellulose is a polymer of hexose and pentose moieties. Starch is readily digestible. Neither cellulose nor hemicellulose is digestible. However, these are metabolized rumen microorganisms. The structure of starch and cellulose are shown the Figure 14-4. Hemicellulose, cellulose, and lignin are collectively considered as fiber in the diet.

Nutrient deficiencies of carbohydrates are considered under energy deficiency, which was described previously.

LIPIDS (FATS AND OILS)

The lipid in animal feed is predominantly triglyceride, which is a chemical produced by combining fatty acids and glycerol: glycerol plus three fatty acids equal triglyceride and three water molecules.

Starch

Cellulose

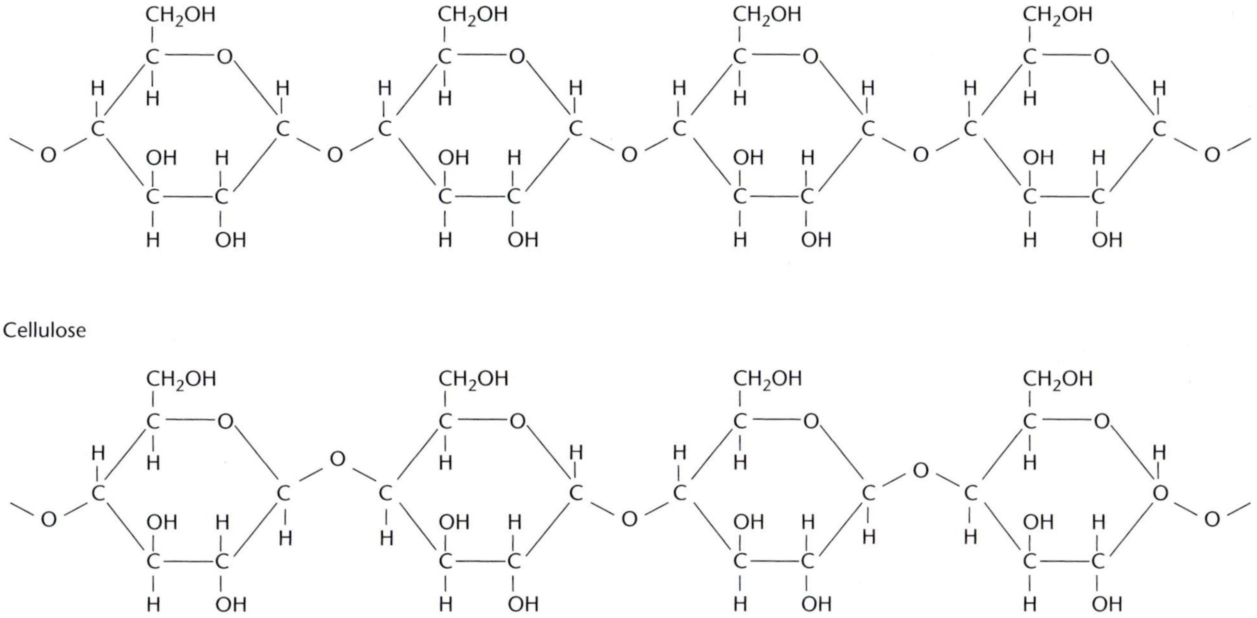

FIGURE 14-4 Structure of starch, a common polysaccharide. Source: Delmar/Cengage Learning.

The chemistry of triglycerides is summarized in Figure 14-5. In addition to triglyceride, there are various phospholipids (e.g., lecithin) together with cholesterol. These are important to parts of the membrane of a cell. Phospholipids consist of a compound of glycerol, two fatty acids, phosphate, and one other compound. There are also lipoproteins, these being lipids that are bound to protein and can function to transport lipids.

Both fats and oils are predominantly triglycerides. They differ in only one way. Fats are solid, and oils are liquid at room temperature. The reason that oils are liquid at room temperature is that the fatty acids in the triglyceride are predominantly unsaturated fatty acids. In fats, the fatty acids are predominantly saturated or monounsaturated fatty acids. Oils are from plants, whereas fats are from animals or are the result of hydrogenation of plant oils.

Fatty acids are hydrocarbon chains ranging from less than 10 carbons to more than 20 carbons long with a carboxylic acid at one end. In saturated fatty acids, there are no double bonds in the hydrocarbon chain, but there are double bonds in unsaturated fatty acids. During storage, oxidation causes unsaturated fatty acids to become rancid with an unpleasant smell.

Addition of fats to animal feeds increases the energy level in the diet. The presence of fat in the diet also facilitates absorption of the fat-soluble vitamins. Fats may also be added to improve the flavor, texture, and palatability of the feed. Added fat reduces any dustiness of the feed. Fats are soluble in ether and other organic solvents. Ether is used in feed analysis to extract the fat from the feed. Therefore, the dissolved substances (fats) are called *ether extract*.

Nutrient deficiencies of fats, excluding essential fatty acids, are considered under energy deficiency, which has been described previously.

FIGURE 14-5 Structure of triglyceride. Source: Delmar/Cengage Learning.

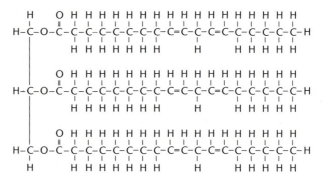

Unsaturated Fat

Saturated Fat

Essential Fatty Acids

Essential fatty acids are required to be present in the feed of nonruminant animals because they cannot be synthesized by the animal. These are unsaturated fatty acids. There are two major groups of essential fatty acids: omega (ω)-3 fatty acids and ω-6 fatty acids (based on the position of a double bond). Examples of ω-3 fatty acids are α-linolenic acid, eicosapentaenoic acid, and docosahexaenoic acid. Examples of ω-6 fatty acids are linoleic acid, γ-linolenic acid, and arachidonic acid.

 PROTEINS

Proteins are a chain or polymer of amino acids. Small proteins are called peptides or polypeptides. Proteins vary widely in shape, solubility, chemical composition, and physical properties. Proteins can range in size from a small peptide of three amino acids through to large proteins, with thousands of amino acids. Biologic functions of proteins include being structural, muscle, skin, connective tissue, enzymes, hormones, receptors, and nutrients in milk and eggs.

More protein is required in the feed for young rapidly growing animals than for mature animals. Protein requirements are also higher during pregnancy and lactation periods.

There are 20 amino acids in animal proteins. All amino acids contain carbon, hydrogen, oxygen, and nitrogen. Some amino acids also contain sulfur. The chemical structure of amino acids is shown in Figure 14-6.

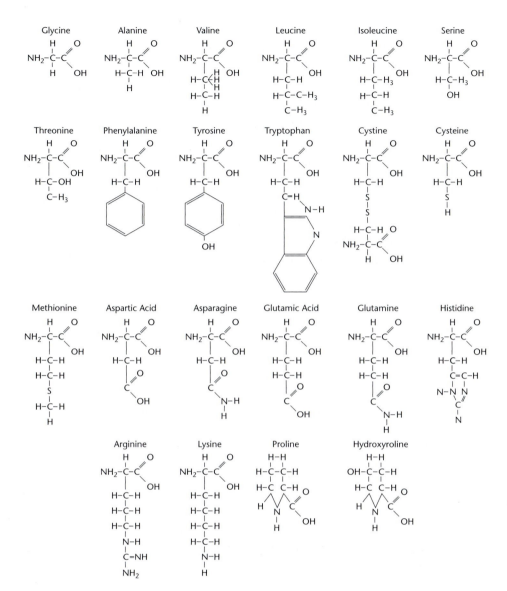

FIGURE 14-6 Structure of amino acids. Source: Delmar/Cengage Learning.

Amino acids contain at least one amino group (NH_2) and one carboxylic acid group (COOH):

$$H$$
$$|$$
$$R-C-NH_2$$
$$|$$
$$COOH$$

The R in the formula represents the side chain of the amino acid. Amino acids can be assigned to various groupings: neutral, acid or basic, or those with either a ring structure or sulfur containing. Natural amino acids with the exception of glycine have an L conformation. Synthetic amino acids such as methionine are a mixture of L and D conformations.

In proteins, the amino acids are combined to give a peptide bond. Formation of a peptide bond is NH$_2$.RCH. COOH + NH$_2$.R'CH. COOH → NH$_2$.RCH. CO.NH.R'CH. COOH

The organs of the body are composed primarily of proteins. These can be proteins with any chemical modification and those that have been chemically or posttranslationally modified. Examples of the posttranslationally modified proteins include the following:

- Phosphoproteins. These are important inside the cell and also in milk protein, including casein, and in yolk proteins.
- Glycoprotein (with carbohydrate added).
- Lipoproteins (with a lipid linked to the protein).

Essential Amino Acids

These are amino acids that must be in the ration of nonruminant animals because the animal either cannot synthesize them or the rate of synthesis is too low for the body's requirements. Nonessential amino acids are still required by the animal but are synthesized in the body from other amino acids. There are 10 essential amino acids for pigs and 14 for poultry (see Table 14-1). Taurine (2-aminoethanesulfonic acid or SO$_3$H-CH$_2$-CH$_2$-NH$_2$) is not an amino acid. It is a metabolite of the amino acid cysteine. It is essential to the health of cats. Cats can be deficient in taurine, and under these circumstances they exhibit retinal degeneration.

Limiting Amino Acids

Nonruminants have requirements for specific amino acids. Essential amino acids are required in the feed in definite proportions. The limiting amino acid is the essential amino acid present in the lowest concentration in the feed (as a percentage of its requirement). Even if other amino acids are present in surplus, they can

TABLE 14.1 List of essential and nonessential but nonetheless important amino acids

ESSENTIAL	NONESSENTIAL
Arginine	Alanine
Histidine	Aspartic acid
Isoleucine	Citrulline
Leucine	Cysteine
Lysine	Cystine
Methionine*	Glutamic acid
Phenylalanine**	Glycine
Threonine	Hydroxyglutamic acid
Tryptophan	Hydroxyproline
Valine	Norleucine
	Proline
Additional Essential for Poultry:	Serine
Alanine	Tyrosine
Aspartic acid	
Glycine	
Serine	

only be used to the extent that the one in shortest supply is available. For example, in pig feeds, lysine is the limiting amino acid, whereas in poultry feeds, methionine may be limiting.

Protein Requirements by Ruminants

Ruminants can meet their amino acid (both essential and nonessential amino acids) needs by digestion of plant proteins and the incorporation of nonprotein nitrogen, such as urea, to microbial proteins in the rumen and subsequent digestion of the microorganism. If proteins in the diet are encapsulated or "protected," they can bypass fermentation and consequent losses in the rumen.

Rumen Bypass Supplements

There are nutritional supplements that are treated such that they are not fermented in the rumen but are digested in the small intestine. These are called *rumen bypass products*. Examples include proteins and amino acids (rumen bypass soybean meal and rumen bypass methionine). The advantage of bypass products is that they can meet the specific nutrient needs of cattle under different physiologic circumstances such as lactation. This is most important in high-producing dairy cattle.

Protein as an Energy Source

Amino acids can be used as an energy source. Many amino acids can be converted to glucose by a process called *gluconeogenesis*, and to fat. This can occur under two circumstances in pigs and poultry:

1. When more amino acids are absorbed than is needed for protein synthesis
2. When the animal is nutritionally deprived and is breaking down protein, for example, in muscle to provide energy

In addition, in carnivores, amino acids are a frequently used source of energy. The nitrogen is "split off" from the amino acids and excreted in the urine as urea. The remainder is used for energy needs or stored as body fat.

Unavailable Protein

When feeds are improperly stored, for instance by being overheated, the protein may be degraded or denatured. The protein is then not digestible and, therefore, is not available to the animal.

Protein Deficiency

When animals do not receive sufficient protein or the amount of essential amino acids is insufficient, there will be severe consequences. Growth will be reduced or cease altogether. In adults, fertility can be reduced. Animals will break down muscle and other proteins in the body to meet the needs of amino acids, even for maintenance (i.e., staying alive).

Sources of Amino Acids

The major sources of amino acids are proteins in soybean meal, corn, and some fish meal. Amino acids are added to diets to bring the proportion of the different amino

Definition

acids to an optimal mix. Corn/soybean diets would be deficient in lysine and/or methionine for optimal growth and utilization of the protein. Animals use proteins until the first limiting amino acid (lysine in pigs and methionine in poultry).

Methionine and its analogue/precursor D,L–2-hydroxy–4 (methylthio)–butanoic acid (sold commercially as Alimet) are produced by chemical synthesis. Other essential amino acids such as lysine, threonine, and tryptophan are produced through fermentation using mutants of *Corynebacterium glutamicum* and recombinant strains of *Escherichia coli*. Lysine is added to pig and poultry corn soybean diets. The lysine is produced from microorganisms by fermentation. An alternative approach is corn bioengineered to have higher lysine contents.

Requirements

The National Research Council publishes nutrient requirements for the feeds of livestock and horses. In practice, the requirements are viewed as a guideline at least by poultry companies in which the feeds will use higher levels of total sulfur-containing amino acids and lysine to increase breast and other white meat components. This is not based on research published in refereed scientific journals but on industry studies.

VITAMINS

Vitamins are specific chemicals that need to be present in small quantities in the feed or food for health. They are not synthesized in the body and may function as coenzymes in metabolic processes. Table 14-2 provides a list of vitamins and what they do.

There are 16 vitamins that are essential in animal nutrition (see Table 14-2). Each vitamin has a specific chemical structure. All are organic chemicals containing carbon, hydrogen, and oxygen. Most vitamins can be readily synthesized chemically and are available in synthetic form.

Vitamins are either water soluble or fat soluble:

- Water-soluble vitamins include vitamins of the B complex (B_1 thiamine, B_2 riboflavin, B_3 niacin, B_5 pantothenic acid, B_7 biotin, folic acid/folate, B_{12} cobalamins) and ascorbic acid (vitamin C). They are poorly stored in the body.
- Fat-soluble vitamins include vitamins A, D, E, and K. These can be stored in the body.

With the exception of ruminants, in which many vitamins are synthesized by the rumen microorganisms, animals require vitamins. The specific requirement for each vitamin varies with species. Another exception is that some animals synthesize vitamin D in the skin when exposed to sunlight. A deficiency of a specific vitamin results in well-characterized symptoms of deficiency. Much of our knowledge regarding the requirements for vitamins came from research using poultry.

Vitamin Deficiencies

Vitamin deficiencies are rare in livestock fed complete feeds because these are formulated to meet the animal's entire nutritional needs. When formulation errors occur, the most common result is very slow growth together with specific symptoms with each vitamin. There are reports of vitamin A deficiency in feedlot cattle, leading to partial or complete blindness and failure to thrive. Ruminants also may show deficiencies of other fat-soluble vitamins. The soluble vitamins are either synthesized by rumen microorganisms or by the animal itself.

TABLE 14-2 Vitamins

NAME	FUNCTIONS	DEFICIENCY SYMPTOMS	SOURCES
Vitamin A (fat soluble) Retinol, Retinoic acid Precursor is Carotene (beta carotene)	Normal eye maintenance; skin tissue maintenance; bone growth; immune functioning; embryonic development; reproduction; Retinoic acid–gene transcription.	Deficiency in humans is night blindness; keratinization of epithelial tissue, lowering disease resistance; reduced appetite; poor growth; weight loss; reproductive problems; paralysis; birth defects. Poultry: reduced egg production and hatchability of eggs.	Good-quality, fresh, green forages; green leafy hay; good-quality grass or legume silage. Exposure to the sun reduces carotene (precursor to vitamin A) content; vitamin premix.
Vitamin D (fat soluble) Plant form: (ergocalciferol) Animal form: (cholecalciferol) Active form 1,25 Dihydroxycholecalciferol	Calcium absorption; calcium and phosphorus metabolism; development of bones; bone growth and remodeling; prevention of rickets.	Deficiency in humans is rickets with abnormal bone development; birth of dead, weak, or deformed young. Mature swine: development of osteomalacia. Poultry: poor feathering and growth, thin-shelled eggs, lower hatchability of eggs.	Sun-cured forages; exposure to sunlight; vitamin premix.
Vitamin E (fat soluble) alpha-tocopherol beta-tocopherol gamma-tocopherol	Antioxidant helping in the absorption and storage of vitamin A; acts in other metabolic functions in the cell; antioxidant.	Muscular dystrophy. Poultry: reduced hatchability of eggs and reproductive problems.	Whole cereal grains; germ or germ oils of cereal grains; green forages; good-quality hay; commercial sources. Seldom deficient in diet unless feed grown on selenium-deficient soils; vitamin premix.
Vitamin K (fat soluble) phylloquinone menadione	Formation of prothrombin necessary for blood clotting; co-factor.	Slow blood clotting; internal hemorrhaging.	Fresh or dry green, leafy feeds, fish meal, liver. Rarely deficient—it is synthesized by bacteria in the rumen of ruminants and large intestine of monogastric animals; vitamin premix.
Thiamine (part of vitamin B complex; also called vitamin B₁)	Coenzyme in energy metabolism.	Deficiency in humans is beriberi; weakness with muscle; nerve and cardiovascular problems; reduced growth rate.	Cereal grains, grain byproducts; brans; germ meals; green, leafy hay; green pastures; brewer's yeast; and milk; vitamin premix.
Biotin (water soluble) (part of vitamin B complex; also called vitamin B₇)	Co–factor of the enzymes involved in carboxylation reactions such as the synthesis of fatty acids.	Dermatitis; loss of hair; cracks in the feet; slow growth. Swine: spasticity of the hind legs. Poultry: reduced hatchability of eggs.	Most grains (wheat and barley are low in available biotin); soybean meal; green forages; yeast; and cane molasses; vitamin premix. It is synthesized by bacteria in the rumen.
Folic acid (folacin) (water soluble) (part of vitamin B complex also called vitamin B₉)	Coenzymes–essential to normal metabolic function of body cells. Involved in the combining of single carbon units into larger molecules. Related to vitamin B₁₂ metabolism.	All species: weakness, slow growth, anemia, birth defects. Young chicks: poor feathering, loss of feather pigment. Breeding hens: lower egg production and hatchability. Turkey poults: nervousness, droopy wings, cervical paralysis. Turkey breeder hens: reduced hatchability.	Green pasture; green, leafy alfalfa hay; and some animal proteins (particularly those made from body organs); vitamin premix.
Vitamin B₁₂ (water soluble) (cobalt containing; synthetic form: cynanocol-balamin)	Coenzyme in a variety of metabolic reactions and is necessary for the maturation of red blood cells.	Deficiency in humans is pernicious anemia. All species: slow growth. Young pigs: lack of coordination in the hind legs. Breeding swine: reduced litter size, higher pig death rate. Breeding chickens: lower egg hatchability.	Vitamin premix; not produced by either animals or plants; bacterially produced in ruminant; found in animal products.
Inositol and Para-aminobenzoic acid (water soluble)	Component of phospholids and signal transduction within cells.	None demonstrated in livestock.	Synthesized in the intestinal tract and are generally not deficient in livestock feeding.

(continued)

TABLE 14-2 Vitamins (continued)

NAME OF	FUNCTIONS	DEFICIENCY SYMPTOMS	SOURCES
Vitamin C (ascorbic acid) (water soluble)	Antioxidant; improve performance in chickens, particularly with heat stress.	Deficiency symptoms have not been observed in farm animals.	Normally farm animals synthesize sufficient amounts in body tissues to meet their needs; synthetic ascorbic acid.
Riboflavin (water soluble) (part of vitamin B complex also know as vitamin B_2)	Coenzymes flavin mononucleotide and flavin-adenine dinucleotide involved in energy and protein metabolism.	Sows: slow growth, poor reproduction, and lower milk production. Young pigs: anemia, diarrhea, vomiting, eye cataracts, stiffness of gait, seborrhea, and alopecia. Calves: lesions around the mouth and alopecia. Young chicks: diarrhea and curled toe paralysis.	Milk; meat scraps; fish meal; green, leafy hay; green pastures; grass silage; and brewer's yeast; vitamin premix.
Pantothenic acid (water soluble) (part of vitamin B complex) also know as vitamin B_5	Part of coenzyme A important for carbohydrate and fatty acid metabolism.	All species: slow growth, loss of hair; and enteritis. Young pigs: stiff legs and lack of coordination when walking (goose stepping). Calves: dermatitis around eyes and muzzle, loss of appetite, diarrhea, weakness, increased susceptibility to respiratory infection, and difficulty in standing. Young chicks: ragged feather development; lesions on mouth, eyelids, and feet. Breeder hens: decreased hatchability of eggs.	Brewer's yeast; cane molasses; dried milk and whey; and fish solubles; vitamin premix.
Pyridoxine (water soluble) (part of vitamin B complex also known as vitamin B_6)	Co-factor in amino acid and essential fatty acid metabolism; production of red blood cells; and in the synthesis of neurotransmitters such as serotonin, dopamine, and norepinephrine, and the hormone epinephrine.	All species: anorexia, slow growth, and convulsions. Chicks: abnormal feathering, and development of nervous symptoms. Hens: lower egg production, poor hatchability, rapid weight loss, and death.	Cereal grains; grain byproducts; rice bran; polished rice; green pastures; green, leafy hay; yeast; and meat and liver meals; vitamin premix.
Niacin (nicotinic acid) (water soluble) (part of vitamin B complex also known as vitamin B_3)	Essential part of enzyme systems involved in lipid carbohydrates, and protein metabolism; needed by all living cells.	Deficiency in humans is pellagra with skin problems. All species: anorexia, slow growth, and unthriftness. Swine: diarrhea, vomiting, and dermatitis. Chicks: inflammation of the tongue, mouth cavity, and upper esophagus; reduced feed consumption; slow growth; poor feather development; and scaly dermatitis of the feet and skin. Turkey poults: hock disorder.	Some available in most feeds; meat and bone meal is a good source; green alfalfa is a fair source; the niacin content of cereal grains is largely unavailable. Surplus: tryptophan in the diet can be converted to niacin. Supplementation often needed in swine rations. Vitamin premix.
Choline (water soluble)	Phospholid as a part of the cell structure; lipid transport; nerve impulse transmission; fat metabolism in the liver; a donor of readily available methyl groups involved in several one-carbon transfer reactions called *transmethylation*. Acetyl choline is an important neurotransmitter.	Slow growth rate; unthriftiness; fattly livers; poor coordination; reproductive problems; lower milk production; and higher death rate in the young. Poultry: perosis (slipped tendons).	Meat scraps; oil meats; brewer's dried yeast; fish meal; distiller's solubles. Also found to a lesser degree in some grains, forages, and dairy products. Vitamin premix or choline chloride.

Commercial Feeds

Vitamins included in commercial feeds for ruminants are A, D, and E. Commercial feeds for pigs and poultry include vitamins A, D, E, and K; riboflavin; niacin; d-pantothenic acid; vitamin B_{12}; and choline chloride.

MINERALS

There are multiple minerals that are essential in animal nutrition (see Table 14-3).

SUPPLEMENTS TO FEED

Probiotics and Prebiotics

Probiotics and prebiotics are beginning to be added to animal feeds and foods. A prebiotic is a nutrient supplement designed to increase the population of

TABLE 14-3 Minerals

MINERALS	FUNCTIONS	DEFICIENCY SYMPTOMS	SOURCES
Calcium (Ca)	Formation of bones and teeth; milk production; forming egg shells; nerve and muscle functioning; maintain acid-base balance in body fluids.	Abnormal bone growth; hypocalcemia (milk fever); rickets; osteomalacia; thin-shelled eggs.	Legume forages; animal-origin protein supplements; fish meat-milk; skim milk; citrus pulp; citrus molasses; calcium carbonate (e.g., ground limestone, oyster shells); calcium phosphate; mineral mix.
Phosphorus (P)	Formation of bones and teeth; milk and egg production; reproduction; conversion of carotene to vitamin A; utilization of vitamin D; component of protein in the soft tissues; component in phospholipids, phosphoproteins, DNA, RNA, and ATP; involved in other metabolic processes.	Rickets; osteomalacia; poor appetite; slow gains; lower milk and egg production; reproductive problems; poor utilization of vitamin D; deficiency of vitamin A; unthrifty appearance.	Wheat bran; wheat middlings; cotton seed meal; linseed meal; meat scraps; tankage; fish meal; dried skim milk; calcium phosphate; much of plant phosphate is bound (nondigestible) as phylate phosphorus; the phosphate can be released by addition of the enzyme phytase to the diet.
Sodium (Na)	Osmotic pressure; maintaining a neutral pH level in the body tissues; muscle and nerve activity.	Slow growth; poor appetite; corneal lesions in the eye; males may become infertile and females may have a delayed sexual maturity.	Salt in block form or as loose salt, or in the mineral mix.
Chlorine (Cl)	Osmotic pressure; acid-base balance; essential for the formation of hydrochloric acid in the digestive juices; muscle and nerve activity.	Slow growth; poor appetite.	Salt in block form or as loose salt, or in the mineral mix.
Potassium (K)	Osmotic pressure; acid-base balance of the body fluids; muscle activity; digestion of carbohydrates.	Poor appetite; lower feed efficiency; slow growth; emaciation; stiffness; diarrhea; decreased milk production in lactating animals.	Forages; grains.
Magnesium (Mg)	Plays a role in activating several enzyme systems in the body; in the proper maintenance of the nervous system; in carbohydrate digestion; and in the utilization of phosphorus, zinc, and nitrates. It is a constituent of bone and is necessary for normal skeletal development.	Decreased utilization of phosphorus; vasodilation (dilation and relaxation of the blood vessels). An acute magnesium deficiency may result in grass tetany.	Magnesium sulphate ($MgSO_4$) or magnesium oxide (MgO) may be mixed with salt or supplement and fed free choice in areas where grass tetany is a risk.

(continued)

TABLE 14–3 Minerals (continued)

MINERALS	FUNCTIONS	DEFICIENCY SYMPTOMS	SOURCES
Sulphur (S)	Essential part of the amino acids cystine and methionine and is important in the metabolism of lipids (as a part of biotin), carbohydrates (as a part of thiamin), and energy (as a part of coenzyme A). Essential part of the amino acid cysteine.	A deficiency of sulphur in the ration will appear as a protein deficiency. Slow growth and a generally unthrifty condition are symptoms of a possible sulphur deficiency. Sheep fed NPN as a nitrogen source may have lower wool production unless the ration is supplemented with sulphur.	Forages, especially legumes, that are harvested in the earlier growth stages should contain enough sulphur for ruminants. If forages are harvested in more mature stages, some sulphur supplementation may be needed to improve the nitrogen utilization. Synthetic DL methionine; 2-Hydroxy methyl 4-(methylthio) butanoic acid (HMTB) (Alimet); animal and fish proteins; grains
Iron (Fe)	Hemoglobin formation; oxidation of nutrients in the cells.	Symptoms of anemia in young pigs include labored breathing (thumps); listlessness; pale eyelids, ears, and nose; flabby, wrinkled skin; and edema of the head and shoulders.	Most grains and forages contain enough iron to meet the needs of older animals. It is recommended that trace-mineralized salt containing iron be fed to ensure against any possible iron deficiency. Poultry and pigs: Mineral premix containing ferrous sulphate; blood meal.
Manganese (Mn)	Utilization of phosphorus; assimilation or iron; reduction of nitrates; in enzyme systems that influence estrus, ovulation, fetal development, milk production, growth, amino acid and cholesterol metabolism, synthesis of fatty acids.	Swollen and stiff joints; abnormal bone development; sterility; delayed estrus; reduced ovulation; abortions; deformed young; young born weak or dead; loss of appetite; slow grains; knuckling over in calves; rough hair coats; pinkeye.	Mineral premix containing manganese oxide or manganese sulfate; trace mineralized salt.
Copper (Cu)	Synthesis of hemoglobin; as an activator in some enzyme systems; for hair development and pigmentation; wool growth; bone development; reproduction; and lactation; growth-promoter in pigs.	Diarrhea; slow growth caused by anemia; swelling of joints; bone abnormalities; abortions; weakness at birth; difficulty in breathing; loss of hair color in cattle, abnormal wool growth in sheep; lack of muscle coordination and possibly sudden death.	Trace-mineralized salt; mineral premix containing copper sulfate.
Zinc (Zn)	Normal development of skin, hair, wool, bones, and eyes; preventing parakeratosis; promoting the healing of wounds; necessary in several enzyme systems, including peptidases and carbonic anhydrase; for protein synthesis and metabolism; required for insulin synthesis; growth-promoter in pigs.	Parakeratosis (rough, thick skin in swine); thickening of skin on the neck, muzzle, and back of ears in cattle; loss of hair; slipping of wool; slow wound healing; poor appetite and slow growth; swelling of hocks and knees; stiff gait; and inflammation of nose and mouth tissues.	Mineral premix containing zinc oxide; trace mineralized salt.
Molybdenum (Mo)	Component of the enzyme xanthine oxidase, which is found in milk and in body tissues. This enzyme is also important for the formation of uric acid in poultry. Molybdenum is involved in stimulating the action of rumen organisms.	No deficiency symptoms noted. Toxicity is of greater concern.	Normal diets do not need supplementation.

(continued)

TABLE 14-3   Minerals (continued)

MINERALS	FUNCTIONS	DEFICIENCY SYMPTOMS	SOURCES
Selenium (Se)	Needed in very small quantities; role in selenoproteins and as antioxidant; important for immune and muscle functioning.	Nutritional muscular dystrophy (white muscle disease) in cattle, sheep, chickens, turkeys, swine, and horses; retained placenta and low fertility in ruminants.	Selenium-deficient areas in the United States include the southeastern coastal states, states along the Great Lakes, New England states, and the coastal northwest. Young calves and lambs raised in selenium-deficient areas are sometimes injected with small amounts of selenium to prevent deficiency symptoms. Mineral mix containing selenite or selenate or organic selenium as selenomthionine; grain or forage from selenium-deficient areas.
Cobalt (Co)	Essential component of the vitamin B_{12} molecule; is used by rumen and cecal bacteria in the synthesis of vitamin B_{12} and in the growth of rumen bacteria.	Poor appetite; general malnutrition; weakness; anemia; slow growth; decreased fertility; lower milk and wool production. Nonruminants may also develop pernicious anemia.	Protein supplements; vitamin-mineral premixes; trace-mineralized salt and mineral supplements.
Iodine (I)	Production of the hormone thyroxine in the thyroid gland.	Development of goiter; weak or dead young at birth; hairlessness at birth; infected navels.	Iodized salt containing 0.007 percent iodine or 0.01 percent potassium iodide will meet the needs of most livestock.
Fluorine (F)	Helps prevent cavities in teeth; possibly slows down osteoporosis in older animals.	Deficiency in the diet is rare and supplementation is not recommended. In livestock feeding, toxic levels are more of a concern than is a deficiency.	Drinking water and forages in most areas contain enough fluorine to meet the needs of livestock.
Chromium (Cr)	Some compounds of chromium may activate insulin for sugar metabolism in the body.	Unknown.	Chromium supplementation of livestock rations in not recommended, because dietary deficiencies have not been demonstrated.
Silicon (Si)	Has been shown to have beneficial effects in the diets of laboratory animals, but the need in livestock feeding is not presently known.	Unknown. Toxic in excessive concentrations.	Silicon supplementation of livestock rations is not recommended, because dietary deficiencies have not been demonstrated.

beneficial microorganisms in the intestinal tract. A probiotic is a "seed" population of beneficial microorganisms (e.g., *Lactobacillus acidophilus*) added to feed or food.

Enzymes

Phytase

Phytic acid (inositol hexakisphosphate) or as a salt in its ionized form, phytate, contains a sugar and six phosphate groups. Much of the phosphate in plant-derived feeds such as grain, including corn, and in soybeans is in the form of phytate. This is an important mechanism for storing phosphate in plants. However, nonruminants do not digest phytate because they do not have the enzyme phytase. Phytate is a problem when feeding to animals because of the following:

• Phytate complexes with minerals such as calcium, iron, zinc, and some amino acids, preventing their absorption.

- Excess nutrients, including phosphate and others, have to be added to the diets.
- There is a large quantity of phosphate in the feces. This potentially leads to environmental pollution with phosphorus in groundwater, rivers, and lakes.

Phytases are enzymes added to pig and poultry diets to break down phytate, making the phosphorus available to the animals. Some phytases are thermostable, and, therefore, the feed can be pelletized. These enzymes are produced by microorganisms in a fermentation process.

Other Enzymes

Other antinutritive components in plant-based feedstuffs include pentosans, which are present in wheat. This compound has the effect of increasing ingesta viscosity and reducing animal growth. Addition of xylanase to animal feeds breaks down the antinutrients and releases nutrients.

TOXIC PLANTS AND FUNGAL TOXINS (MYCOTOXINS)

Mycotoxins

Fungi produce mycotoxins. These are often metabolites, and they are found in moldy grain. Mycotoxin contaminants have negative effects on animals consuming the grain. Examples of mycotoxins include aflatoxins, ergot mycotoxins, fumonisins, and zearalenone.

Aflatoxins

Aflatoxins are the most commonly occurring mycotoxins in feedstuffs, including corn worldwide. Examples include B1, B2, G1, G2, M1, and M2. Aflatoxins are produced by strains of *Aspergillus flavus* and *Aspergillus parasiticus*.

Ergot Mycotoxins

Ergot is a fungus, and can be found on wheat, barley, sorghum, and some other grasses. It produces a mycotoxin, with a number of toxic alkaloids, including ergotamine. This is a tripeptide chemically linked to a psychoactive compound ergoline. These mycotoxins act in the central nervous system causing severe reactions, such as convulsions, hallucinations, and a burning or prickly sensation in the hands or feet. Because of its effect of mimicking the neurotransmitter dopamine, ergot alkaloids inhibit prolactin secretion and stop lactation.

Endophyte Fungi

Endophyte fungi in tall fescue grasses produce mycotoxins causing reproductive problems in pregnant horses (abortion, dystocia, fetal death, and lactation failure) and fescue toxicosis in cattle. There are several alkaloids, including diazaphenanthrene, pyrrolizidine, and ergot, in endophyte-infected fescue.

INTERESTING FACTOID

Consumption of grain contaminated with ergot by people causes delusions and convulsions. Ergotism was known as Holy Fire or St. Anthony's Fire in Europe in the Middle Ages. Ergotism may have been involved in strange events in history, such as the following:

- The Salem witch trials.
- The failure of Peter the Great and his Cossack army to capture the great city of Constantinople after both the men and their horses consumed rye contaminated with ergot.
- The story of a sailing ship found with no crew: the *Mary Celeste*, which became the *Marie Celeste* in a short story by Arthur Conan Doyle, who wrote the Sherlock Holmes stories. It has been suggested that the crew jumped overboard, suffering from hallucinations from consuming ergot-contaminated grain.

Fumonisins

Fumonisins are a common mycotoxin on corn. Ingestion of animal feed containing fumonisins can be fatal in horses at concentrations of more than 10 ppm. Fumonisins reduce the efficiency of growth in pigs (<100 ppm) but have severe negative effects on the health of pigs (at concentrations >100 ppm). For chickens, concentrations of fumonisins need to exceed 200 ppm to even affect production efficiency.

Bioengineered Bt corn has been reported to have less fumonisins and other mycotoxins. This is thought to be due to the reduced damage by the European corn borer larvae and other organisms. There is consequently less damaged tissue for mold to infest (see fig. 14-7).

Zearalenone

Zearalenone is another mycotoxin with marked estrogenic activity and can cause infertility, for instance, in pigs. A possible source of zearalenone is clover.

Avoiding Mycotoxin Contamination

To avoid mycotoxin contamination of food and feed, grain is tested for mycotoxin concentrations to ensure that contamination is below acceptable limits. The addition of specific clays to feed contaminated with mycotoxins make them unavailable by adsorption onto the clay. In addition, enzymes have been developed that deactivate specific mycotoxins.

Toxic Plants

There are many plants that produce chemicals that are toxic to animals. These toxic compounds are called *phytotoxins*. Examples of toxic plants include the following:

FIGURE 14-7 Comparison of Bt (two on left) and non-Bt (two on right) ears of corn showing fungal infestation. Kindly provided by Dr. Gary Munkvolt, Seed Science Center, Iowa State University.

- Various hemlock species, which are the most toxic plants in the United States. Cattle are particularly vulnerable.
- Leaves of the yew tree that are potentially eaten by cattle, sheep, and goats, or by a dog penned near the tree.
- The leaves of rhubarb that are potentially eaten by cattle, sheep, and goats.
- Red maple leaves if eaten by horses.
- Wild black cherry leaves can be consumed by all livestock and companion animals.
- Bulbs such as iris bulbs, with dogs probably the most likely to consume these.
- Rhododendron and azalea leaves can be consumed by all livestock.
- *Veratrum californicum* is found in rangelands in the western states. It is very toxic to grazing livestock, and with teratogenic effects in sheep in utero, including a single eye (cyclops), lack of a pituitary gland, and greatly delayed parturition or failure to initiate parturition.

 # ANIMAL FEEDS, FOODS, AND THEIR FORMULATION

The nutritional requirements of animals are met by the provision of a balanced diet formulated to meet their nutritional needs. Nutritionists formulate animal feeds or foods using three criteria:

1. The set of nutrient requirements as recommended for the species (of a subset of species such as beef and dairy cattle, meat and egg-producing chickens) by the National Research Council. These recommendations are based on the available research information and a consensus of a panel of eminent nutritionists. This is often supplemented by the latest research findings.
2. The relative cost of different potential constituents of the diet. This is known as *least cost formulation*.
3. Chemical analysis of the various constituents of the diet.

Energy

Corn is the most widely used energy source for pigs and poultry feed in the United States. Moreover, corn is a constituent supplying energy in a feed concentrate supplement for cattle. This is why much of the nation's livestock industry is located in the corn belt in the Midwest. Sorghum/milo grain, oats, barley, and wheat are the other commonly used energy feeds. Other sources of energy can be citrus and other pulps, spoiled fruits and vegetables, and outdated sell date human foods.

Protein

The sources of protein to be in the feed include the following:

- Plant-derived protein
- Animal-derived protein
- Bacterial and other microorganism-derived protein and/or specific amino acids, such as lysine, tryptophan, and threonine

- Chemically synthesized amino acids or their precursors (D or L methionine and hydroxy methyl-thio butanoic acid sold commercially as Alimet)
- Nonprotein nitrogen (the microorganisms in the rumen that can synthesize amino acids from simple chemicals such as urea)

Plant-Derived Protein

Plant-derived protein sources include soybean meal, corn gluten, distillers' dried grains, canola meal, cottonseed meal, linseed oil meal, peanut oil meal, and brewers' dried grains. Soybean meal is the most common source of protein. Plant proteins can be used as the only protein supplement for ruminants. To improve performance, ruminants receive supplements containing a protein source.

Legumes

Legumes are plants that can fix atmospheric nitrogen (N_2) to ammonia and nitrate, and then incorporate these into amino acids, protein, and other nitrogenous compounds. Legumes have nodules in their root system containing nitrogen-fixing bacteria living in a symbiotic relationship. The bacteria fix the nitrogen from the air in soil and make it available for use by the plant. Examples of legumes are alfalfa, beans, clovers, peas, and soybeans. Legumes are often high in protein and require little or no nitrogen fertilization.

The seeds of legumes are used in animal feed; for example, soybean meal in pig and poultry feed; clover or alfalfa hay for cattle, sheep, and horses; and peas in dog food. Soybeans (*Glycine max*) are native to Asia. They are now the major protein source for animal feed in North America and the rest of the world. The major producers of soybeans are the United States and Brazil.

Ruminant Feed

Ruminants receive three major types of feed: forage; concentrates, predominantly grain diets with high energy/high nitrogen; and salt containing trace elements.

Forage

Forage is high-fiber feeds, including hay, silage, pasture, and rangeland.

Hay

Hay is a storage form of plant materials for ruminants and horses. It can be dried clover or dried alfalfa (see figs. 14-8 and 14-9) or dried grasses (e.g., timothy).

Silage

Silage is a storage form of food for ruminants. The high-moisture crop (e.g., corn with the stalks and leaves, as well as chopped kernels containing the grain, cob, and husk) is harvested at a time when there is the maximal amount of digestible nutrients per acre. It is then chopped (to say one-quarter-inch long) and placed in a silo, pit, or enormous plastic bag in such a way that air is excluded. The plant material ferments because of the presence of lactic acid bacteria such as *lactobacillus plantarum*. These microorganisms are either present on the plants or added as a silage inoculant. Fermentation is anaerobic, that is, in the absence of oxygen. When silage is exposed to air, spoilage will begin. Sealing silage in an anaerobic environment

FIGURE 14–8 Alfalfa plants. Kindly provided by Limin Kung, Department of Animal and Food Science, University of Delaware.

FIGURE 14–9 Alfalfa being harvested to hay. Kindly provided by Limin Kung, Department of Animal and Food Science, University of Delaware.

is important. During silage formation, a liquid is produced. Corn silage contains about 40% grain on a dry weight basis and, therefore, is a high-energy feed suitable for high-producing dairy cows. Silage is an example of a high-moisture feed.

Grain Diets, Including Concentrates

Grains

Grains are a major ingredient of feeds for cattle, pigs, poultry, and many companion animals. They are ground and may be fed as a meal (powder) or may be extruded as pellets to improve the palatability of the feed. Grains in animal feeds include corn or maize; sorghum, also known as milo; barley, wheat, and, for horses, oats; and rice, for example, in dog food.

Animal Proteins

Animal proteins can contain a better balance of essential amino acids than is the case of plant proteins. Moreover, they contain more protein: frequently over 47% crude protein. Commonly used animal proteins in livestock feeds include fish meal, blood meal, bonemeal, protein derived from rendering, dried skimmed or whole milk, and feather meal.

Animal proteins can be used for balancing rations for pigs and poultry, and as supplements for cattle. Given the transfer of the prions from rendered animal protein sources as the method of transfer of the bovine spongiform encephalopathy, protein from ruminants is prohibited from use in the diets of cattle, sheep, and any other ruminant livestock.

Cat food and, to a lesser extent, dog food is formulated using a mix of animal protein, including meat, fish, and by-products, together with some grain.

Use of By-Products of Corn Milling in Animal Feed

The National Corn Growers Association (www.ncga.com) plays a strong role in nationally promoting the use of corn. Corn grain (or kernels) has the following constituents: starch (62%), water (13%), protein (8%), nondigestible fiber (8%), oil/fat (7%), and ash/minerals (2%).

The corn kernel or grain has the following components: the germ, which is high in oils; starch; gluten, which is high in protein; and bran, which is the hull and other fiber.

The goal in corn milling is to separate the components. Two approaches are used: wet milling and dry milling. In both, the corn is first cleaned to remove foreign materials.

Wet Milling

The first step in wet milling entails soaking or "steeping" the kernels in water with sulfur dioxide. This results in the kernels becoming swollen and softened for the milling process. The soluble components, such as the starch, move into the steeping liquor and will be later concentrated. The whole process is summarized in Figure 14-10.

Dry Milling

The corn is subject to grinding to produce a product with a small particle size. This can be used directly in livestock feed. Digestion of the corn is facilitated by

Steps in the Wet Corn Milling Process

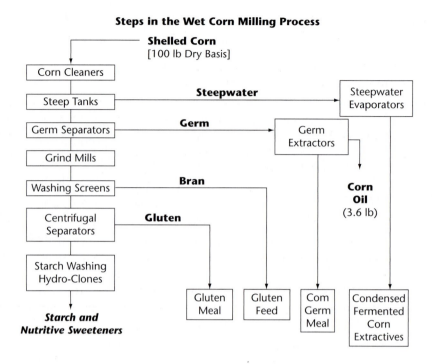

the small particle size of the ground corn. An alternative use for the ground corn is as a feedstock fermentation of the sugars of the starch to ethanol. The solids remaining after fermentation are *distillers' grains*, and the materials dissolved in the fermentation fluid are the *solubles*. The latter can be dried.

The by-products or co-products in corn milling that are included in livestock feed include the following:

- Distillers' grains with solubles (DDGS) are a by-product of dry milling, and are included in the feeds for poultry, pigs, and cattle. DDGS have only 10% water (90% dry matter) with about 25% crude protein and 0.8% phosphate. They can be used at about 25% in pig or poultry diets or to replace concentrates in feeding cattle. This by-product is becoming increasingly important as the ethanol production from cornstarch increases (see the section "The Impact of Bioenergy on Animal Feed" that begins on the next page).

- Wet distillers' grains for cattle feed.

- Corn gluten feed is a co-product of wet milling manufacturing and is used in cattle feed. It is the part of the commercially shelled corn remaining after extraction of most of the starch. Corn gluten feed is a source of protein (20–25%), is low in starch and oil but high in digestible fiber. It is lower in energy, protein, and fat than DDGS, but the availability of phosphate is higher.

- Corn gluten meal is a co-product of wet milling manufacturing. It is the dried residue from corn after the removal of most of the starch and germ, and the separations of the bran. It is used as a protein source in poultry and livestock feeds. It also serves as a good source of xanthophylls (golden yellow).

THE IMPACT OF BIOENERGY ON ANIMAL FEED

Corn is processed by either dry milling or wet milling. In either case, the starch is used as the chemical feedstock for fermentation to produce ethanol. The ethanol is then used as a fuel as a 10% blend with gasoline. Because of the greater efficiency in ethanol production, there is a tendency to increase dry milling. One of the by-products or co-products of dry milling is DDGS. The solubles are the result of drying >75% of the solids of the whole stillage remaining after fermentation and distilling the ethanol. They are blended with the distillers' dry grains. The DDGS can be dried to form dry DDGS. The dry DDGS can be used in feeds for poultry, pig, and cattle. A number of ethanol plants do not have driers, and, therefore, the DDGS is marketed as a wet DDGS. This is then used in either dairy or beef cattle operations. With the increasing production of ethanol, it is likely that the availability of DDGS will exceed 10 million metric tons by 2010.

With the use of sorghum, or barley, wheat, or blends for ethanol production, there is variation in the composition of DDGS. The beverage alcohol industry also produces grain DDGS (whiskey distilleries) or brewers' grains (beer manufacturing) that can be used in livestock feed.

Impact of Ethanol on the Livestock Industry

Ethanol production for fuel in the United States is increasing. According to the U.S. Department of Energy and Renewable Fuels Association, in 1980, 200 million gal were produced; in 1985, 600 million gal; in 1992, 1 billion gal; in 1998, 1.4 billion gal; and in 2004, 3.4 billion gal. Ethanol production is likely to exceed 6 billion gal in 2007 or 2008.

The increasing use of corn for the production of ethanol is having a number of effects that will impact the livestock industry:

- Increased price of corn, which is a major constituent of poultry feed.
- Increased availability of by-products of ethanol production, such as DDGS. (ethanol co-products are being used in almost half of beef cattle and pig operations in the midwestern United States).
- An increase in acreage planted to corn and a consequent decrease in the soybean acres planted. In 2007, the U.S. Department of Agriculture reported that growers plant on 12.1 million acres more than in 2006, with a concomitant decrease in soybean acres. This, in turn, is likely to impact the price of soybeans.

Other Issues with Imported Food and Feedstuffs

The United States now imports critically important feed ingredients, including amino acids and vitamins. Government payments to U.S. farmers increase the production of feed grains and depress the world prices. Chinese authorities acknowledged that ingredients exported to make pet food in the United States contained melamine, a chemical that the U.S. Food and Drug Administration suspects led to scores of pet deaths. A similar problem was reported for milk.

Definitions

Primary product or *principal product* is what an industrial process is designed to produce.

By-product is a material of some value produced along with the primary or principal product in an industrial process.

Co-product is when two or more products, often of equal or similar value, are produced in an industrial process.

ISSUES *for discussion*

1. Given that corn, soybeans, and other components of livestock feed can be used for people, in a hungry world, what is the justification of using these for livestock?
2. It is sometimes said that if we all became vegetarians, all grain could be used to feed the hungry. What is the difference between a consumer choosing to eat corn-fed beef, or pork or chicken, and choosing to drive to work or fly for vacation?
3. What are the impacts on the environment, particularly water, of crop production for livestock? How can they be alleviated?
4. What would be the impact on your community if there were a sudden switch away from field crops going to livestock?
5. What impact does ethanol production from corn have on the price of corn, the costs for animal production, and the price of land?
6. Is subsidizing ethanol use in fuel good for the American public? Is it good public policy? (It might be interesting to review the available literature on the World Wide Web regarding the energy efficiency of ethanol production.) What is the impact of ethanol on U.S. energy independence?
7. What might be the impact of increasing ethanol production on alleviating world hunger?
8. Should poultry receive animal-derived protein, including protein from poultry by-products, in their diet? Should corn that can be used for food or feed be used to produce a fuel?

ISSUES *for discussion*

1. If U.S. agriculture wants to export to other countries, should we reciprocate and allow imports of agricultural products?
2. What are your opinions on whether or not the United States imports feed ingredients or other agricultural products?
3. Should there be changes in the procedures to improve safety?
4. What are the risks in one country becoming the source of crucial feed ingredients?
5. What is the impact of U.S. government policy and payments to grain producers on the following?
   - Livestock producers in the United States and elsewhere
   - U.S. price of food
   - The ability of farmers in developing countries to compete without an "even playing field"

REVIEW QUESTIONS

1. Name and give examples of the three general forms of livestock feed.
2. What is a nutrient?
3. What are the five groups of nutrients?
4. Name and describe the terms associated with descriptions of energy in feeds.
5. Describe the chemical composition of carbohydrates.
6. Which carbohydrates are easily digested, and which are not?
7. What is a hexose sugar?
8. Name the most common simple hexose sugar in an animal.
9. How are polysaccharides formed?
10. What is the most important polysaccharide in an animal's body?
11. What parts of the plant store the most easily digested carbohydrates?
12. Discuss the digestion of fiber.
13. What is basal metabolism?
14. What are the processes in an animal that require energy?
15. What problems occur as a result of a deficiency of energy in the feed?
16. Describe the chemical composition of fats.
17. Why do fats have more energy value than carbohydrates?
18. What is the difference between a saturated and unsaturated fatty acid?
19. Name three essential fatty acids.
20. Why are fats added to the rations of livestock?
21. What is a protein? Describe its chemical composition.
22. What are nonprotein nitrogen compounds?
23. What is crude protein, and how is the crude protein content of a feed determined?
24. What is digestible protein?
25. Compare the utilization of protein and nonprotein nitrogen by ruminant and nonruminant animals.
26. Define amino acids, and describe their chemical composition.
27. What is the difference between essential and nonessential amino acids?
28. Define protein quality, and discuss protein quality as it relates to formulating rations for ruminant and nonruminant animals.
29. What is a limiting amino acid, and how does this relate to feed formulation?
30. What are the functions of protein in an animal?
31. What are some signs of protein deficiency in the animal's diet?
32. What may cause protein in the feed to be unavailable?
33. Define a protein supplement.
34. What is a vitamin?

35. How does a vitamin differ from carbohydrates and proteins?

36. Which vitamins are soluble in water, and which are soluble in fat or fat solvents?

37. Which vitamins are commonly synthesized in the rumen?

38. How does the solubility of vitamins affect the need for supplying them in the diet?

39. How may vitamins be supplied other than through natural feed sources?

40. Name the major minerals needed by livestock.

41. Name the trace minerals needed by livestock.

42. Describe the general functions of minerals in nutrition.

43. What are some general symptoms of mineral deficiency in the diet?

44. How are minerals added to an animal's feed?

45. What are the functions of water in an animal?

46. What is net energy of a feed?

47. What are prebiotics and probiotics?

REFERENCES AND FURTHER READING

Cheeke, P. E. (2005). *Applied animal nutrition: Feeds and feeding* (3rd ed.). Upper Saddle River, NJ: Prentice Hall.

Kellems, R. O., & Church, D. C. (2002). *Livestock feeds and feeding* (5th ed.). Upper Saddle River, NJ: Prentice Hall.

Mycotoxins

Maloy, O. C., & Inglis, D. A. *Ergot*. Retrieved July 20, 2009, from http://pnw-ag.wsu.edu/smallgrains/Ergot.html

Munkvold, G., Osweiler, G., & Hartwig, N. *Moldy grains, mycotoxins and feeding problems*. Retrieved July 20, 2009, from http://www.oardc.ohio-state.edu/ohiofieldcropdisease/Mycotoxins/mycopagedetrimental.htm

Olukosi, O. A., Sands, J. S., & Adeola, O. (2007). Supplementation of carbohydrases or phytase individually or in combination to diets for weanling and growing-finishing pigs. *Journal of Animal Science, 85*, 1702–1711.

Probiotics

MayoClinic.com. *Lactobacillus acidophilus*. Retrieved July 20, 2009, from http://www.mayoclinic.com/health/lactobacillus/NS_patient-acidophilus

Schroeder, W. (1997). *Corn gluten feed: Composition, storage, handling, feeding and value*. North Dakota State University Extension Publication AS-1127. Retrieved July 20, 2009, from http://www.ag.ndsu.edu/pubs/ansci/dairy/as1127w.htm

CHAPTER 15

Animal Genetics and Breeding

OBJECTIVES

This chapter will consider the following:

- The impact of genetics of livestock
- Fundamentals of genetics and genomics (DNA, genes including dominant and recessive genes, chromosomes, including comparative genomics, gene mapping, genome sequencing, gene ontology, single nucleotide polymorphisms [SNPs], and epigenetics)
- Applications of animal genetics and breeding (heritability, quantitative trait loci [QTL], and candidate genes, with information on the genetics of animal color)
- Cloning
- Transgenic animals

Breeds are considered, albeit briefly, under the different species.

INTRODUCTION: THE IMPACT OF GENETICS ON LIVESTOCK

During the past 65 years, there have been huge increases in the efficiency of animal production. Two examples will be considered in some detail, namely, the 3.5-fold increase in milk production/yield per cow, and the 3-fold increase in the rate of growth in chickens.

Impact of Genetics on Dairy Production

There have been huge increases in average yield of dairy cow. This is clear from the changes in yield between 1940 and 2005: in 1940, 5.1 thousand lb per cow per year (or 2.3 × 1,000 kg per cow per year); and in 2005, 18.3 thousand lb per cow per year (or 8.3 × 1,000 kg per cow per year).

The steady increase in yield has been due to the application of scientifically researched management approaches, including the following (in order of impact):

- Quantitative genetics used to improve markedly the genetic merit of cows
- Improved nutrition
- Improved animal health
- Improved management
- The availability of bovine somatotropin (also known as growth hormone)

With the widespread use of artificial insemination (used in the United States since the 1930s), it has been possible for farmers to purchase semen from superior bulls. The genetic merit of a bull is evaluated based on the production/yield of its daughters, the butter fat and protein in the milk, the longevity of milking, mastitis resistance, and the pregnancy rate of the daughters. Improved milk production is achieved by the use of bulls with high genetic merit. These pass on production traits with high heritabilities. In addition, culling lower-producing cows improves average production by the herd and prevents inferior genetics being passed on. A corollary to the increased yield of milk per cow has been a decline in the number of dairy cows: in 1940, 24 million dairy cows in the United States; and in 2005, 9 million dairy cows in the United States.

Genetic Selection to Improve Poultry Growth

There have been tremendous improvements in the rate of growth of meat chickens. In 1945, in a "classic" text by Brody, he reported that it took 3 months (13 weeks) to grow a 2.6 lb (1.2 kg) chicken (average growth rate 0.1 kg per week). Today, an 8.1 lb (3.7 kg) chicken is routinely grown in 8 weeks, with an average growth rate of 1 lb (0.45 kg) per week.

Gerry Havenstein, Peter Ferket, and colleagues at North Carolina State University examined whether it was improved nutrition or improved genetics that was making the difference (see Table 15-1). The vast majority of the improvement was due to the effect of breeding or genetics. Not only was there a very large genetic effect on growth and feed efficiency but also more of the carcasses could be processed for meat.

QUOTATION

THAT WAS MISAPPLIED BY THE AUTHOR BUT APT FOR ANIMAL BREEDING

All animals are equal but some are more equal than others.

George Orwell (1903–1950), British author and journalist in the book *Animal Farm* that was a parody about totalitarian regimes

TABLE 15-1 The relative impact of the application of genetics and nutritional research and development on the growth rate and feed efficiency of chickens ("broilers").

GENETICS	NUTRITION	WEIGHT AT 42 D OF AGE (KG)	FEED-GAIN RATIO
2001	2001	2.90	1.63
2001	1957	2.27	1.92
1957	2001	0.64	2.14
1957	1957	0.59	2.34

Source: Data are from Havenstein et al. (2003).
Note: Researchers at North Carolina State University compared the growth rate and feed efficiencies of commercial chickens with a randomly bred population mimicking the commercial chickens of 50 years ago. The birds were fed commercial feed formulated to either today's standards or those of 50 years ago.

In a similar way in turkeys, Gerry Havenstein together with colleagues at North Carolina State University and Ohio Agricultural Research and Development showed that the large improvement in performance of turkeys was due to genetic selection (see Table 15-2).

 ## FUNDAMENTALS OF ANIMAL GENETICS AND GENOMICS

Heredity is the transfer of traits or genes from each of two parents to the next generation. We have known for thousands of years that both parents contribute characteristics to the offspring, regardless of whether the offspring is a plant, animal, or person. The study of heredity is genetics. There has been tremendous progress in our understanding of genetics in the last 150 years, with the following developments:

- The development and validation of the concept of the gene (dominant or recessive) for a single trait stemming from the work of Gregor Mendel.
- The identification of chromosomes with genes arranged along a specific order.
- The chemical basis of genetics is DNA encoding specific proteins.
- The mechanism established that RNA (ribose nucleic acid) copies of specific sequences of DNA are made (transcribed). These RNA molecules are then used to make specific proteins in what is called *translation*.
- Recently, the sequence of the genome (all the genes) of a number of species, including humans, cattle, horses, and chickens, has been reported.

Heredity is based on the transfer of genes from each parent. The unit of heredity or inheritance is the gene. Genes are arranged along chromosomes in a

TABLE 15-2 The relative impact of the application of genetics and nutritional research and development on the growth rate and feed efficiency of turkeys

GENETICS	NUTRITION	AGE AT 8.5 KG (D)	FEED-GAIN RATIO AT 8.5 KG
2003	2003	84	1.63
2003	1966	98	1.92
1966	2003	168	2.14
1966	1966	168	2.34

Source: Data are from Havenstein et al. (2007).

Definitions

An allele is a specific locus on a chromosome usually encoding a gene.

A *candidate gene* is one of the genes suspected in underlying the expression of a trait.

A *chromosome* consists of a long DNA molecule encoding multiple genes.

Epigenetics are heritable changes in gene expression that are not encoded by DNA.

Epistasis is where two or more genes interact.

A *gene* is the unit of heredity and encodes a single protein. Genes on the same chromosome tend to stay together except that there is crossing over at a single point on the chromosome (see fig. 15-1). Linkage between genes has facilitated the development of gene maps.

Gene expression is the synthesis of RNA from the DNA template of the gene. The level of gene expression varies with physiologic and developmental state. Gene "knock-outs" have been produced in mice with a specific gene missing or not being expressed. These are very useful genetic models for researchers. Gene maps show the sequence of genes along a chromosome.

Genomics is the study of the entire genome (structural genomics) or comparisons between the genome of different species (comparative genomics) or transcription (functional genomics).

Genotype is the full collection of genes/DNA in a specific individual.

Heritability is the proportion of the total variation (genetic and environmental) in phenotype that is due to gene effects.

Histones are proteins in close proximity to DNA in the chromosomes. They can influence gene expression. Genes on different chromosomes segregate independently during meiosis.

Phenotype is the biologic characteristics of an individual animal, and is based both on the genotype and environmental factors.

Quantitative trait loci (QTL) are regions of DNA linked to the genes underlying a trait.

Selection index is an index of net merit weight traits based on their economic importance, heritability, and any genetic correlations existing in them.

Sex-linked genes are nonreproductively related genes that are located on the X chromosome in mammals or Z in birds. Therefore, a male mammal (or female bird) will only have one copy of the gene, whereas a female mammal (or male bird) will have two.

Transcription is the synthesis of RNA from the DNA template.

Translation is the synthesis of a polypeptide chain or protein from the RNA template.

specific order. DNA is the chemical coding for the genes. The genetic code of an animal (its DNA) is called the *genome*. Associated with the DNA in the chromosomes are proteins called *histones*. Every cell (except gametes) has pairs of chromosomes with each member of the pair having the same genes, although there are likely to be some differences in the exact structure for DNA for many genes. Gametes have half the number of chromosomes (one from each pair). When a male and female gamete unites, a zygote is formed with genes from each parent. Not only is inheritance of characteristics determined by the sequence of the DNA but also there can be modifications of the DNA (such as DNA methylation) or associated proteins (changes to the histone proteins) that change the activity of the gene.

Deoxyribonucleic Acid (DNA)

DNA is a double helix composed of two strands of an alternating deoxyribose (a five-carbon sugar molecule) and phosphate. The side strands are linked together by bases; two are purines (adenine and guanine), and two are pyrimidines (thymine and cytosine) (see fig. 15-1). Adenine connects only to thymine across the helix, and cytosine connects only to guanine. The unit of DNA is the nucleotide, which is comprised of one deoxyribose, one phosphate, and one base. There are four nucleotides: adenosine monophosphate (A), cytidine monophosphate (C), guanosine monophosphate (G), and thymidine monophosphate (T).

The Genetic Code

The genetic code consists of a series of codons. Each codon is three nucleotides, and in all animals, plants, and other living organisms, codons code for a specific amino acid. The characteristics of the genetic code include the following:

- Codons are three nucleotides or bases in a specific sequence.
- Codons do not overlap.
- Multiple codons exist for the same amino acid (termed *degeneracy*).
- Polypeptide chains are initiated by the codon for methionine (AUG).
- The three codons—UAA, UAG, and UGA—are stop codons, causing polypeptide chain termination.
- The code is universal with the same codons coding for specific amino acids in all living organisms.

Chromosomes

There are pairs of chromosomes. Genes are arranged in a line in chromosomes (see fig. 15-2). The number of chromosomes varies considerably from species to species, as can be seen in Table 15.3.

Gene

The gene is the unit of heredity. Genes are composed of DNA. Upstream of the open reading frame of a gene is a noncoding region of the DNA. This contains regulatory elements such as a promoter region. Binding of a transcription factor

DNA ladder separates to form two identical DNA ladders

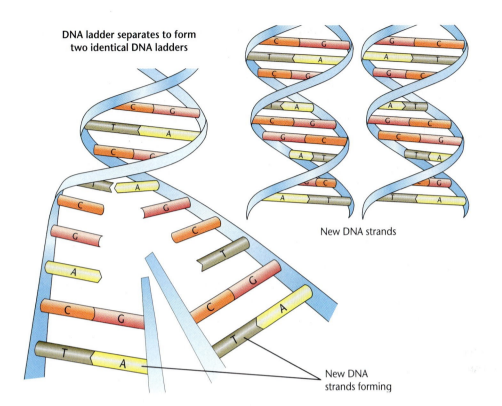

New DNA strands

New DNA strands forming

FIGURE 15-1 Structure of DNA showing helix, nucleotides, and replication. Source: Delmar/Cengage Learning.

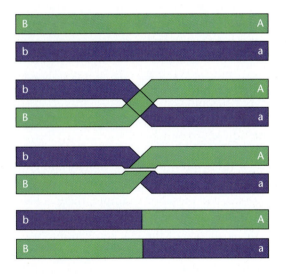

FIGURE 15-2 *A*, New combinations of genes are formed when chromosomes crossover and split. Source: Delmar/Cengage Learning. *B*, Chromosome at metaphase of meiosis showing centromere and where crossing over can occur.

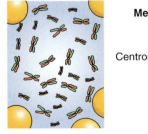

Metaphase chromosomes

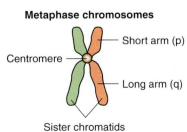

Centromere

Short arm (p)

Long arm (q)

Sister chromatids

TABLE 15-3 Chromosome number in different domestic animals

Buffalo	60
Cat	36
Catfish (Channel)	58
Cattle	60
Chicken	78
Dog	78
Donkey	62
Guinea pig	16
Goat	60
Horse	64
Pig	38
Rabbit	44
Sheep	54
Turkey	82

to the promoter greatly increases gene transcription of the open reading frame of the gene. The transcribed RNA is composed of exons and introns, with the latter spliced out (removed) to generate the messenger RNA. There is translation of the messenger RNA to generate polypeptides or proteins: binding of transcription factor → gene → transcription of the open reading frame → RNA (e.g., exon intron exon) → splicing to produce messenger RNA → translation → protein

Microsatellites

Microsatellites or simple sequence repeats (repeated motifs of one to six bases) are widely used for gene mapping

Genome Sequencing

With the completion of the sequencing of the human and mouse genomes, there has been tremendous progress sequencing the genomes of domestic animals, such as cattle, chickens, horses, and pigs. For example, the chicken genome has a haploid content of 1.1×10^9 base pairs (bp) of DNA encoding 16,715 protein encoding genes. In 2002, the sequence of the chicken genome was first assembled. In 2007, the first draft of the horse genome sequence (2.7 billion DNA bp) was reported and placed in public databases that are freely available (www.uky.edu/Ag/Horsemap/). The DNA that was sequenced came from a Thoroughbred mare named Twilight, from Cornell University (see fig. 15-3).

Comparative Genomics

This involves comparing the genomes of different species. An example is the comparison of the positions of genes along the chromosome. This is identical in a species. Closely related species have very similar sequences of genes along the chromosome. More distantly related species have more differences.

FIGURE 15-3 Image of Twilight, the horse. Researchers have sequenced the genome of this Thoroughbred mare from Cornell University in Ithaca, NY. Source: http://www.genome.gov/pressDisplay.cfm?photoID=20008.

Genetic Changes or Mutations

There are a number of possible changes to the genome of an animal. Examples are listed below. Most changes are deleterious or have no effect. If they are a severe problem, they are unlikely to be passed on to the next generation because they may result in death as, for instance, in the case of embryonic lethal mutations. Genetic changes or mutations include deletion of a gene or part of a gene; gene duplication; gene inversion; gene translocation; and point mutation of a single nucleotide leading to SNPs (see the section on SNPs later in the chapter for more details).

Epigenetics

Epigenetics are heritable changes in gene expression that are not encoded by DNA. These include covalent modification of histones and methylation (adding a methyl group typically at specific cytosine residues—CpG dinucleotide site) of the DNA. This is epigenetic programming and is largely not reversible in the life of an animal. Methylation has several functions, including tissue-specific gene expression, imprinting genes, X-chromosome inactivation, and immobilization of mammalian transposons. During gametogenesis, there is epigenetic programming with methylation and reprogramming with demethylation (removal of methyl groups). Epigenetic effects or defects can be induced by environmental factors. For instance, feed restriction during gametogenesis has been shown to lead to defects in the offspring.

Gene Ontology

It is critical that scientists use a common vocabulary to describe the functions of gene products. The functional genomic and bioinformatics community uses gene ontology to annotate what genes do.

Telomeres

Telomeres are multiple TTAGGG tandem repeats at the end of chromosomes in at least higher vertebrates. The telomeres stabilize the chromosomal ends and protect the genes in the subtelomeric regions. Telomeres shorten with progressive cell cycles, either in the animal or in tissue culture, leading to the view that shortened telomeres are related to aging. Addition of telomeric repeats can be accomplished by the enzyme, telomerase, a RNA-dependent DNA polymerase. High levels of telomerase activity are observed in malignant tissues and in fetal tissues, both where there are high rates of cell division.

 TYPES OF GENE ACTION

Chromosomes and, therefore, genes are homologous pairs. A homozygous gene pair is one that carries two genes for a trait. For example, a polled cow might carry the gene pair PP. A horned cow must carry the gene pair pp. For a cow to have horns, it must carry two recessive genes for the horned trait.

A heterozygous gene pair is one that carries two different genes (called *alleles*) for the trait. For example, a polled cow might carry the gene pair Pp. This cow is polled because the P gene is dominant, but it carries a recessive gene for the horned trait. If this cow is mated to a bull with a gene pair Pp, some of the calves will be polled, and some will have horns. There are six basic types of genetic combinations possible for a single gene pair:

- Homozygous $\times$ homozygous (PP $\times$ PP) (both dominant)
- Heterozygous $\times$ heterozygous (Pp $\times$ Pp)
- Homozygous (dominant) $\times$ heterozygous (PP $\times$ Pp)
- Homozygous (dominant) $\times$ heterozygous (recessive) (PP $\times$ pP)
- Heterozygous $\times$ homozygous (recessive) (Pp $\times$ pp)
- Homozygous (recessive) $\times$ homozygous (recessive) (pp $\times$ pp).

Genes may be additive, dominant, partial dominant, overdominant, or recessive. The phenotypes are shown in Table 15-4.

 APPLICATIONS OF ANIMAL GENETICS AND BREEDING

Many traits that are heritable are based on multiple genes. This is polygenic or quantitative inheritance. An example of this is growth rate that can be described by a normal or Gaussian curve (see fig. 15-4). Quantitative trait loci are regions of DNA linked to the genes underlying a trait. Similarly, a candidate gene is one of the genes suspected in underlying the expression of a specific trait. Animal breeders use

TABLE 15-4 Comparison of the phenotypes when the gene is dominant, recessive, additive, partially dominant, and overdominant

GENOTYPE	PHENOTYPE
Dominant (DD) versus recessive (dd)	
DD	++++
Dd	0
dd	0
Additive	
AA	++++
Aa	++
Aa	0
Partial dominant	
BB	++++
Bb	+++
Bb	0
Overdominant	
BB	++++
Bb	++++++
Bb	0

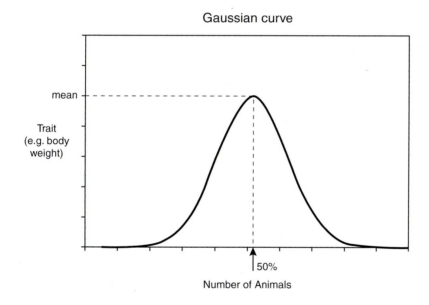

Gaussian curve

FIGURE 15-4 Traits show a normal probability or Gaussian distribution with a bell-shaped curve.

quantitative genetics together with marker-assisted selection (MAS) to improve animal breeds, or to produce hybrids or crosses between breeds or different grandparent stock.

Heritability

Many traits are due to multiple genes. Heritability is the proportion of the total variation (genetic and environmental) in phenotype that is due to gene effects: phenotype equals additive effects plus dominance effects plus epistatic effects plus environment.

A heritability estimate expresses the degree to which a trait is passed on from parent to offspring. Some traits are highly heritable, such as those related to carcass quality; some are moderately heritable, such as growth rate; and others, such as those related to reproduction, have a low heritability.

Breeding Systems

There are three breeding systems used with domestic animals:

1. Straightbreeding or purebred breeding. Advantages of this are that only a purebred animal is eligible for registry in the particular breed association or equivalent. There is greater homozygosity. This can lead to the disadvantage of purebreds with either inbreeding (mating of closely related animals) or linebreeding (animals closely related genetically but several generations apart), occurring with the risk of genetic diseases resulting from homozygosity of recessive genes. Purebred breeding tends to be a specialized business.
2. Crossbreeding, or producing hybrid offspring, is when two or more different breeds are crossed. There can be two-breed crosses, three-breed crosses, and rotation breeding. The advantage of crossbreeding is that there is hybrid vigor or heterosis. This is where the offspring is superior to the average of the parents. Beef cattle, pigs, and sheep are frequently terminal crossbred to produce market animals. The terminal cross is between the parents of animals that ultimately will be slaughtered for meat.
3. Synthetic inbred lines for grandparent stock each may be based on one or several breeds. There are frequently four such grandparent lines. Intense selection is performed on the grandparent lines. The breeding program is designed to maximize selection pressure and heterosis.

Marker-Assisted Selection (MAS)

MAS is the use of quantitative genetics coupled with genomics (such as candidate genes). It further increases the accuracy of breeding value estimation, and is particularly useful for traits that are expensive and/or difficult to measure, such as meat quality and animal health. An example is the estrogen receptor pig fecundity test. This is based on polymorphism in the estrogen gene. Females with two copies of one specific allele produce more offspring than with an alternate allele. This is obviously significant to breeding. Jeremy Rifkin, an activist opposed to genetically modified crops, has stated guarded support for MAS, as have environmental organizations such as the Sierra Club. Moreover, organic production can use varieties developed using MAS.

Single Nucleotide Polymorphisms (SNPs)

SNPs (pronounced "snips") are when a single base pair is different between animals of the same species. These SNPs account for much of the genetic differences between individuals and will be increasingly used by animal breeders.

The Genetics of Animal Color

The color in fur/hair in domestic animals is due to the natural polymer melanin. This is made in the animal from the amino acid tyrosine. The two chemical forms of melanin are eumelanin (giving black or brown colors) and pheomelanin (giving an orange/red). Melanin granules can be deposited in each hair shaft in

---

GENETICS OF CAT COLOR

There are at least nine genes that control the color of cats (see Table 15-5 and fig. 6-3 [Chapter 6] for examples of different colorations in cats):

1. Agouti (A- or aa). The agouti allele is dominant (A-), and only cats homozygous for non-Agouti allele (aa) show the lack of pattern.
2. Black or brown (B, b, or bl). These have differing amounts of eumelanin deposited in the hair. B is dominant, giving a black color; bb or bbl gives brown; and blbl gives a cinnamon color.
3. Color (C, cb=cs, c). The Siamese cat has the allele, cs, at the albino gene. There is a temperature sensitivity to the tyrosinase gene, essential for the manufacturing of melanin by the animal. It is inactive at higher temperatures. At the extremities such as the ears or legs (cooler), there is active enzyme and, therefore, melanin. This leads to pigmented extremities. Cats with cc are complete albinos.
4. Dilution (D, d). In cats with dd, there is dilution of colors.
5. Inhibitor (I, I). It affects the synthesis of yellow/red pigment (pheomelanin).
6. Orange (O, o). This is a sex-linked trait with the orange gene on the X chromosome. It is two alleles: the dominant orange (O) and the nonorange (o). In the presence of the orange allele, there is not full expression of the black locus:
   - Females are OO (orange colored), Oo (tortoiseshell), or oo (black colored).
   - Males are either O- (orange colored) or o- (black colored).
7. Spotting (S, s). This affects melanocyte production and migration. Codominant expression as the SS cats have a harlequin or "van" appearance, with color on the top of the head and tail. Ss cats are bicolor with white patches on their coat.
8. Tabby (Ta, Tm, tb). Tm is mackerel tabby or tigre, and tbtb as a blotched (classic) tabby. Ta is a ticked tabby.
9. White (W, w). It has a reduced melanocyte number and, therefore, melanin production. W cats are white, often with blue eyes.

TABLE 15-5 Effect of B (black), O (orange), and D (dilute) genes on color in cats

PIGMENT	BLACK GENES	COLOR OF CAT IF DENSE D GENOTYPE	COLOR OF CAT IF DILUTE DD GENOTYPE
Eumelanin (black)	B-	Black	Blue
	bb	Chocolate brown	Lilac
	B'b'	Cinnamon	Fawn
Pheomelanin (orange/red	OO (females)	red/orange	Cream
	O- (males)		

Note: Applications of animal genetics and breeding.

different arrangements. In many species, including cats and dogs, there is a banding of pigmentation along the hair. This is referred to as *agouti*.

Molecular Genetic Control of Coat, Hair, and Skin Color

There is increasing knowledge of the molecular control of coloration. There are two coat color pigments, both forms of melanin. The two forms are eumelanin, a black pigment; and pheomelanin, a red pigment. The color of the skin or hair of mammals is due to the presence of one or both forms of melanin by the melanocytes, together with the amount of the pigment and number of melanocytes.

In the wild-type or natural nondomesticated state, both melanin types are usually synthesized, but with mutations in several genes, melanocytes produce a single type of pigment. For instance, the melanocortin receptor 1 (MC1R) is the switch between producing eumelanin or pheomelanin. Activation of the MC1R by a form of melanocyte-stimulating hormone (α-MSH) leads to the synthesis of the black pigment, eumelanin, because of increases in the amount of the key enzyme tyrosinase. Mutations of the *MC1R* gene can result in it being switched on continuously without the ligand or not respond to the ligand. An example of the former is the classic dominant extension (*E*) coat color locus that produces a uniform black coat color and is an allele of the *MC1R* gene. Recessive alleles of the *MC1R* gene increase the amount of the red pigment. The agouti (*A*) locus encodes a protein, agouti-signal peptide. This induces production of the red pigment pheomelanin by being an α-MSH antagonist (competing with α-MSH for binding to the MC1R). Other genes affecting coloration include tyrosinase-related protein, β-defensin (also binds to the MC1R), and *KIT*, which encodes the mast/stem cell growth factor receptor that is essential for migration and survival of neural crest-derived melanocyte precursors.

 ## THE MSH: MELANOCORTIN RECEPTOR SYSTEM

There are five different melanocortin receptors:

1. MC1R is present on melanocytes and is activated by the ligand α-MSH, causing increased synthesis of eumelanin. It is antagonized by agouti-signaling protein, preventing production of the black color and leading to synthesis of the red pigment. β-Defensin prevents the effect of agouti-signaling protein and, therefore, causes black coloration.
2. MC2R is present on cells in the adrenal cortex. These receptors are activated by the adrenocorticotropic hormone, stimulating the production of the stress hormone cortisol in mammals and corticosterone in birds and rodents.
3. MC3R (ligand γ-MSH).
4. MC4R is present on neurons in the feeding center in the brain. Activation of the MC4R by its ligand, β-MSH, decreases feeding. The agouti-signaling protein is an antagonist for the MC4R.
5. MC5R.

Proopiomelanocortin (POMC) is a precursor of multiple biologically active peptides, many of which bind to melanocortin receptors. These peptides include the following:

- α-MSH produced in the hypothalamus and intermediate lobe of the pituitary gland
- β-MSH produced in the hypothalamus and intermediate lobe of the pituitary gland
- γ-MSH produced in the hypothalamus and intermediate lobe of the pituitary gland
- Adrenocorticotropic hormone produced in the anterior lobe of the pituitary gland
- B-Lipotropic hormone produced in the anterior lobe of the pituitary gland
- B-Endorphin produced in the hypothalamus and intermediate lobe of the pituitary gland

The gene-encoding POMC encodes multiple peptides. The anterior pituitary processes POMC to adrenocorticotropic hormone and β-lipotropic hormone. While the intermediate lobe of the pituitary and neurons processes POMC to α-MSH, β-MSH, γ-MSH, and β-endorphin

Nutritional Genomics

It is becoming increasingly possible to tailor the diets of an individual animal and person to its specific genetic code. This is called *nutritional genomics*.

Genetic Counseling

In human health care, genetic counselors work with families in which there have been birth defects or genetic disorders, or in which there is the risk of inherited diseases. In livestock, animals known to pass on genetic diseases are normally not used in breeding programs. In companion animals, there have been breeding programs to reduce the incidence of genetic disease.

Animal Genetics Companies

Animal breeding is increasingly performed by large multinational companies, including Genus, with divisions in cattle (Genus/ABS) and pigs (PIC). Other major genetic companies include Alta Genetics (cattle), Hypor and Monsanto (pigs), Cobb-Vantress (poultry owned by Tyson), Hubbard-ISA (broiler poultry), Lohman-Hy-Line International (poultry) containing Avigen, Lohman and Hy-Line, Hendrix Poultry, Hubbard-ISA (laying hens), Nicholas Turkeys, BUTA, and Hybrid turkeys (turkeys).

Heritage Breeds

Commercial poultry are produced from synthetic breeds or varieties. It is likely that this trend will continue into pigs and, potentially, other livestock. However, there is some demand for meat from the original breeds and varieties. Examples of heritage turkey varieties include the Bourbon Red, Blue Slate, Narragansett,

Royal Palm, and Spanish Black. These may be used in gourmet foods and "high-end" restaurants.

 ## CLONING OR NUCLEAR TRANSFER

Cloning entails using nuclear transfer from early embryo cells (cleavage stage embryos or inner cell mass cells from the very early embryo or cultured inner cell mass cells, morula cells or 16- to 20-cell stage cells) or from a somatic cell into an activated enucleated oocyte. This produces a totipotent blastocyst that will go to term. The resulting progenies are identical to the donor of the nucleus with the exception of the mitochondrial genes in the recipient oocyte. Thus, a cloned animal will have virtually all the genetic characteristics of the donor. Cloning has been developed to produce domestic livestock and companion animals. This was first developed in the 1980s for cattle and then pigs by a team led by Neal First at the University of Wisconsin. The earliest example of animal cloning using non-embryo-derived cells was from a team in Scotland, reporting in 1997 that it had produced the sheep, Dolly. She was the result of nuclear transfer from a somatic (mammary) cell line into an oocyte. The first cloned cat was reported from Texas A&M University in 2002. She was named cc for "carbon copy." She is not identical to her mother because of environmental factors. The first cloned horse (named Prometea) was reported in 2003 by a French/Italian team. A cloned dog (Snuppy) was reported from a South Korean team in 2005.

Cloning has the potential to benefit animal agriculture with improved growth, feed efficiency, disease resistance, and reproduction. Cloning of pets provides owners of a much-loved animal the ability for that cat to be "re-created." However, based on cat cloning, the cloned offspring will be different. Cloning of horses has the potential to ensure the continuation of the genetics of a prized horse.

Meat and milks from cloned livestock are safe to eat. The Food and Drug Administration in December 2006 concluded the following: "edible products from healthy clones that meet existing requirements for meat and milk in commerce pose no increased food consumption risk(s) relative to comparable products from sexually-derived animals." For cloning to be successful, there must be proper epigenetic reprogramming. The genome reprogramming includes global reduction in DNA methylation. This is a normal process in development completed by the time of implantation. Abnormalities with cloned livestock include placental abnormalities, markedly increased fetal growth and high birth weight, and difficulties in delivery and losses due to spontaneous abortion or perinatal death.

Transgenic Animals

Figure 15-5 summarizes how transgenic cattle are produced. A further advantage of nuclear transfer from either a somatic or embryonic cell line in vitro is that a transgene is added to the genome. Thus, a transgenic sheep, goat, or cow could produce a human protein in large quantities, and then this could be used as a drug.

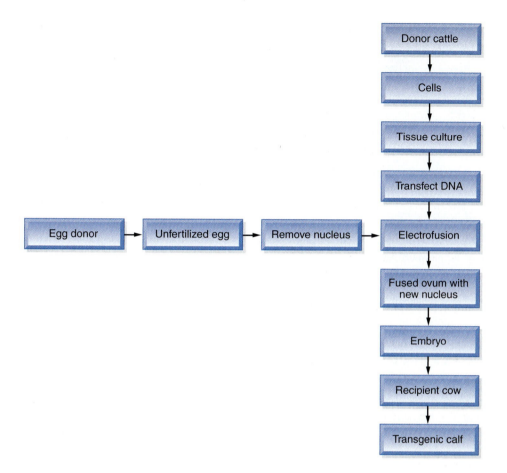

FIGURE 15–5 Schematic diagram illustrating how transgenic cattle (or sheep) are produced.

ISSUES *for discussion*

1. The U.S. government has expended large amounts of money in sequencing the genome of domestic animals. Provide pros and cons (arguments for and against) for the activity. Good investment or waste of taxpayers' money?
2. Animal geneticists are largely educated/trained in state-supported/funded universities. Is this a good use of public funds to support private industry, that is, animal breeding companies or large producers?
3. Production traits such as growth have been the focus of animal breeders. What traits would you consider should be the focus of geneticists in the future?
4. What are the ethical issues in genetic improvement of livestock?
5. Discuss whether it is ethical to produce clones or transgenic animals.

Students are encouraged to research the issues in questions 6–9 on the Web prior to the classroom discussion.

6. What are the ethical issues of breeding of dogs?
7. To what extent are dog breeders regulated? Should there be changes to the extent of regulation?
8. Should dogs be bred if they pass on genetic diseases?
9. What are puppy mills?

REVIEW QUESTIONS

1. What is a chromosome, DNA, a codon, inheritance, a genotype, a phenotype, and an SNP?

2. How many pairs of chromosomes do each of the following animals have: cats, cattle, catfish, horses, donkeys, pigs, sheep, goats, chickens, turkeys, and rabbits?

3. What is a gene?

4. What are exons and introns?

5. What is a gene map?

6. What is the genome?

7. What is the open reading frame of a gene?

8. What is transcription, and how is it controlled?

9. What is translation?

10. Does genetic counseling have a place in the breeding of domestic animals? What is marker-assisted selection?

11. What is the difference between a genotype and phenotype?

12. What are dominant, recessive, additive and partially dominant, and overdominant genes?

13. What is epistatic gene interaction?

14. What is a sex-linked trait?

15. What are linkage and crossover?

16. What is a heritability estimate, and how is it used to improve livestock through breeding?

17. Define straightbreeding and crossbreeding.

18. What is purebred breeding?

19. What is an advantage of a purebred animal?

20. What are the disadvantages of a purebred animal?

21. Define inbreeding.

22. Why is crossbreeding performed?

23. What is heterosis?

24. What is a synthetic inbred line, and how is it used in poultry breeding?

25. What is a cloned animal?

26. What is the difference between a cloned animal and a transgenic animal?

REFERENCES AND FURTHER READING

Buchanan, D. S. (2005). Gene action, types of. In W. G. Pond & A. W. Bell (Eds.), *Encyclopedia of animal science* (pp. 456–458). New York: Marcel Dekker.

Buchanan, D. S. (2005). Genetics: Mendelian action. In W. G. Pond & A. W. Bell (Eds.), *Encyclopedia of animal science* (pp. 463–465). New York: Marcel Dekker.

Havenstein, G. B., Ferket, P. R., Grimes, J. L., Qureshi, M. A., & Nestor, K. E. (2007). Comparison of the performance of 1966- versus 2003-type turkeys when fed representative 1966 and 2003 turkey diets: Growth rate, livability, and feed conversion. *Poultry Science, 86,* 232–240.

Havenstein, G. B., Ferket, P. R., & Qureshi, M. A. (2003). Carcass composition and yield of 1957 versus 2001 broilers when fed representative 1957 and 2001 broiler diets. *Poultry Science, 82,* 1509–1518.

Havenstein, G. B., Ferket, P. R., & Qureshi, M. A. (2003). Growth, livability, and feed conversion of 1957 versus 2001 broilers when fed representative 1957 and 2001 broiler diets. *Poultry Science, 82,* 1500–1508.

Pond, W. G., & Bell, A. W. (2005). *Encyclopedia of animal science.* New York: Marcel Dekker.

Rohrer, G. A. (2005). Gene mapping. In W. G. Pond & A. W. Bell (Eds.), *Encyclopedia of animal science* (pp. 459–462). New York: Marcel Dekker.

Rohrer, G. A. (2005). Genomics. In W. G. Pond & A. W. Bell (Eds.), *Encyclopedia of animal science* (pp. 469–471). New York: Marcel Dekker.

CHAPTER 16

Animal Reproduction

OBJECTIVES

This chapter will consider the following:

- The reproductive characteristics of males and females
- Reproductive organs, including their anatomy and functioning
- Sex determination
- Embryonic/fetal development of the reproductive organs
- Reproductive hormones
- Production of the gametes, including spermatogenesis, oogenesis, and meiosis

 ## INTRODUCTION TO REPRODUCTION

Reproduction is fundamental both to the survival of species and to agriculture. The efficiency of reproduction is an important key to animal breeding. Maximal fertility is a goal of the livestock producer and for a species.

 ## REPRODUCTIVE OR SEXUAL CHARACTERISTICS

Male Sexual Characteristics

The reproductive or sexual characteristics of a male can be summarized as follows:

- Production of very large numbers of the haploid male gametes (spermatozoa) by the primary reproductive organ, the testes
- Production of the male hormones, that is, androgens, particularly testosterone, which act to stimulate the growth of the penis, ducts, and glands, and the development of the secondary sexual characteristics and male behavior, such as mounting, libido, and aggression
- Penis, duct (vas deferens and urethra in mammals), and accessory organs, such as epididymis, ampulla, prostate, seminal vesicles, and bulbourethral glands (In poultry and other birds, there are no glands or a true penis.)
- Secondary sexual characteristics such as the comb, wattle, and spurs in roosters; snood and color in tom turkeys; pheromones in pigs, goats, and sheep; and antlers in deer
- Behaviors, including mounting, aggression, and crowing (rooster)
- Libido, which is the desire to mate

Female Sexual Characteristics

The reproductive or sexual characteristics of a female can be summarized as follows:

- Production of the female gametes (ova) by the primary reproductive organ, the ovaries (In poultry there is a single ovary.)
- Production of estrogens, particularly estradiol

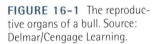

- Female reproductive tract (In mammals, it is the fallopian tubes or oviducts, uterus or uterine horns, cervix, and vagina. In poultry and most other birds, it is a single oviduct.)
- Secondary sexual characteristics, and lack of male characteristics
- Behaviors, including sexual and maternal
- Libido, which is the desire and willingness to mate

 REPRODUCTIVE ORGANS

In both males and females, the primary reproductive organs are the gonads, that is, the ovaries in females and testes in males. There must also be accessory organs that allow the spermatozoa and ova to meet and to provide the environment necessary for the development of the conceptus. In addition, there are behaviors that lead to mating and that increase the chances of the mating leading to a successful next generation.

Male Reproductive Organs

The reproductive systems of mammals show strong similarities. The bull, boar, and rooster reproductive systems are shown in Figures 16-1 to 16-3, respectively. The dog and horse penises are illustrated in Figure 16-4. The composition of livestock and poultry semen is summarized in Table 16-1.

Testes

The testes are the primary reproductive organs, with spermatozoa produced in very large numbers in the seminiferous tubules (see fig. 16-5). In addition, the testes have specific cells in the interstitium between the seminiferous tubules,

FIGURE 16–1 The reproductive organs of a bull. Source: Delmar/Cengage Learning.

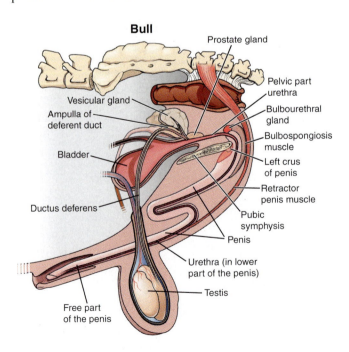

Boar

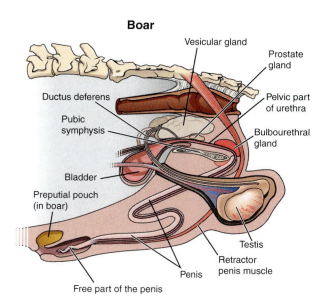

Vesicular gland

Prostate gland

Ductus deferens

Pelvic part of urethra

Pubic symphysis

Bulbourethral gland

Bladder

Preputial pouch (in boar)

Testis

Retractor penis muscle

Penis

Free part of the penis

FIGURE 16–2 The reproductive organs of a boar. Source: Delmar/Cengage Learning.

the Leydig cells. These produce the male hormone testosterone. The testes are held outside the body in a bag-like structure, the scrotum. This enables the temperature of the testes to be maintained at a lower temperature (3–5°C) than the body, without which spermatogenesis becomes very inefficient, and the animal is largely infertile. In cold weather, the testes are brought close to the body because

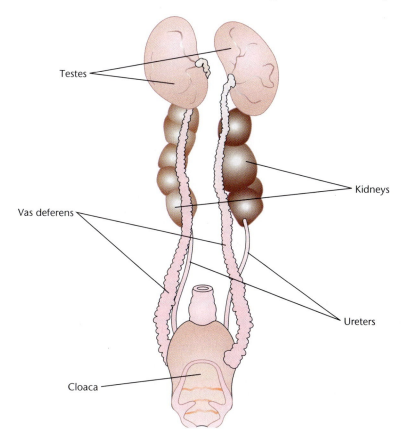

Testes

Kidneys

Vas deferens

Ureters

Cloaca

FIGURE 16–3 The reproductive organs of a male chicken (rooster). Source: Delmar/Cengage Learning.

FIGURE 16-4 Comparison of the reproductive system of the dog and horse.

Dog

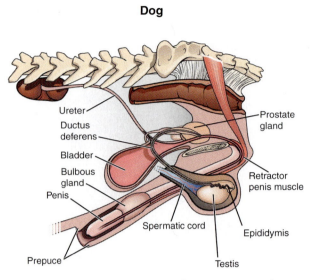

A. The reproductive organs of a dog showing the penis (essential for coitus), muscles for erection and ejaculation, and prostate gland, which is an accessory gland adding secretions to the semen. Source: Delmar/Cengage Learning.

Stallion

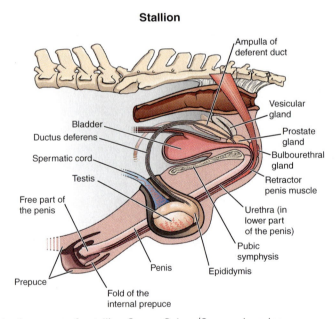

B. The reproductive organs of a stallion. Source: Delmar/Cengage Learning.

of contraction of the cremaster muscles connecting the testes to the abdominal cavity and the tunica dartos muscles in the wall of the scrotum.

Reproductive Tract and Accessory Organs

The functions of the reproductive tract and accessory organs are to transport the spermatozoa during ejaculation and add secretions. Semen contains large

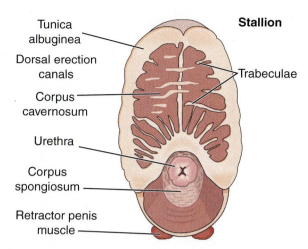

Stallion

- Tunica albuginea
- Dorsal erection canals
- Corpus cavernosum
- Urethra
- Corpus spongiosum
- Retractor penis muscle
- Trabeculae

C. Cross section of the penis of the horse showing the urethra (urine and semen pass through this), and the corpus spongiosum and cavernosum, which become filled with blood during erection. Source: Delmar/Cengage Learning.

numbers of spermatozoa together with secretions from the accessory glands, namely, the epididymis, ampulla (not present in pigs), seminal vesicles (very large in pigs), the prostate gland, and bulbourethral (or Cowper's gland). The compositions of semen from a number of domestic animals are summarized in Table 16-1. The secretions protect the spermatozoa. The reproductive ducts are the vas deferens, including two, with one from each testis/epididymis, and the urethra.

TABLE 16-1 Semen composition in livestock and poultry species

	CATTLE (BULL)	HORSE (STALLION)	PIG (BOAR)	CHICKEN (ROOSTER)
Volume of ejaculate (mL)	6	80[a]	175	0.3
Spermatozoa concentration[b] (billion per mL)	1.4	0.225	0.25	5
Nutrients for spermatozoa metabolism				
Fructose[c] (g/100 mL)	530	2	9	4
Citrate[c] (g/100 mL)	700	30	173	0
Sorbitol (g/100 mL)	75	40	12	5
Glyceryl phosphoryl choline[d] (g/100 mL)	300	70	175	20
Other constituents				
Protein (g/100 mL)	7	1	4	2.4
Chloride (g/100 mL)	250	450	345	150
Sodium (g/100 mL)	225	260	585	350
Calcium (g/100 mL)	40	26	6	10

Source: Adapted from Hafez and Hafez (2000).
[a]*The ampulla is an important accessory organ with secretions added to semen in the horse. There is no ampulla in pigs.*
[b]*Percent motile sperm >60% and percent normal appearance >75%.*
[c]*Produced largely in mammals by the seminal vesicles.*
[d]*Produced in the epididymis.*

Definitions

Androgen is a male sex hormone, such as testosterone.

Artificial insemination (AI) is when semen is placed into the cervix.

Conceptus is the production of conception being the embryo/fetus together with the surrounding membranes(s).

CL (plural *corpora lutea*) produces progesterone that in turn allows and maintains pregnancy.

Estrous is an adjective describing the estrous cycle (in U.S. English), whereas *estrus* is a noun denoting the time of heat.

Function of the gonads is to produce gametes and sex hormones.

Gonad is the ovary in females and testis in males.

Libido is the desire to mate.

Monoestrous is where there is a single estrus followed by an absence of ovarian activity.

Polyestrous is where an animal will show repeated estrus separated by 16–23 days depending on the species.

Pregnancy recognition is when the conceptus signals its presence to the mother to allow pregnancy to continue.

The *primary reproductive organ* is the gonad.

The *primordial germ cell* migrates to the site destined to be the gonad. Cells become gonocytes in the developing gonad and ultimately become gametes.

Progesterone is the hormone that is essential to the maintenance of pregnancy.

FIGURE 16-5 *A*, Section of bull testis showing seminiferous tubules 0.2 mm (200 μm) in diameter and interstitial tissue, spermatogonial divisional activity, and spermatid heads are visible nearing release into the lumen. Source: Delmar/Cengage Learning. Photo by Keith Schillo. *B*, Micrograph of the microanatomy of the boar testis. Note how the Leydig cell, which produces testosterone, is separated from the seminiferous tubule (where spermatogenesis is occurring) by a membrane. Also note how the developing spermatozoa are in close proximity with other cells. Image courtesy of J.J. Ford, USMARC, Agricultural Research Service, U.S. Department of Agriculture. *C*, Schematic diagram of the microanatomy of the seminiferous tubule producing spermatozoa. Source: Delmar/Cengage Learning.

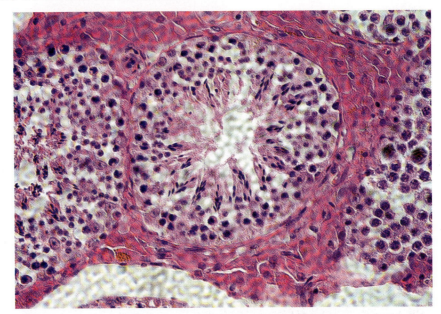

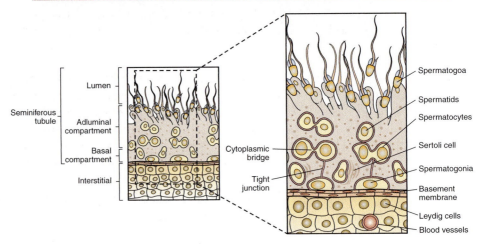

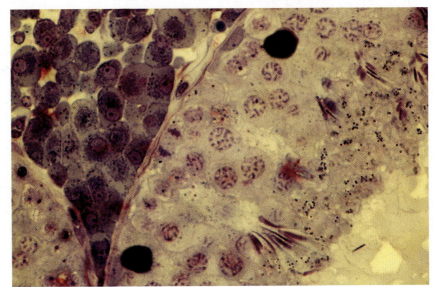

The epididymis is a coiled structure also in the scrotum. The functions of the epididymis include the (1) storage of quiescent but live spermatozoa before ejaculation, (2) maturation of the spermatozoa, and (3) addition of a nutrient to semen for the spermatozoa to use for movement. The epididymis is between the testes to the vas deferens.

Seminal vesicles produce secretions that are added to the spermatozoa. These contain proteins to stimulate capacitation of the spermatozoa after ejaculation and gelling or clotting of the semen. In addition, there are significant quantities of the sugar fructose together with citrate. These are to provide energy for the spermatozoa. There are also prostaglandins, which stimulate contractions of the female reproductive tract. The seminal vesicles are very large in the boar. There is a tremendous need for these secretions to produce the average of 0.35 pt (175 mL) of ejaculate in the boar, with some boars producing about 1 pt of ejaculate.

The prostate gland produces secretions that are added to the spermatozoa. The secretions are slightly alkali (to neutralize the lactic acid present because spermatozoa are stored in the epididymis) and contain low concentrations of proteins.

The vas deferens (two) and urethra are the ducts through which spermatozoa pass along during ejaculation. The urethra connects the bladder and vas deferens with the tip of the penis. Urine from the bladder also passes through it. During ejaculation, semen is expelled into the vagina (or cervix in some species).

The secretion from the Cowper's gland acts to clean the urethra.

Penis

The penis becomes erect at times of sexual arousal such that it can be inserted into the vagina. The corpus cavernosum (plural corpora cavernosa) is a spongy tissue surrounding the urethra in the penis (see fig. 16-4). During sexual arousal, the corpora cavernosa becomes rigid due to it being engorged with blood because there is much less blood draining from the penis. In bulls, rams, and boars, the sigmoid flexure and the retractor muscle extend the penis from the sheath (see figs. 16-1 and 16-2). In dogs, cats, and most primates, there is a bone in the penis that facilitates erection. This is the os penis, or baculum (see fig. 16-4).

Nervous Control of Erection and Ejaculation

Nerves control both erection and ejaculation. The parasympathetic nervous system and nitric oxide–releasing nerve terminals lead to dilation of blood vessels in the penis. The discovery of the role of nitric oxide as the chemical mediator of penile erection led to the development of drugs such as sildenafil (Viagra) for erectile dysfunction (impotency). Ejaculation is controlled by the sympathetic nervous system.

Poultry Male Reproduction

Poultry have testes within the body cavity but lack a penis, epididymis, and accessory organs such as prostate, Cowper's glands, and seminal vesicles (see fig.

Reflex ovulators show estrus but require mating to ovulate.

Sertoli cells are in the seminiferous tubules of the testis. They function as "nurse cells" for developing spermatozoa.

Testis (plural *testes*) is the primary male reproductive organ because it produces spermatozoa. The testes also produce the male reproductive hormone testosterone.

Vas deferens connects the epididymis with the urethra. It provides an environment suitable for survival of spermatozoa and the additions of the secretions of the accessory glands resulting in semen.

16-3). Chicken semen has low concentrations of nutrients (see Table 16-1). After ejaculation, some of the spermatozoa are stored in glands in the vaginal wall.

Castration

It has been known for several thousand years that after the removal of testes, there is a change in the behavior of the animal. Castration causes a considerable reduction in aggression. In companion animals after castration, there are less undesirable behaviors, such as spraying in cats.

Female Reproductive Organs

The reproductive system of the female functions to produce ova; to provide sites for insemination, fertilization, and embryonic/fetal development; to give a route from the site of insemination to that of fertilization; to produce female sex hormones and the overlapping group of hormones associated with maintenance of pregnancy; and to cause parturition or birth. The female reproductive systems of the cow and dog are summarized in Figures 16-6 and 16-7, respectively. The reproductive systems of other female mammals are similar to that of the cow.

Ovary

There are two ovaries whose principle function is to produce the ova. The ova develop in follicles associated with other cells a developing fluid-filled lumen. The follicle develops from a primary, to a secondary, to a tertiary follicle. When the follicle is fully mature (Graafian follicle), it produces estradiol. This stimulates the release of the luteinizing hormone (LH) and then ovulation of the ova surrounded by the thin membrane, the zona pellucida. One or multiple follicles can ovulate at the same time. The remnants of a follicle after ovulation first form a corpus hemorrhagicum (bloody or red body), and this develops into the CL (or yellow body) (see fig. 16-7). The CL produces the hormone progesterone. After functioning in the estrous cycle or pregnancy, the CL breaks down by the process of luteolysis for the corpus albicans (or white body).

Mammalian Reproductive Tract

The ovum is caught by the infundibulum of the oviduct. The oviduct functions to move the ovum to the uterus and as the site for fertilization. After fertilization, the ovum is called the zygote or conceptus. The uterus (plural *uteri*) is the site where the embryo implants, and the fetus develops. The walls of the uterus are composed of a glandular layer, the endometrium, and layer of smooth muscles, the myometrium (see fig. 16-8).

There are several types of uteri depending on whether and the extent to which the two uteri or uterine horns fuse:

- The duplex uterus has two uterine horns with two distinct lumens and two cervices. This is found in rodents and rabbits.
- The bipartite uterus has two uterine horns and a fused uterine body but is divided by a partial septum and a single cervix. This is found in cats, dogs, and pigs, and is well suited for the production of litters of offspring.

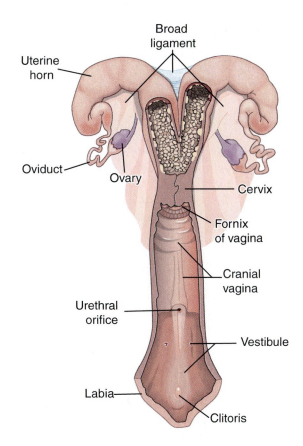

FIGURE 16-6 The female reproductive system of a cow. Source: Delmar/Cengage Learning.

Bitch

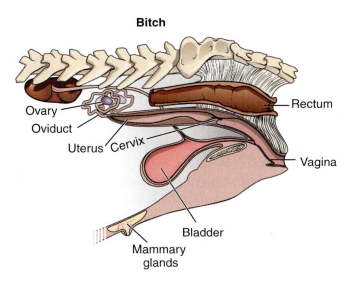

FIGURE 16-7 The female reproductive system of a dog. Ovaries produce ova. Fertilization occurs in the oviduct. The conceptus develops in the uterine horn. The cervix is closed to protect the uterus and developing fetus from bacterial invasion. Spermatozoa pass through the cervix after coitus. The cervix opens prior to parturition (whelping). The penis is placed into the vagina during mating or coitus. Source: Delmar/Cengage Learning.

- The bicornuate uterus is very similar to the bipartite uterus, but without the septum. It is found in cattle.
- The simplex uterus has no uterine horns and a single large uterine body. It is found in humans and other primates.

FIGURE 16-8 Cross section of the uterine horn (pig) showing generalized structure, including glandular endometrium and muscular myometrium.

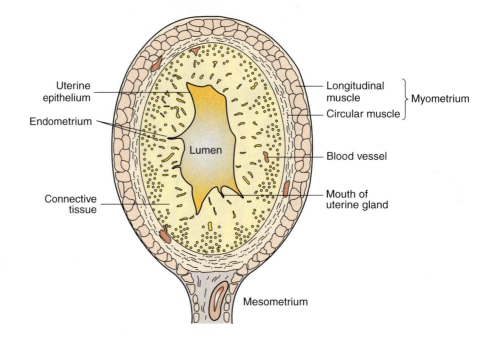

The cervix is the barrier between the uterus and vagina. It is normally constricted because of connective tissue and muscles with a mucus plug to prevent foreign materials such as pathogens entering the uterus. During estrus, there is some relaxation, and before parturition, the connective tissue breaks down to facilitate delivery of the neonate.

It is relatively relaxed to allow the passage of sperm into the uterus; during pregnancy, it remains tightly closed, sealing the uterus from the vagina. The vagina is the site for insemination, and when expanded, becomes part of the birth canal during the birth process. The vulva is the external opening. It consists of the labia majora (of the same embryonic origin as the scrotum in the male), the labia minora, and the clitoris, which is erectile tissue with the same embryonic origin as the penis in the male (see fig. 16-7).

Avian Female Reproductive Tract

In poultry, there is a single ovary and reproductive tract, known confusingly as the *oviduct* (see fig. 16-9). The left ovary and left oviduct develop, whereas the right ovary and oviduct regress. In the mature hen, the ovary is large with large yellow or yolk-filled follicles with the largest, the F1 follicle, up to 1 in (2.5 cm) in diameter. There are also small yellow and even smaller white follicles together with post-ovulatory follicle(s). There is no CL because poultry do not become pregnant. During follicular development, the ovum takes up yolk precursors that are synthesized by the liver under estrogen stimulation. The oviduct is made up of five regions:

- Infundibulum. It functions by engulfing the ovulated ovum, where fertilization occurs. The ovum spends 15–30 minutes here.
- Magnum. It functions by adding concentrated egg white proteins and membranes to the egg. The ovum spends 2–3 hours here.

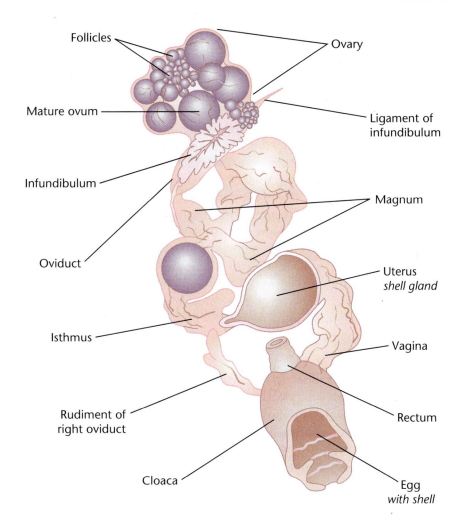

Follicles

Ovary

Mature ovum

Ligament of infundibulum

Infundibulum

Magnum

Oviduct

Uterus
shell gland

Isthmus

Vagina

Rudiment of right oviduct

Rectum

Cloaca

Egg
with shell

FIGURE 16–9 Reproductive system of the chicken (or turkey). Note the single ovary with large yolk-filled follicles and a single duct (the oviduct). Source: Delmar/Cengage Learning.

- Isthmus. It functions by adding fluid to egg white. The ovum spends 1–2 hours here.
- Uterus. It functions by adding calcium carbonate ($CaCO_3$) shell to the egg with calcium coming as ionized calcium (Ca^{++}) from the blood and the carbonate (CO_3^{++}) coming from bicarbonate (HCO_3) also in the blood. The ovum spends 20–26 hours here.
- Vagina. It functions by expelling the egg due to muscular contractions (this is called *oviposition*), and is the site for insemination and spermatozoa storage in vaginal ducts or glands. The ovum spends only a few minutes here.

Ovulation occurs about 45 minutes after oviposition of the previous egg, leaving time for the spermatozoa to migrate to the infundibulum. The oviduct completes its development during sexual maturation, largely under the influence of the ovarian hormone estradiol. Poultry can be molted by a variety of techniques, including changes in day length and diet. This results in the ovary and oviduct undergoing regression. After molting, the ovary and oviduct regenerate, and subsequent egg production is much higher than before the molt.

DEVELOPMENT OF THE REPRODUCTIVE ORGANS

Development of the Gonads

Primordial Germ Cells

The embryonic cells destined to become gametes are the primordial germ cells. These migrate to the embryonic gonad early in fetal development. Once they arrive in the gonad, they are called *gonocytes*, which develop into either oogonia or spermatogonia. In the male, primordial germ cells are enclosed in seminiferous tubules.

After puberty, the spermatogonia undergo multiplication (mitosis and then meiosis) and differentiation to form spermatozoa. In contrast, oogenesis occurs only during fetal development. The oogonia proliferate through a series of mitotic divisions and then enter meiosis, becoming dictyotene oocytes in primordial ovarian follicles. Mitosis is occurring in cell clusters called oocyte nests or germ line cysts. These oocytes then separate and become central to the primordial follicles. Postnatally, many oocytes are destroyed by the process of programmed cell death.

Sex Determination

Sex is genetically determined. Mammals having XY (or much less commonly XXY) sex chromosomes will develop testes. In contrast, XX or X0 individuals develop as a female with two ovaries. Therefore, not only is the presence of a Y chromosome required for testis development, but its absence is also needed for the development of the ovaries. The usual situation in mammals is that males have an X together with a Y chromosome that is small and gene poor. Females have two X chromosomes. The principal gene controlling the sex of a mammal is the *Sry* gene (sex-determining region Y gene) on the Y chromosome. A mammal with a *Sry* gene in virtually all cases develops testes.

Sry gene

The *Sry* gene (sex-determining region Y gene) is located on the short arm of the Y chromosome. In rats but not domestic animals, there are multiple copies of the *Sry* gene. This gene is essential to the development of the testis and, therefore, sex determination. The *Sry* gene product protein is a transcription factor that begins to be expressed in the genital ridge. This factor induces coelomic epithelial cells adjacent to the genital ridge to differentiate into Sertoli cells. The transcription factor causes the indifferent or bipotential gonad to develop into a testis. In rare cases, the *Sry* gene is on an X chromosome. Under these circumstances, the testis will develop. Ovaries develop in animals with a mutation in the *Sry* gene that prevents the production of a functional protein.

Sex Determination in Birds

Female birds have a W and Z chromosome, and males have two Z chromosomes. The W chromosome is small, like the Y chromosome in mammals. The sex-determining genes in birds are not known, but at least some are probably on the W chromosome.

In addition, it is thought that the number of Z chromosomes is an important factor in avian sex determination. This is the Z-dosage model. In ZZW chicken embryos, the right gonad develops into testis and the left into an ovotestis. In triploid ZZZ chickens, there is normal male sexual development, but a low number of and abnormal sperm are produced. There are ZO or WO chickens, but this is embryonically lethal.

Ductal Development

In mammals and birds, two sets of genital ducts develop. These are the Wolffian and Müllerian ducts. This development of two pairs of ducts occurs in both males and females. The Wolffian duct develops, when stimulated by androgens such as testosterone, into the vas deferens and associated accessory glands. The Müllerian duct regresses because of the anti-Müllerian hormone, which is produced by the testes. In female mammals, the Müllerian duct develops into the oviduct, uterus, cervix, and (posterior) vagina under the influence of estrogens, whereas the Wolffian duct regresses. In birds, only one Müllerian duct develops; the other regresses.

Descent of the Testes into the Scrotum

Testicular descent is essential for the normal growth of the testes and the process of spermatogenesis. The testes descend down a ligament, the gubernaculum, to the scrotum. Fetal Leydig cells produce a hormone called *insulin-like factor-3*, which is also known as *relaxin-like factor*. Release of the insulin-like factor-3 induces massive growth of the gubernacula (plural of *gubernaculum*). In "knockout" mice with the *Insl3* gene disrupted, the testes are retained in the body cavity (bilateral cryptorchidism).

REPRODUCTIVE HORMONES

The overall hormonal control of reproduction is summarized in Figure 16-10. The reproductive hormones include gonadotropin-releasing hormone (GnRH), gonadotropin inhibitory hormone (GnIH), luteinizing hormone (LH), follicle-stimulating hormone (FSH), testosterone, estrogens, progesterone, prostaglandins, and oxytocin.

Gonadotropin-Releasing Hormone (GnRH)

GnRH is a modified peptide with 10 amino acid residues. It has been also called *LH-releasing hormone* in the past, but because the peptide stimulates the release of both LH and FSH, the term GnRH is preferable and is now used. GnRH is the primary regulator of reproduction, stimulating the release of two of the reproductive hormones of the anterior pituitary gland: LH and FSH.

Gonadotropin Inhibitory Hormone (GnIH)

GnIH inhibits the release of both gonadotropins, LH and FSH, in both mammals and birds. It is an "RF amide" peptide with Arg-Phe-NH2 at the C terminus. The RF peptides are hormones in invertebrates.

GENETICS OF SEX DETERMINATION

The genetics of sex determination in mammals, including dogs, cats, cattle, pigs, and horses, is males are XY and females XX; males are heterogametic, producing X and Y spermatozoa; females are homogametic, producing X ova only; and the critical gene for sex determination is the *Sry* gene located on the Y chromosome.

The genetics of sex determination in birds, including chickens, ducks, turkeys, and parrots, is males are ZZ and females ZW; males are homogametic, producing Z spermatozoa only; and females are heterogametic, producing W and Z ova.

FIGURE 16–10 Schematic
diagram of the hormonal
control of reproduction in
1. male and 2. female with
control of the production
of estradiol in 2a and
progesterone in 2b.

1. MALE

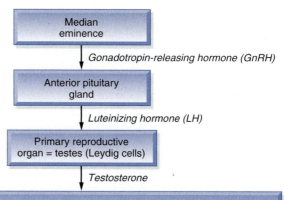

2. FEMALE

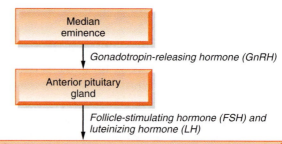

a.

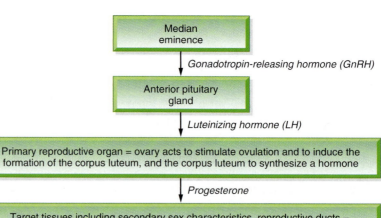

b.

Luteinizing Hormone (LH)

LH is a glycoprotein hormone that is produced by the anterior pituitary gland. The functions of LH are to stimulate the Leydig cells in the testis to produce testosterone and other androgens in the male, induce ovulation in the female, cause the development of the CL or luteinization in the female, and stimulate the CL to produce progesterone in the female.

Follicle-Stimulating Hormone (FSH)

FSH is also a glycoprotein hormone that is produced by the anterior pituitary gland. Its functions are to stimulate spermatogenesis and follicular development in the male and the female, respectively.

Testosterone

Testosterone is the male sex hormone. It is a steroid, and like all steroids it is synthesized from cholesterol. Testosterone acts to stimulate much of maleness, that is, penis size, ducts, and glands that are important to semen and ejaculation, secondary sex characteristics, and the brain to increase male behavior. Testosterone is used to enhance cattle growth (see Chapter 17).

Estrogens

Estrogens, such as estradiol, are the female sex hormones. They are steroids produced by the follicle before ovulation. Estrogens act to stimulate much of femaleness, that is, development of the female ducts, including the oviducts, uterus, cervix, and vagina; mammary development; and stimulating female behavior. Estradiol is also used to enhance cattle growth (see Chapter 17).

Progesterone

The major role for progesterone is maintaining pregnancy. It is a steroid produced by the CL (a part of the ovary) and by the placenta in some animals.

Prostaglandins

Prostaglandins are modified fatty acids that play critically important roles in reproduction. One prostaglandin, prostaglandin F_2^{α}, causes the breakdown of the CL and stimulates contractions of the uterus. Another, prostaglandin E_1, causes the laying of the egg in the chicken.

Oxytocin

Oxytocin is a peptide hormone produced by the posterior pituitary gland. It causes uterine contractions during the birthing process and milk to be let down from the mammary gland.

GAMETOGENESIS

With the need to produce large numbers of spermatozoa, spermatogenesis occurs throughout the life of the male animal, except in the seasonal breeders, when it is restricted to the breeding season. In contrast, in the female, the number of oocytes in the ovary is established at birth. After birth, there is the loss of oocytes after ovulation or follicular atresia and by cell death. Cell division requires duplication of the genome by mitosis (or meiosis). In addition, there are nuclear mitochondria replication factors released from the nucleus to stimulate mitochondrial multiplication.

Meiosis

Meiosis is division of cells during gametogenesis to produce gametes, that is, spermatozoa and ova. In meiosis or reduction division, the daughter cells end up with only half the chromosome—one of each pair—and are referred to as *haploid*. During meiosis, there is crossing over or recombination within a pair of chromosomes. Meiosis is summarized in Figure 16-11.

The process of gametogenesis is summarized in Figure 16-12. In the female, most of the process of gametogenesis is completed during fetal development, with meiosis arrested at prophase I before birth. Meiosis is stimulated to resume by the hormone that induces ovulation, namely, LH. In the male, meiosis begins

FIGURE 16–11 The process of reduction division or meiosis. Source: Delmar/Cengage Learning.

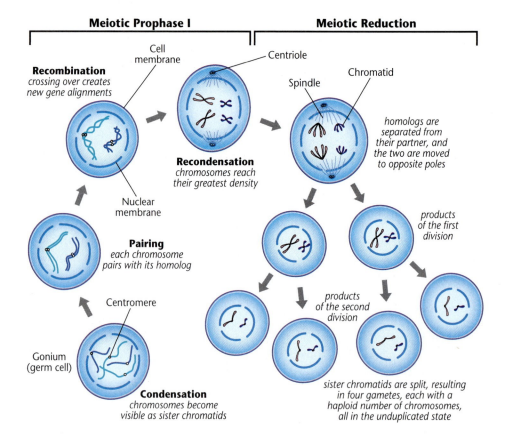

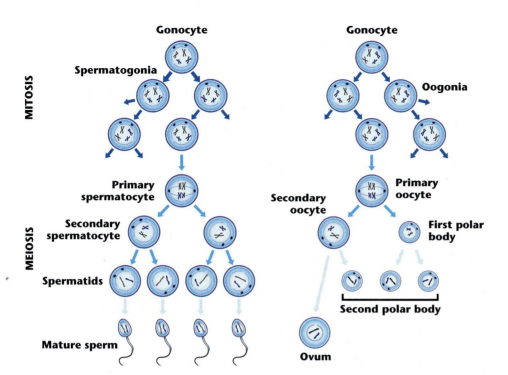

FIGURE 16-12 A schematic representation of gametogenesis. Source: Delmar/Cengage Learning.

at puberty. Spermatogenesis requires the hormone FSH and testosterone. To produce motile spermatozoa, spermatogenesis involves three distinct processes:

1. Repeated mitosis producing large numbers of spermatogonia
2. Meiosis going from primary spermatocytes to spermatids
3. Metamorphosis of the spermatid with a loss of cytoplasm and the development of the tail or flagella, producing the light motile spermatozoa (see fig. 16-13)

The Sertoli cells play a critically important role in spermatogenesis, acting as nurse cells for the developing spermatozoa (fig. 16-5). To ensure that all the developing spermatozoa in a given section of seminiferous tubules are at the same stage of development, there is cell-to-cell communication across gap junctions.

DEVELOPMENT OF THE FOLLICLE

The ovum develops in the ovary in a specific structure, the follicle. Figure 16-14 illustrates the development of the follicle. Initially, there are large numbers of ova, each surround by the single layer of granulosa cells in primary follicles. There is development of a membrane around the ovum called the *zona pellucida*. This helps to ensure that only spermatozoa from the same species fertilize the egg after ovulation. The primary follicle develops into a secondary follicle, with large increases in the number of granulosa cells and the emergence of a theca layer of cells.

Under stimulation of a hormone (FSH or follicle-stimulating hormone), the secondary follicle develops into a Graafian or preovulatory follicle. There are continued increases in granulose cells with the appearance of a fluid-filled lumen or cavity in the follicle. The follicle protrudes or sticks out from the surface of the ovary. The granulosa and theca cells together produce the estrogen estradiol.

FIGURE 16-13 *A,* Structure of spermatozoa from various domestic animals showing the head, which contains the nucleus with deoxyribonucleic acid and the acrosome containing enzymes that aid fertilization; mitochondria-rich midpiece; and flagella, with its whip-like movement as it moves the spermatozoa. *B,* Bull spermatozoa. Bull sperm is about 0.08 mm (80 µm) in length. Courtesy of Harold Hafs, Rutgers, The State University of New Jersey.

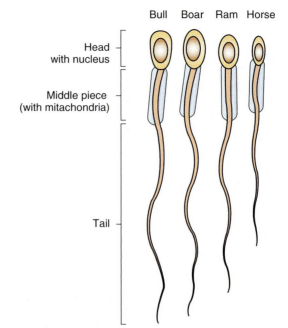

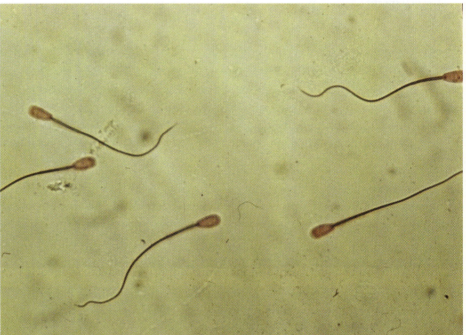

After ovulation and, therefore, release of the ovum, the remnants of the follicle go on to form the CL (see fig. 16-15).

 ## ESTROUS CYCLE

Estrous cycles are found in the vast majority of mammals. The estrous cycle is marked by a period of estrus or heat, with this being the only time that the female will allow coitus or mating; this is known as being receptive. In contrast, coitus can occur at any time during the human menstrual cycle. Poultry have ovulatory

FIGURE 16-14
Development of the follicle in
the ovary.

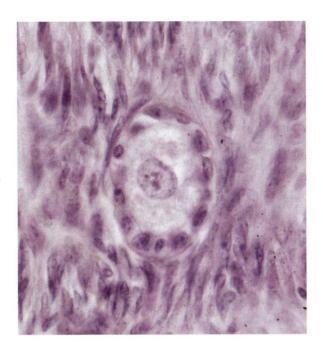

A. Primary follicle. The oocyte is surrounded by a single layer of (cuboidal) cells and a basement membrane. The cells around the oocyte are destined to become granulosa cells. Bovine oocytes normally are about 0.1 mm in diameter. Most primary follicles are located near the surface of the ovaries. Courtesy of Harold Hafs, Rutgers, The State University of New Jersey.

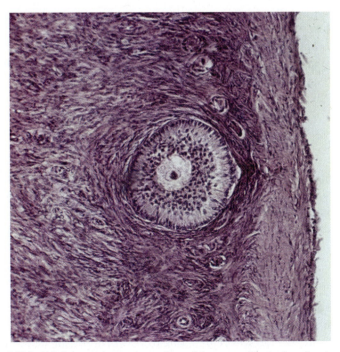

B. Secondary follicle. This is located near the surface of the ovary. The tunica albuginea covering the surface of the ovary is on the *right*. The oocyte is usually about 120 μm in diameter, has a small darkly staining nucleus, and is surrounded by several layers of small darkly staining granulosa cells. Courtesy of Harold Hafs, Rutgers, The State University of New Jersey.

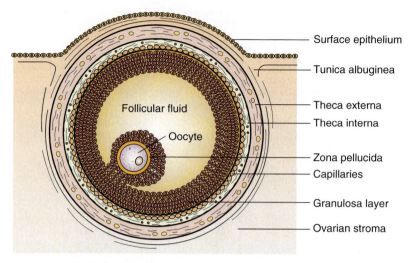

C. Pre-ovulatory or Graafian follicle. The follicle is bulging from the surface of the ovary before ovulation. The preovulatory follicle contains the ovum surrounded by the zona pellucida and cumulus oophorus (granulose cells), a fluid-filled cavity or lumen, and the granulosa and theca internal cells that together produce the estrogen hormone, estradiol.

cycles with ovulations occurring approximately every day, with the egg laid 1 day later. Mating occurs at any time, but spermatozoa are stored in vaginal glands and are then released to fertilize the ovulated ovum. For more details on estrous cycles, see the "Interesting Factoid" sidebar on this page.

Biology of the Estrous Cycle

The estrous cycle can be divided into two phases: the follicular phase, when the follicle is growing and then ovulated; and the luteal phase, when there is a CL functioning. Figure 16-15 shows the anatomic changes in the ovary of a cow. Figure 16-16 illustrates the hormonal changes during the estrous cycle.

The Follicular Phase

In this phase, the follicle(s) develops and grows bigger (fig. 16-14). The preovulatory follicle or Graafian follicle produces estradiol. This induces the behavioral changes of estrus. In addition, the estradiol stimulates the preovulatory increase in the hormone, LH (estradiol acting on the hypothalamus to release GnRH, and this in turn stimulates the pituitary gland to release very large amounts of LH). The LH then causes ovulation.

The Luteal Phase

The remnants of a follicle form the CL, which produces the hormone progesterone, which in turn allows and then maintains pregnancy. If the animal does not become pregnant, the CL will regress due to effects of prostaglandin $F_{2\alpha}$ from the uterus. After CL breakdown, the next follicles develop, leading to a preovulatory follicle producing estradiol.

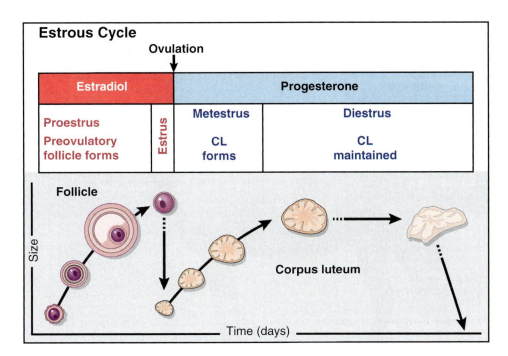

FIGURE 16-15 Schematic view of the changes in the ovary during the estrous cycle. The follicle develops, growing larger with more cells and the appearance of a fluid-filled cavity. After ovulation, the remnants of the follicle form the CL. Source: Delmar/ Cengage Learning.

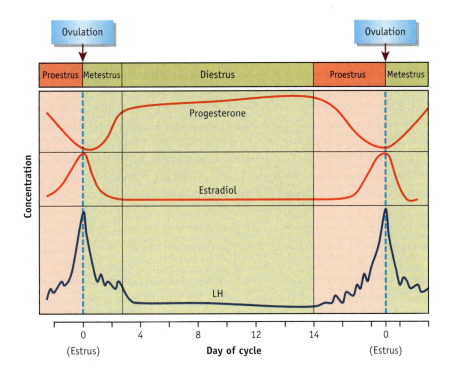

FIGURE 16-16 Schematic view of the hormonal and other changes with time during the estrous cycle of a sheep. The follicle develops producing estradiol, and this in turn stimulates a large increase in the release of LH from the pituitary gland. The LH induces ovulation. After ovulation, the CL produces progesterone to prepare for pregnancy, were it to occur. Source: Delmar/ Cengage Learning.

Types of Estrous Cycles

There can be either long or short estrous cycles, respectively about 16–23 days, or 4 or 5 days between estruses (see Table 16-2). In reflex ovulators such as cats, rabbits,

TABLE 16-2 The length of the estrous cycle in various species

SPECIES	LENGTH OF ESTROUS CYCLE (D)
Cattle	21
Horses	21–23
Pigs	21
Sheep	16–17
Rats and mice	4 or 5

and llamas, only LH will be released, and, therefore, ovulation will occur only if there has been coitus. Some species such as cattle show estrous cycles and are called *polyestrous*, others such as sheep and horses show cycles in the breeding season and are called *seasonally polyestrous*, and others such as the dog show a single estrus and are called *monoestrous* with a long period of anestrus (no "heats").

MATING

In domestic mammals, mating or coitus involves the male mounting a receptive female, that is, a female at estrus or heat. The erect penis is inserted into the vagina in mammals, with ejaculation into the posterior vagina or cervix, depending on species. In many mammals such as the pig, the ejaculated semen coagulates into a gel. Subsequently, the spermatozoa are gradually released. Semen coagulates because of specific proteins produced by the seminal vesicles and released at ejaculation.

The spermatozoa move into the oviduct to fertilize the ovum. Movement is due to the tail of the spermatozoa, together with both muscular and cilia-caused movement of fluid in the female tract. After ejaculation, the spermatozoa undergo a change such that they become capable of fertilizing the ovum. This is a process called *capacitation* and is triggered by specific protein(s) from seminal vesicles.

After mating, there is an inflammation response in the female reproductive tract. In mammals, spermatozoa retain their ability to fertilize for only about 24–30 hours after mating. The egg cell lives about 12 hours after it is released if it is not fertilized. In mammals, the oviduct can act as a short-term sperm reservoir, storing the sperm before the arrival of the ovum. Fertilization is the union of the spermatozoa and the ova to produce a zygote. In reflex ovulators such as the cat, stimulation of the female reproductive tract during coitus results in the release of LH and ovulation. If the resultant ovum is not fertilized in these species, they become pseudopregnant.

In poultry, there are some differences in mating. Capacitation is not required. Mating can occur at any time. There is not a true penis, but intromission does occur. Spermatozoa are stored for up to 2 weeks in duct- or gland-like structures in the wall of the vagina.

Artificial Insemination (AI)

AI involves placing a sample of semen into the female reproductive tract (by the cervix, into the cervix, or into the uterus) manually. AI is used with the vast majority of dairy cows, sows, and turkeys. It is of growing importance in other species. The rate of pregnancy achieved by AI in cattle is reported to be between 70% and 80% but is lower in poorly managed herds.

The advantages of AI include the following:

- Use of genetically superior males.
- Dilution of semen allows many more females to be bred to a genetically superior male.
- Ease of record keeping, that is, it allows paternity to be readily established.
- It reduces the risks of injury to the female during coitus or to people working with the animals.
- It reduces the risks of the spread of venereal diseases.
- Semen is readily transported within the United States and internationally.

AI requires the following:

- Collection of semen
- Evaluation of the semen for the number of live spermatozoa and any abnormalities
- Dilution of the semen
- Storage of the diluted semen, either stored at 4°C (or 40°F) or frozen in liquid nitrogen
- Insemination of the female

The semen is collected from the male into an artificial vagina, or manually or by the electro-ejaculation. The semen sample is evaluated for the number of live, dead, and abnormal spermatozoa. Semen can be diluted such that it can be used for more females and to maintain the viability of spermatozoa, particularly in species in which semen is frozen.

Sperm Sorting

Spermatozoa can be separated based on the differences in DNA content (e.g., Y-containing sperm have less DNA) using DNA-binding fluorescent dye/stains and a cell sorter (or flow cytometer) that sorts sperm one at a time very rapidly. The rate of sorting is 10–20 million spermatozoa per hour. Inseminating sperm that have been through a cell sorter results in 90% of the offspring of the gender selected. For instance, this will allow more daughter dairy cows to be produced.

PREGNANCY OR GESTATION

Gestation is the time during which an animal is pregnant. The duration of gestation varies with species (see Table 16-3).

During pregnancy, the conceptus develops first into an embryo surrounded by membranes and then a fetus in the uterus. The early development of the cattle conceptus is shown in Figure 16-17, and that of the pig conceptus in Figure 16-18.

TABLE 16–3 Gestational lengths in various livestock and companion animal species

SPECIES	GESTATION LENGTH (D)
Cattle	285
Horse	336
Sheep	147
Goat	150
Pig	114

(continued)

FIGURE 16–17 Schematic illustration of early embryonic development showing morphologic changes that occur between formation of the zygote and hatching the blastocyst.

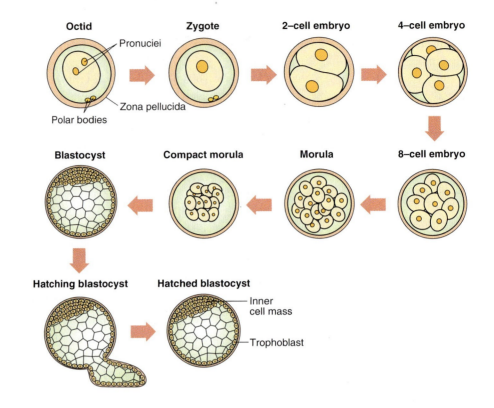

TABLE 16–3 Gestational lengths in various livestock and companion animal species   (continued)

SPECIES	GESTATION LENGTH (D)
Dog	52
Cat	60
Mouse	20
Rat	22
Guinea pig	65
Hamster	16–17

INTERESTING FACTOID

There are three layers of an embryo: endoderm, mesoderm, and ectoderm. The endoderm goes on to form the gastrointestinal tract and liver. The mesoderm develops into the cardiovascular system, kidneys, muscles, bone, and adipose tissue, and the dermis of the skin. The ectoderm goes on to form the nervous system, including the brain and the epidermis of the skin.

There is a continuum in development in utero with a number of distinct processes, including the following:

- Cell division being immediately after fertilization (fig. 16-17).
- The fertilized ovum or zygote, or preimplantation embryo hatching from the membrane surrounding it, the zona pellucida (fig. 16-18).
- Differentiation with the formation of embryonic/fetal membranes and the major layers: endoderm, mesoderm, and ectoderm (fig. 16-18).
- Implantation, which requires shedding of the zona pellucida, correct orientation of the blastocyst, adhesion or sticking, and, finally, invasion of the endometrium.
- Embryonic development with organ formation initiated. Once all the organ systems are developed, bone formation has started, and implantation has occurred, the embryo is called the *fetus*.
- Growth and development of the fetus.

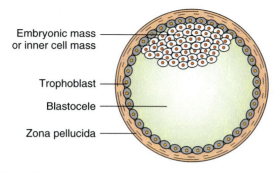

Embryonic mass
or inner cell mass

Trophoblast

Blastocele

Zona pellucida

A. Schematic showing section through a blastocyst about 6 days after fertilization.

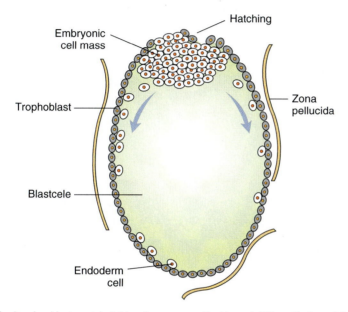

Hatching

Embryonic
cell mass

Trophoblast

Zona
pellucida

Blastcele

Endoderm
cell

B. Schematic showing blastocyst hatching from zona pellucida and differentiation of the endoderm beginning about 8 days after fertilization.

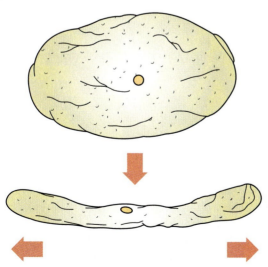

C. External view of blastocyst showing marked lateral (side) expansion beginning from about 8 days after fertilization.

FIGURE 16–18
Development of the fertilized ovum or conceptus in the pig starting at the blastocyst stage. Cleavage, morula, and blastocyst formation together with hatching from the zona pellucida are similar to that in cattle.

Within the embryo, the endoderm goes on to form the intestines, pancreas, and liver. The mesoderm goes on to form the muscles, cardiovascular system, kidneys, and inner layer of the skin (the dermis). The ectoderm forms the outer layer of the skin (the epidermis) and the nervous system, beginning with the neural tube.

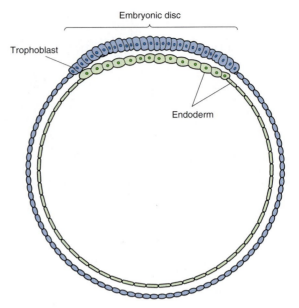

D. Schematic showing complete differentiation of the endoderm at about 9 days after fertilization.

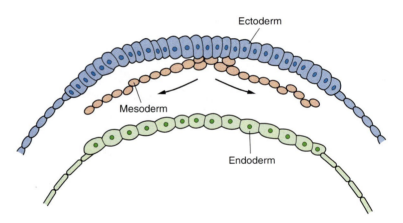

E. The third layer, mesoderm, begins to form at about 9 days after fertilization.

Fertilization

The zona pellucida is required for sperm-egg recognition with the spermatozoa binding to specific zona pellucida proteins. After fertilization, changes in the zona pellucida proteins do not allow further sperm binding and prevent polyspermy, that is, fertilization by more than one spermatozoon. Meiosis of mammalian eggs is arrested at the metaphase of the second meiotic division. Fusion of the spermatozoa with the ovum rapidly leads to resumption of meiosis after changes in intracellular calcium. The second polar body is extruded.

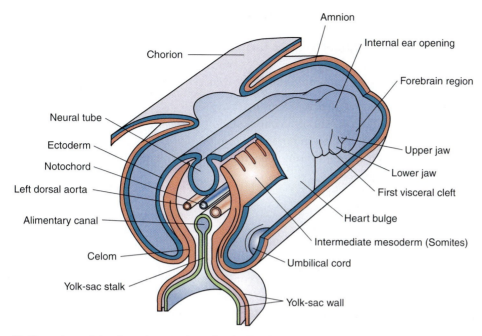

F. The embryo of the pig at about 15 days after fertilization.

Embryonic/Fetal Membranes

The embryo/fetus develops surrounded or partially surrounded by embryonic/ fetal membranes. The membranes are the chorion, made up of ectoderm and mesoderm; amnion, made up of ectoderm and mesoderm; allantois, made up of endoderm and mesoderm; and yolk sac, made up of endoderm and mesoderm.

Between the membranes are fluid-filled cavities protecting the embryo/fetus from mechanical shocks and injuries. The majority of these fluid-filled cavities are theamniotic and allantoic cavities. The relative growth of the membranes, cavities, and embryo/fetus in the pig is shown in Figure 16-19.

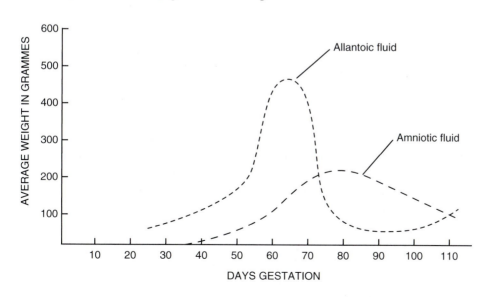

FIGURE 16-19 Changes in the weights of the fetal pig, the surrounding membranes, and the fluid in the allantoic and amniotic cavity. Source: Delmar/Cengage Learning.

Placenta

The placenta is the site of contact between the developing embryo/fetus and the mother. Nutrients, respiratory gases, and wastes are exchanged across the placenta. Placentas are categorized into various types based on the following:

1. The embryonic/fetal membranes involved
   - Choriovitelline (embryonic chorion and yolk sac), which is found early in pig development plus in horses
   - Chorioallantoic (embryonic chorion and allantois), as in cattle, sheep, dogs, and cats
2. Their anatomy, for example
   - Diffuse, e.g., horses
   - Cotyledonary, e.g., cattle
   - Zonary, e.g., dogs
   - Discoid, e.g., humans
3. The number of maternal and embryonic/fetal layers
   - Epitheliochorial, with six layers between maternal (epithelium, connective tissue, and endothelium) and fetal (chorionic epithelium, mesenchyme, and endothelium) blood, e.g., horses
   - Syndesmochorial, with the loss of maternal epithelium, e.g., cattle
   - Endotheliochorial, with the loss of maternal epithelium and connective tissue
   - Hemochorial, with the loss of all three maternal layers, e.g., humans

In ruminants, there is a unique placenta in which gaseous exchange and nutrients pass from mother to fetus. This arrangement is called a *placentome* or *cotyledon*. There are between 60 and 80 "placentomes" or cotyledons. The importance of all the cotyledons being fully functioning and required can be readily demonstrated experimentally. Removal of some cotyledons results in reduced growth of the fetus. Each is made up of a cotyledon (its cells coming from the conceptus or zygote) and a caruncle (its cells being of uterine origin). Within the cotyledon there are a series of giant binucleate cells. These are derived from the trophoblast and produce specific proteins associated with pregnancy, including placental lactogen, prolactin-related protein I, and pregnancy associated glycoproteins. These giant binucleate cells play a critical role in implantation and placental functioning.

Pregnancy Recognition

It is critical that the conceptus signals its presence to the mother to allow pregnancy to continue. Otherwise, the mother will reject the conceptus, and pregnancy will terminate. Successful pregnancy requires communication between the conceptus and the uterus. In cattle, the trophectoderm of the conceptus releases a signal protein, interferon t, that inhibits the production of prostaglandin F2α from the uterus. Otherwise, the prostaglandin F2α will pass to the CL (in the ovary), causing it to regress and ceasing to produce progesterone, the hormone essential to pregnancy. In pigs, conceptus-generated estrogens prevent prostaglandin F2α from reaching the CL. There are a host of other proteins that play critical roles in allowing pregnancy. One is granulocyte-macrophage colony-stimulating factor. The mother becomes immune tolerant to the placental trophoblast cells invading the uterus despite these being "foreign." Hormones play a critically important role in pregnancy (*see* boxed text). An example is relaxin.

HORMONES OF PREGNANCY

Progesterone, which is produced by the placenta in sheep and by the CL in cattle, maintains pregnancy.

Placental lactogen, produced by the placenta, induces mammary development.

Prolactin-related proteins are produced by the placenta, and their role is unknown. *Pregnancy associated glycoproteins*, which are produced by the placenta, are involved in implantation.

Relaxin is produced by the CL, and prepares the cervix and pelvis for parturition or the birth process.

Estradiol is produced by the placenta and prepares the uterine muscles for the birth process.

Sex ratio. The sex ratio of offspring would be expected to be 1:1. However, there can be sex ratio skewing. There is evidence from studies in sheep, cattle, and pigs that females with higher body condition scores or receiving improved nutrition produce more male offspring.

Relaxin has a number of critically important roles in pregnancy and parturition. In most domestic animals, relaxin is produced by the CL. An exception is the horse, in which the placenta is the main source of relaxin. Relaxin reduces contraction of the uterus, therefore reducing the frequency of myometrial contractions. This facilitates the maintenance of pregnancy. In preparation for parturition, relaxin has marked effects on the pubic ligament, uterus, and cervix. Relaxin induces relaxation of the pubic ligament and, therefore, enlarges the birth canal. It promotes softening and extensibility of the cervix due to changes in the collagen fibers and hydration. Without relaxin, labor takes much longer, and the number of pigs born alive greatly decreases.

PARTURITION

Parturition is the process of giving birth to the neonate. There are three stages of parturition. In phase 1, the uterus contracts until the cervix is completely dilated. The end of phase 1 is marked by "transition." The uterus contracts, delivering the neonate through the birth canal in phase 2. In phase 3, the uterus contracts, delivering the placenta and fetal membranes.

Before parturition, there is a buildup of the uterine muscles (such as the myometrium) and a thinning of the cervix. The fetus moves into the position to facilitate a safe birth, the front feet and head close to the cervix. Parturition is signaled by the fetus. There is increased production of estradiol and other estrogens, and decreased production of progesterone. Together, this sensitizes the myometrium. There is an increasing release of oxytocin from the posterior pituitary gland.

REPRODUCTIVE PROBLEMS

Reproductive problems are either reduced fertility or sterility, which is a complete lack of any fertility. These conditions can be either permanent or temporary. Causes for infertility are the following:

- Anatomic, such as cryptorchidism or freemartinism.
- Genetic, such as Turner's syndrome in which there is only one X chromosome and no Y chromosome.

Definitions

Cryptorchidism is the failure of the testes (one or both) to descend down the inguinal canal into the scrotum. If only one testis is retained in the body cavity, this is referred to as *unilateral cryptorchidism*; if two, *bilateral cryptorchidism*. This causes sterility or, if only one descends, reduced fertility. Another problem is that the retained testis/testes can become cancerous.

Dystocia is difficult birth.

Stillbirth is the delivery of a dead fetus. Frequently, the fetus dies during the birth process. The difference between a stillborn calf and a neonatal death is the latter refers to when the calf has taken at least one breath before dying.

TABLE 16-4 Relationship between difficult calving or dystocia and percentage of deliveries of calves by dairy cows

DAIRY CALF DELIVERY	PERCENTAGE OF DELIVERIES	DYSTOCIA SCORE
Unassisted birth	63	1
Some assistance with the birth	26	2
Severe dystocia	11	3

Source: Lombard et al. (2007).

- Nutritional, such as a nutrient deficiency, e.g., low calcium in the diet. This will stop egg production in hens.
- Environmental, such as heat stress.
- Toxicologic, as with toxic materials in the feed.
- Infectious disease related, including viral, bacterial, and protozoan diseases that can adversely affect fertility.

Dystocia

Dystocia is difficult birth. Table 16-4 summarizes the incidence of dystocia by the percentage of deliveries. There is a much higher incidence of dystocia in calf heifers than in older cows. Delivery of large calves or twins is also associated with dystocia. As dystocia becomes more severe, there is an increasing chance of stillbirth (birth of a dead fetal calf) or neonatal mortality.

Brucellosis

Brucellosis is known as *contagious abortion* in cattle. It will cause abortion most frequently in the fourth and fifth months of pregnancy but also it will impair fetal development. It is a bacterial disease, caused by bacteria of the genus *Brucella*, that can affect cattle, sheep, goats, deer, pigs, and people. Not only are there effects in the female reproductive tract, but also brucella can infect the epididymis of the male and be transmitted as a venereal disease that is spread during coitus or mating.

Leptospirosis

Leptospirosis, which is a disease caused by spirochaete bacteria of the genus *Leptospira*, causes a series of problems, including fever, and liver and kidney problems in cattle, pigs, horses, dogs, and people. It can also cause abortion. Rats are a frequent source of the infectious organism.

Cryptorchidism

Cryptorchidism is heritable. It is found in livestock, such as pigs and, to a lesser extent, cattle and sheep. It is also found in companion animals, including horses, dogs, and, more rarely, cats. It is particularly common in dogs, with an incidence rate of over 10% depending on the breed.

Intersex

Freemartinism is found in cattle and sheep. It is a situation in which 90% of the females are completely infertile, with an incompletely formed reproductive tract

MISMATING IN DOGS

The best way to prevent a bitch from becoming pregnant as the result of a mismating to an "undesirable" dog is for the owner to ensure that the female does not have the opportunity to mate. This involves confinement during estrus such that males cannot get to the female or having her on a leash when walking and actively preventing coitus. A pregnancy can be terminated by surgery, ovariohysterectomy, or spaying during pregnancy. However, if you want to allow for a future pregnancy, there are alternative veterinary approaches that involve the use of drugs and hospitalization. Dopamine agonists, either cabergoline or bromocriptine, are drugs that suppress prolactin secretion. When these are administered, prolactin concentrations in the blood decline, and the CL regresses. In dogs, the CL requires the presence of prolactin. The dopamine agonist is given in combination with prostaglandin $F_{2\alpha}$ to induce uterine contractions and also exert a luteolytic effect destroying the CL. The progesterone antagonist mifepristone (RU486) has been shown to be effective in terminating pregnancy in clinical trials.

It is important to know if the bitch is pregnant. Simply finding her with a male does not automatically indicate that they have mated. The absence of detectable spermatozoa in the female tract suggests that mating has not occurred, but this does not exclude mating completely.

and very small ovaries. Freemartinism is observed when there are twin fetuses, one male and the other female, and there is incomplete separation of circulatory systems of the two fetuses and, therefore, mixing of the blood at the fused placenta. There can be mixing of the cells of the two fetuses resulting in a XX/XY chimera. Moreover, hormones, including anti-Müllerian factor and androgens, impair uterine development. There is little effect of freemartinism on the fertility of the male twin.

 ## REPRODUCTIVE MANIPULATION

Management can be used to manipulate reproduction. For instance, changing the day length is important to bring some animals into reproduction condition. Photoperiodism also can give an annual pattern of a physiologic function, such as reproduction in some small rodents (hamsters) and in birds (e.g., turkeys and, to a lesser extent, chickens). Photoperiodism is when an animal is genetically programmed to respond to either an increase in day length (in the spring) or a decrease in day length (in the fall). This is often followed by a time of year when the animal is no longer responsive to changes in day length.

Drugs Used to Control Reproduction

There are a variety of products available to assist with or control animal reproduction.

Progesterone

Progesterone implants or inserts (e.g., the EAZI-BREED CIDR Insert) are used in both beef and dairy production (see fig. 16-20). These are molded from elastic rubber containing progesterone (1.4 g). They are placed intravaginally for

FIGURE 16-20 Intrauterine insert used for estrus synchronization.

7 days, releasing progesterone continually. This mimics the luteal phase of the cycle. When they are removed, progesterone levels in the blood decrease, and estrus/ovulation is triggered within 3 days. This facilitates AI, estrous detection, embryo collection, and transfer, and can be used to overcome anestrus. Moreover, it can be used in a feedlot to ensure that the heifers do not become pregnant. The effectiveness of this system is enhanced when prostaglandin F_2^α (Lutalyse or dinoprost tromethamine) is used to cause the breakdown of a CL if there is one. A synthetic progestin (altrenogest Regu-Mate) is used to suppress estrus in mares. When the drug is withdrawn, the mare comes into estrus in a reasonable, predictable manner.

Melengestrol Acetate

This is a progesterone-like chemical that is orally active. Commercially, melengestrol acetate is placed in the feed to suppress the estrous cycle in beef heifers. Moreover, cattle receiving melengestrol acetate show increased growth rates.

Prostaglandin F_2^α (Lutalyse)

Prostaglandins are widely used in veterinary practice with domestic animals. Their use is based on their luteolytic effect (i.e., causing the CL to break down), and because they stimulate contractions of the uterus. Prostaglandins also have the following uses:

- To synchronize estrus in cattle, pigs, and horses.
- To induce parturition in pigs. Prostaglandin F_2^α induces parturition with 90% of the gilts and sows farrow in a 24-hour period. This reduces labor costs and facilitates cross fostering of the newborn pigs. Prostaglandin F_2^α must be used or ordered by a licensed veterinarian.
- To induce abortion in dogs and horses.
- To terminate pseudopregnancy in mares.
- To control the time of estrus in mares and with difficult-to-breed mares.
- To remedy both retained placenta and chronic endometritis.

Caution should be used with prostaglandin F_2^α or Lutalyse because they are skin soluble, that is, readily absorbed across the skin. Moreover, they cause abortion in pregnant animals and miscarriage in women.

Gonadotropin-Releasing Hormone (GnRH)

GnRH can be used along with prostaglandin F_2^α for estrous synchronization. This involves four injections:

- GnRH to induce ovulation and CL formation, if there is a follicle at the right stage
- Prostaglandin F_2^α after 7 days to induce luteolysis.
- GnRH to induce ovulation after another 2 days (day 9) to induce ovulation
- Insemination on day 10

Follicle-Stimulating Hormone (FSH), or Follitropin

Daily injections of this are used to cause large numbers of follicles to develop for superovulation in donor cattle for embryo transfer.

Oxytocin

Oxytocin is used in cattle to assist with difficult births and to cause milk letdown.

Dopamine Agonists (Cabergoline or Bromocriptine)

The CL in the dog is maintained by prolactin. If prolactin levels can be caused to decline, then there will be luteolysis and the end of pregnancy.

Pregnancy Detection Diagnostic Kits

In Dogs

In dogs, pregnancy detection kits can be used to determine if a bitch is pregnant. One such diagnostic kit detects the presence of the pregnancy hormone relaxin. Relaxin is detectable in the serum or plasma 22–27 days after mating, when there is implantation. The diagnostic kit enables distinguishing between pregnancy and pseudopregnancy. An alternative approach is to measure the hormone progesterone.

In Cattle

The presence of a placental protein named *pregnancy-specific protein B* can be used to determine whether or not a cow is pregnant. An alternative approach is to measure the hormone progesterone in milk. These approaches complement physically testing for pregnancy by feeling the uterus.

Contraception in Wildlife

The traditional approach to preventing overpopulation in managed wildlife has been to reduce the population by programs of hunting or culling designed to maintain an optimal population. A nonlethal method of population control is contraception or reduced fertility. Given that males inseminate multiple females, contraceptive efforts have focused on the female. An approach that has been successful is immunization against GnRH and, thereby, preventing ovulation.

1. What is the primary reproductive organ?

2. What is libido?

3. Giving examples, what are secondary sexual characteristics?

4. What are the roles for the following: scrotum, testis, Leydig cells, Sertoli cells, gubernaculum, epididymis, seminal vesicles, prostate gland, Cowper's gland, and vas deferens?

5. Why is the temperature of mammalian testes kept lower than the body temperature?

6. What is an os penis?

7. What is the role of the corpus cavernosa in achieving an erection?

8. Why is castration performed?

9. Why is castration more difficult in poultry?

10. How is the reproductive system of male and female poultry different from that of their mammalian counterparts?

11. What are the differences between gametogenesis in males and females?

12. What is the difference between a spermatid and a spermatozoon?

13. What is the function of the corpus luteum?

14. What causes ovulation?

15. What is estrous behavior, and what causes it?

16. What is the difference between a duplex uterus and a bipartite uterus?

17. What happens to the ovum in the infundibulum, magnum, isthmus, uterus, and vagina of a hen?

18. What are the embryonic origins for the uterus and vas deferens?

19. What are the homologues to the labia majora and clitoris in a male?

20. What genotype normally ensures male or female phenotypes in mammals?

21. How is this different in poultry?

22. What is the gene required to produce a male?

23. What are the roles of GnRH, LH, FSH, estradiol, testosterone, and progesterone in reproduction?

24. What are the roles of progesterone, estradiol, relaxin, prostaglandin $F_{2\alpha}$, and oxytocin in reproduction?

25. What is the length of the estrous cycle in sheep, cattle, and horses? Why is it not appropriate to consider the length of the estrous cycle in dogs?

26. What is a reflex ovulator?

27. How long are spermatozoa normally fertile in mammals? Why is this different in poultry?

28. What is the success rate for artificial insemination?

29. What is the length of pregnancy of three species of domestic animals?
30. What is hatching of the zygote in development?
31. What are the different types of placenta?
32. How does the mother "know" physiologically that she is pregnant?
33. What is the role of the placenta?
34. What are the phases or stages of parturition?
35. What causes parturition?
36. What are the muscles that contract during parturition?
37. Give examples of problems with reproduction.
38. How is reproduction manipulated by drugs or management?

REFERENCES AND FURTHER READING

Bloomfield, F. H., Oliver, M. H., Giannoulias, C. D., Gluckman, P. D., Harding, J. E., & Challis, J. R. (2003). Brief undernutrition in late-gestation sheep programs the hypothalamic-pituitary-adrenal axis in adult offspring. *Endocrinology, 144,* 2933–2940.

Cooper, D. W., & Larsen, E. (2006). Immunocontraception of mammalian wildlife: Ecological and immunogenetic issues. *Reproduction, 132,* 821–828.

Grant, V. J., & Chamley, L. W. (2007). Sex-sorted sperm and fertility: An alternative view. *Biology of Reproduction, 76,* 184–188.

Hafez, B., & Hafez, E. S. E. (2000). *Reproduction in farm animals* (7th ed.). Ames, IA: Blackwell.

Johnson, G. A., Burghardt, R. C., Bazer, F. W., & Spencer, T. E. (2003). Osteopontin: Roles in implantation and placentation. *Biology of Reproduction, 69,* 1458–1471.

Lombard, J. E., Garry, F. B., Tomlinson, S. M., & Garber, L. P. (2007). Impacts of dystocia on health and survival of dairy calves. *Journal of Dairy Science, 90,* 1751–1760.

Marrable, A. W. (1971). *The embryonic pig: A chronological account.* London: Pitman Medical.

Pond, W. G., & Bell, A. W. (2005). *Encyclopedia of animal science.* New York: Marcel Dekker.

Reece, W. (2004). *Dukes' physiology of domestic animals* (12th ed.). Ithaca, NY: Cornell University Press.

Rosenfeld, C. S., & Roberts, R. M. (2004). Maternal diet and other factors affecting offspring sex ratio: A review. *Biology of Reproduction, 71,* 1063–1070.

Sisson, S., Grossman, J. D., & Getty, R. (1975). *Sisson and Grossman's the anatomy of the domestic animals* (5th ed.). Philadelphia: Saunders.

Swenson, M. J. (1979). *Dukes' physiology of domestic animals* (9th ed.). Ithaca, NY: Cornell University Press.

Wallace, J. M., Aitken, R. P., Milne, J. S., & Hay, W. W. (2004). Nutritionally mediated placental growth restriction in the growing adolescent: Consequences for the fetus. *Biology of Reproduction, 71*, 1055–1062.

Whittow, G. C. (2000). *Sturkie's avian physiology* (5th ed.). New York: Academic Press.

CHAPTER 17

Animal Growth and Development

OBJECTIVES

This chapter will consider the following:

- Hyperplasia (increases in cell number) and hypertrophy (increases in cell size)
- Methods to measure growth
- The growth of some of the major organ systems (bone, muscle, and fat/adipose tissue)
- The hormones controlling growth
- Effect of gender on growth (sexual dimorphism of growth)
- Nutrition and growth
- Genetics and growth

Objectives continue on the next page.

 ## INTRODUCTION

Growth is a complex of biologic processes leading to a mature animal. It is critically important to the production of meat by livestock and the performance of horses. The cellular basis of growth encompasses two distinct processes:

1. Hyperplasia, or the increase in the number of cells because of cell division and cell differentiation. For a specific organ or tissue, there can also be cell recruitment.
2. Hypertrophy, or the increase in the size of individual cells.

Cell division is a critically important part of growth and development. The process of cell division is mitosis (see fig. 17-1). In mitosis, the daughter cells have the complete set of DNA in the parent cell. There is replication of the DNA and duplication of the chromosomes.

Much of the cell division component of growth occurs before birth. However, there are exceptions with the fat or adipose tissue and with muscle fibers incorporating satellite cells. When nutrition is adequate, growth follows a sigmoidal curve (see fig. 17-2)

There is a progression in the growth of organs, with the growth of bone preceding that of muscle, and the growth of adipose tissue following (see fig. 17-3).

 ## MEASUREMENT OF GROWTH

Growth is measured in different ways in different species. We are all familiar with the situation in humans where we measure height. Sometimes with children, there is a place in the house where the height on a birthday is marked on a wall. Similarly in horses, we measure growth by the height in hands at the withers. For livestock, poultry, and aquaculture species, together with cats and dogs, we measure growth by the increase in body weight or average daily gain (ADG). However, both of these are "crude" measures of growth because they provide little information on the growth of muscle, fat, and other organs.

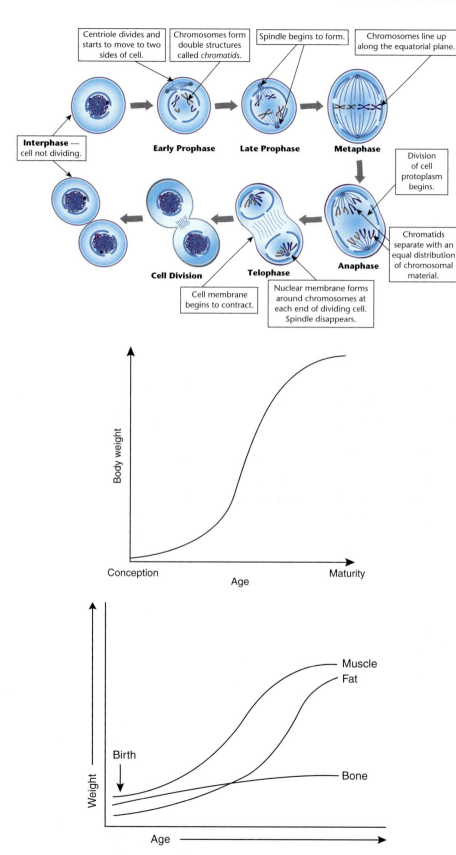

FIGURE 17-1 The process of cell division or mitosis. Source: Delmar/Cengage Learning.

Centriole divides and starts to move to two sides of cell.

Chromosomes form double structures called *chromatids*.

Spindle begins to form.

Chromosomes line up along the equatorial plane.

Interphase — cell not dividing.

Early Prophase

Late Prophase

Metaphase

Division of cell protoplasm begins.

Chromatids separate with an equal distribution of chromosomal material.

Cell Division

Telophase

Anaphase

Cell membrane begins to contract.

Nuclear membrane forms around chromosomes at each end of dividing cell. Spindle disappears.

FIGURE 17-2 A sigmoidal curve describes growth of an animal from conception to maturity. Source: Delmar/Cengage Learning.

Body weight

Conception

Age

Maturity

INTERESTING FACTOID

The timing of the growth of organs is first, bone; second, muscle; and third, adipose tissue.

FIGURE 17-3 The growth of bone, muscle, and fat occurs on a different time sequence, with bone growing before muscle, and adipose tissue (fat) growth last. Source: Delmar/Cengage Learning.

Weight

Birth

Muscle

Fat

Bone

Age

Definitions

Carcass analysis is used to find out if there are changes in the amounts of specific tissues, such as individual muscles.

Growth hormone (GH) is the major pituitary hormone required for growth. This protein hormone is also called *somatotropin* (ST)

Hyperplasia is the increase in the number of cells in an individual or organ due to cell division.

Hypertrophy is the increase in the size of individual cells in an organ or individual.

Marbling is the adipose tissue (fat) found in muscle. It increases the percent fat of the meat but improves its eating qualities.

Puberty and growth. Growth ceases sometime after puberty in many species of animals. There is no increase in height in people or horses after puberty, when the long bones stop growing. There can be increases in adipose tissue and muscle after puberty.

There are techniques that are used to provide more specific information on growth, including the following:

- Serial carcass analysis. Animals are slaughtered at different ages, and the carcass composition is determined by dissection. The increase in muscles and other organs during growth is calculated. Serial carcass analysis may be followed by chemical analysis to determine the amounts of protein, fat, water, and ash, and, therefore, the increase in these chemical constituents during growth.
- Use of isotopes (radioactive and nonradioactive) to measure the rate of synthesis of protein or fat. Whole-body 40K provides a nonlethal method for determining whole-body muscle because the natural potassium is found in intracellular water, which makes up about 90% of muscle.
- Noninvasive methods such as ultrasound, electrical impedance, and magnetic resonance imaging to determine the composition of a live animal.
- Nitrogen retention. This is a good method for estimating muscle growth. It involves accurately determining the amount of nitrogen consumed by an animal, the amount in the feces, and, therefore, the amount absorbed: nitrogen absorbed equals nitrogen consumed in feed minus nitrogen in the feces.
- The nitrogen retained is the nitrogen absorbed less than the nitrogen excreted in the urine: nitrogen retention equals nitrogen absorbed minus nitrogen excreted.
- Back fat probe. Back fat is a very good indicator of the overall amount of fat in an animal. To measure fat depth, a small incision in the back of a pig, steer, or heifer is made after use of a local anesthetic. A narrow ruler is then inserted.

 THE HORMONES CONTROLLING GROWTH

The principal hormone controlling growth is growth hormone, or GH. This protein hormone is released from the anterior pituitary gland under the control of releasing factors from the hypothalamus. GH stimulates growth of bone and muscle indirectly by enhancing the production of insulin-like growth factor (IGF)-I from the liver. This, in turn, stimulates growth. Figure 17-4 summarizes

FIGURE 17–4 The hypothalamic pituitary (ST)–IGF-I–growth axis. GHRH = GH-releasing hormone; SRIF = somatotropin release-inhibiting factor.

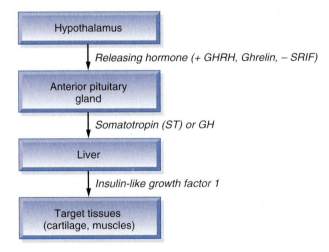

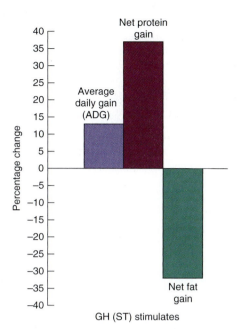

FIGURE 17–5 ST stimulates growth in pigs. Source: Delmar/Cengage Learning.

Postnatal is defined as occurring after birth. For example, in the majority of animals, postnatal growth occurs due to hypertrophy.

Prenatal is defined as occurring before birth. For example, differentiation and cell division occur prenatally.

Sexual dimorphism is when a characteristic of an animal (unrelated to reproduction) shows differences between males and females. An example is the sexual dimorphism in growth rate and mature adult size.

the hypothalamic-pituitary (GH)–IGF-I–growth axis. The effect of GH on the growth of pigs is illustrated in Figure 17-5. Another hormone important for growth is thyroid hormone thyroxine and its active form, triiodothyronine.

SEXUAL DIMORPHISM OF GROWTH

There is considerable variation in growth rate in a given species, or strain or breed within a species. On average, the male grows faster than the female and will achieve a large mature body size. This is the case regardless of whether we consider people, livestock, poultry, horses, dogs, or cats. Examples of sexual dimorphism of growth are included in Table 17-1.

One of the major factors causing males to growth faster than females is the hormone testosterone. This hormone and related chemicals are referred to as *anabolic androgens.* They are used to enhance growth in female cattle and illegally by some athletes.

Although the phenomenon of the male being bigger and growing faster than the female seems universal, there are less well-known species, including some reptiles, in which the female is larger and grows faster than the male.

NUTRITION AND GROWTH

It is probably self-evident that the full expression of the maximum growth rate requires optimum nutrition. If required nutrients, including energy, are not at the required levels, growth is slowed or becomes nonexistent. Growth rate can be increased above the normal maximum in "catch-up" growth. This is when an

INTERESTING FACTOID

What is the difference between GH and ST? The simple answer is NONE! Endocrinologists—people who study hormones—use the term *growth hormone* (GH), as do physiologists. Biochemists and chemists use the term *somatotropin* (ST). Interestingly, when bST (*b* is for bovine) was marketed for dairy cattle, the pharmaceutical industry used the term *ST*, and the opponents of its use called it *growth hormone(s)*, or *GH.*

TABLE 17-1 Examples of sexual dimorphism of growth based on mature body weight in pounds

SPECIES	MALE	FEMALE
German shepherd dog	64	54
Cat	8.6	6.0
Bison	1,200	950
Cattle	1,500–2,000	1,050–1,500
Pigs	440–500	750–800
Turkeys	55–60	25

animal has been on poor nutrition, and optimal nutrition is then provided. When the optimum diet is restored, there is very rapid growth.

GENETICS AND GROWTH

In livestock and poultry, there has long been selection of animals with superior growth rates. Progress has been accelerated by the approaches of quantitative genetics. This is now being supplemented by the use of genomics and candidate gene identification to augment selection.

STRESS AND GROWTH

When an animal is stressed, the adrenal cortex produces a stress hormone. The type of stress is over hours or days rather than seconds. Examples include crowding, transportation, and gross nutritional deficits such as starvation. The adrenal stress hormone is a steroid and is synthesized from cholesterol. The stress hormone is chemically related to cortisone, which we use as an anti-inflammatory drug. There are two related adrenal cortical stress hormones in different species: cortisol in people, cattle, sheep, horses, and pigs; and corticosterone in poultry, other birds, and rodents.

Cortisol (or corticosterone, i.e., the other stress hormone) acts in an animal's body to reduce growth. There is reduced growth of the skeleton and breakdown of the muscle proteins. For instance, in horses, injections of cortisol have been shown to lead to nitrogen retention, which is a very good indicator of growth and particularly muscle growth, decreasing to virtually zero. This is different from the "fight-or-flight" situation when animals release epinephrine (or adrenaline) from the adrenal medulla into their bloodstream for a very short-term effect (seconds or minutes). You see the fight-or-flight reaction when watching a scary movie.

Disease and Growth

As might be expected from common sense, disease is associated with reduced rates of growth. The major animal diseases are discussed in detail in Chapter 20.

ABNORMALITIES IN GROWTH

Dwarfism

Dwarfism is the failure to grow at the appropriate rate and to achieve normal mature weight. It has several different causes, such as the following:

- GH deficiency. This is seen in dogs. For instance, there are German shepherd dogs with inherited pituitary dwarfism and growth retardation. Levels of both GH and IGF-I are less than 10% of normal.
- IGF-I deficiency. In people, Laron's dwarfism is where there are low rates of growth but normal secretion of GH. What is thought to be happening is that the levels of mediating hormone, IGF-I, are very low. A similar situation is found in miniature cattle and dwarf chickens. The GH control of IGF-I release is uncoupled because of problems with the GH receptor.

Hypothyroidism

Thyroid hormones are essential for growth and development. In some places, there is a high incidence of congenital hypothyroidism in horses, with new-born foals showing musculoskeletal deformities. This may be due to high levels of nitrate (NO_3) in the dam's feed or other goitrogenic plants, that is, plants that are toxic to animals by affecting the thyroid gland. These interfere with the synthesis of thyroglobulin and, therefore, thyroxine. In dogs, there is adult-onset hypothyroidism.

Runting in Pigs

There are growth-impaired baby pigs, which continue to show lower rates of growth through weaning and after weaning. It is thought that this is due to the uterine environment restricting intrauterine and postnatal growth.

AGING

Aging has little relevance to livestock because they will either go to market or be culled at an age well before they can be considered as old aged. However, veterinarians and animal owners are increasingly concerned about aged dogs, cats, and horses. These show many of the same conditions as in aged people, ranging from mental deterioration, reduced amounts of muscles, diabetes and other endocrine diseases, and cardiovascular disease.

Dogs' and Cats' Age

It is often said that a dog's age in human years is seven times the dog's age. This is an oversimplification. A better "rule of thumb" is 12 dog or cat years per human year for the first 2 years, then 4 dog or cat years per human year for each year thereafter (see Table 17-2).

TABLE 17-2 Comparison of a dog or cat's age to "human years"

DOG OR CAT YEARS	HUMAN YEARS
½	10
1	15
2	24
4	32
8	48
12	68
16	88

Table 17-2 approximates to the situation for small dogs. Large dogs age faster. For instance, based on a chart developed by Fred L. Metzger, DVM, a 10-year-old dog is equivalent to the following in "human years":

- If 0–20 lb, 56 years old
- 21–50 lb, 60 years old
- 51–90 lb, 66 years old
- 90 lb or greater, 78 years old

There is increasing interest in maintaining the health of the older dog or cat.

Aged Horses

About 15% of the equine population in the United States is older than 20 years of age. These older horses are often referred to as *geriatric horses*. The old horses have a greater incidence of some diseases such as colic, dental disease, tumors, lameness, and pituitary disease, but not of others, including laminitis, allergies, respiratory tract disease, or thyroid disease. Aged horses show an impaired stress response with no increase in the release of the stress hormone cortisol when placed on a treadmill for an exercise test. In contrast, in young and middle-aged horses, the concentration of cortisol is rapidly increased and stays high for at least 1 hour.

 ## GROWTH STIMULATORS/PROMOTANTS

In Cattle

In cattle, hormonal implants are used extensively to improve the growth and feed efficiency. The implants contain estrogen, including estradiol or zeranol, a fungal compound or mycotoxin with estrogen activity in cattle; androgen, including testosterone or synthetic androgen trenbolone; and progesterone.

Implants for heifers and steers frequently contain both an estrogen and androgen, whereas those for young bulls contain predominantly estrogens. Use of these implants results in a 15–20% increase in growth rate (particularly muscle) together with improvement in feed efficiency. Along with improved performance, some decreases in carcass quality are observed. Another advantage of these is that they are size neutral, working just as well for a large- or small-scale producer.

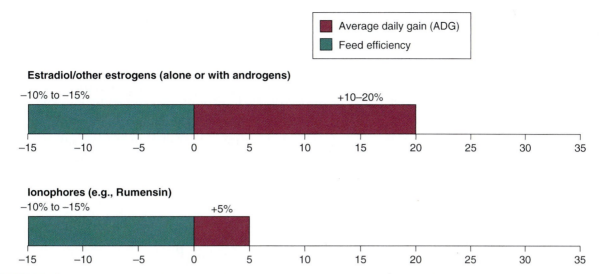

Estradiol/other estrogens (alone or with androgens)

−10% to −15%   +10–20%

Ionophores (e.g., Rumensin)

−10% to −15%   +5%

FIGURE 17–6 Stimulators of growth used in the beef cattle industry. Source: Delmar/Cengage Learning.

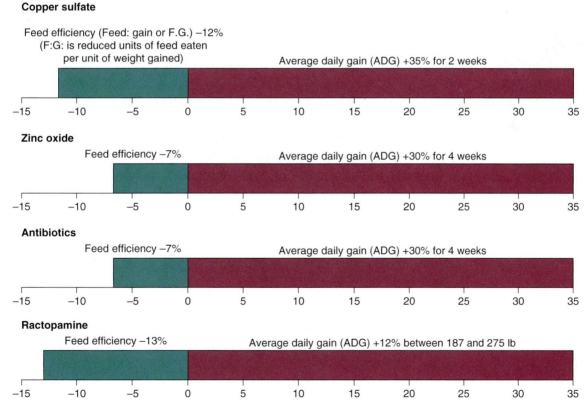

Copper sulfate

Feed efficiency (Feed: gain or F.G.) −12%
(F:G: is reduced units of feed eaten
per unit of weight gained)   Average daily gain (ADG) +35% for 2 weeks

Zinc oxide

Feed efficiency −7%   Average daily gain (ADG) +30% for 4 weeks

Antibiotics

Feed efficiency −7%   Average daily gain (ADG) +30% for 4 weeks

Ractopamine

Feed efficiency −13%   Average daily gain (ADG) +12% between 187 and 275 lb

FIGURE 17–7 Stimulators of growth and the efficiency of growth in pigs. Examples are from individual research studies. Source: Delmar/Cengage Learning.

These implants have received Food and Drug Administration (FDA) approval after a stringent approval process based on scientific risk assessment. Although they have been approved in the United States, Canada, and other countries, they are prohibited in the European Union. Between 1954 and 1979, a synthetic estrogen, diethylstilbestrol, was used to improve performance in cattle. In addition, it was used in human medicine for estrogen replacement, for prostate cancer treatment, as a postcoital birth control (an off-label use not approved by the FDA), and to reduce miscarriages. In the case of the latter, its use was associated with an increased incidence of vaginal and breast cancer for the children. Because of its teratogenic effects (adverse effects on the fetus), diethylstilbestrol is no longer used in human medicine or as an animal growth promotant.

β-Adrenergic Agonists as Growth Stimulators/Promotants

Elanco Animal Health has commercialized the β-adrenergic agonist ractopamine to improve the performance of pigs and cattle. This drug has been approved by the FDA. Ractopamine acts by increasing the deposition of muscle protein and reducing the amount of triglyceride/fat deposited. It acts by increasing the breakdown of triglyceride and reducing the synthesis of both fatty acids and triglyceride.

In pigs, ractopamine is marketed as Paylean 9. Addition of Paylean 9 to the pig feed has the following advantageous effects: increased yield by 0.5%, improved feed efficiency by 14.3%, and increased growth rate by 19.4%. However, there may be reductions in palatability of the meat. In cattle, ractopamine is marketed as the feed ingredient Optaflexx. It similarly increases yield, growth rate, and feed efficiency.

Antibiotics as Growth Stimulators/Promotants

Antibiotics have been used in poultry and pig industries as antimicrobial growth promoters for some 50 years. The European Union has banned antimicrobial growth promoters based on the view that they are responsible, at least partially, for the development of antibiotic resistance. It is not clear the extent to which this is the case, and what is the relative significance of this compared with the high level of prescribing antibiotics to people (so-called "over-prescribing") and the failure of patients to complete the full treatment and consume all the prescribed drug.

1. What is the difference between hypertrophy and hyperplasia?
2. When does hyperplasia predominantly occur?
3. How do the number of cells in an organ increase? (Hint: It is more than cell division.)
4. When does bone, muscle, and adipose tissue growth occur?
5. What type of curve describes the growth of a domestic animal?
6. What is carcass analysis? How do researchers use serial carcass analysis?
7. Why is marbling important for meat?
8. Is growth hormone the same as somatotropin?
9. What techniques are used to measure growth?
10. What is ADG?
11. What is the basis of measuring nitrogen retention in an animal?
12. Why is a back fat probe used?
13. What is sexual dimorphism of growth?
14. Are male animals always bigger than females?
15. How does nutrition influence growth?
16. What is "catch-up" growth?
17. How does an animal's genetics influence its growth?
18. How do animal breeders use genetics to improve the growth rate of animals or breeds?
19. Does stress affect growth? If so, how?
20. Give examples of stresses that lead to the production of the adrenal stress hormones.
21. Is stress the same as the "fight-or-flight" response? If no, how do they differ?
22. What is the mature body weight of an animal? Why does growth of the long bones stop after puberty?
23. Do animals develop and/or age at a different rate than people?
24. Give examples of the effects of aging in dogs, cats, and horses.

REFERENCES AND FURTHER READING

Arango, J. A. Cundiff, L. V., & Van Vleck, L. D. (2004). Comparisons of Angus, Charolais, Galloway, Hereford, Longhorn, Nellore, Piedmontese, Salers, and Shorthorn breeds for weight, weight adjusted for condition score, height, and condition score of cows. *Journal of Animal Science*, *82*, 74–84.

Gerrand, D. E., & Grant, A. L. (2003). *Principles of animal growth and development*. Dubuque, IA: Kendall Hunt.

Scanes, C. G. (2003). *Biology of growth of domestic animals*. Ames, IA: Blackwell.

Lactation

INTRODUCTION TO LACTATION AND ITS IMPORTANCE

The common feature to all mammals is the production of milk by mammary glands of the mature female to provide nutrition for the newborn. This provides a tremendous advantage to the young mammal. Figure 18-1 shows a young calf suckling at the teat of its mother. It is thought that the mammary gland evolved from sweat glands. Although all mammalian species produce milk, mammals are not the only animals that produce milk. In pigeons and doves, both parents produce milk. This is fat-filled cells sloughed off from the crop, part of the upper gastrointestinal tract. The crop milk is then regurgitated to the young. The hormone prolactin stimulates the production of the crop milk.

Milk contains a mixture of proteins (particularly casein, a milk phosphoprotein), lactose (the milk sugar), lipids/fats (both triglyceride and phospholipids), minerals in solution (e.g., sodium, potassium, calcium, magnesium, or chloride), or complexed with the proteins (such as phosphorus), and water. The composition of milk varies markedly among different species, as can be seen in Table 18-1.

The first milk produced is the colostrum. This has a much higher concentration of protein because of the high concentrations of maternal antibodies (both immunoglobulin G and M). The intestines of many species are permeable to proteins in the first day or few days of life. These antibodies pass into the circulation of the newborn, providing it with passive immunity against potentially invading disease-causing pathogens. Species in which antibodies from colostrum pass from milk through the gut into the circulation include cattle, horses, pigs, dogs, and probably cats. Babies and newborn rabbits have already received passive immunity by placental transfer of antibodies, and, thus, gut impermeability to proteins is not a problem. Antibodies in colostrum and the continuing lactation also can bind to pathogens in the gastrointestinal tract and potentially aid the young animal.

FIGURE 18-1 Hereford cow with calf suckling from udder/mammary glad. Courtesy of Getty Images/Photodisc.

TABLE 18-1 Comparison of the composition of milk from different mammalian species

	PROTEIN (G %)	FAT (G %)	CARBOHYDRATE (G %)	WATER (G %)
Aquatic mammals (whales, seals, porpoises)	10–11	41–49	0.1–1.3	40–45
Cow	3	3.8	4.8	87
Dog	9	10	2.7	75
Donkey	2	1	6	90
Human	1	4	7	88
Horse (mare)	6–7	1–2	7	89
Pig (sow)	12–13	7–8	4–5	80–82
Rats and mice	9–12	13–15	3	68

 ## ANATOMY OF THE MAMMARY GLAND

Mammary glands in different mammals vary in number and position. The number of mammary glands shows marked differences with species. In virtually all mammalian species, there is an even number of mammary glands. Examples of differences in the number of mammary glands are two in humans, goats, and sheep; four in cattle; eight to 10 in dogs; and 12–16 in pigs. In the mare, there are four mammary glands, but only two teats. Each teat has two streak canals from two mammary glands.

Mammary glands are always located on the ventral or front surface. There are three positions for mammary glands: anterior or thoracic, as found in humans, other primates, and elephants; intermediate or abdominal; and posterior or inguinal, as with cattle, sheep, goats, and mares. In litter-bearing species such as rats, dogs, cats, and pigs, there are a series of mammary glands from anterior to posterior.

The mammary gland is composed of the mammary gland proper together with the nipple or teat (see fig. 18-2). Milk passes out of the mammary gland through the teat, which also acts to prevent pathogens invading the ducts of the mammary gland. There is no nipple in the most primitive of mammals, the monotremes, such as the duck-billed platypus.

The mammary gland is made up of lobes, and with each lobe is a series of lobules. Each lobe and lobule is surrounded by connective tissue. There are several tissue types in a mammary gland: alveoli, ducts, connective tissue surrounding the lobes and lobules, and adipose or fat tissue.

The mammary gland consists of a series of branching ducts leading from the teat to the many milk-producing alveoli (singular, alveolus):

- Teat cistern, one per mammary gland, which is absent in humans
- Udder or gland cistern, one per mammary gland, which is absent in humans
- Galactophores or cistern ducts, approximately eight per mammary gland
- Interlobar ducts (multiple, branching)
- Intralobar ducts (multiple, branching)
- Interlobular ducts (multiple, branching)

Definitions

Alveolus (plural *alveoli*) is the site of milk production.

Casein is the major phosphoprotein in milk.

Colostrum is the first milk produced and is high in maternal antibodies, giving the neonate (newborn animal) passive immunity against potentially invading disease-causing pathogens.

Galactopoiesis is the maintenance of milk production.

Lactogenesis is the initiation of milk production.

Lactose is milk sugar, which is a disaccharide of glucose and galactose.

Lobular-alveolar system is the system for milk production in the alveolar, and its transport through the ducts in the lobes and lobules.

Mammary alveolar cells are exocrine cells producing a secretion (milk). The alveolus leads into a duct and, therefore, to the outside of the animal.

Mammogenesis is the development of the mammary gland.

Myoepithelial cells are smooth muscle cells surrounding the alveolus. Contraction of these stimulated by the hormone oxytocin causes milk letdown, that is, milk being squeezed from the alveolus down the ducts to the udder and teat. Stages in the development of lactation are mammogenesis, lactogenesis, and galactopoiesis.

FIGURE 18-2 Anatomy of the mammary gland of a cow. Source: Delmar/Cengage Learning.

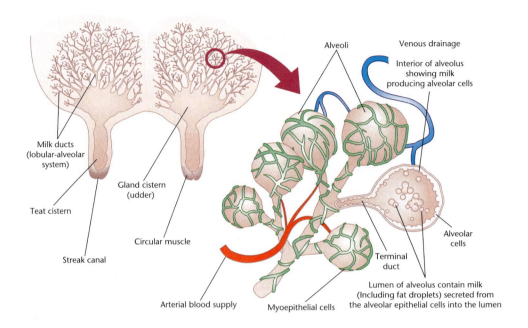

- Intralobular (multiple, branching)
- Terminal duct (one per alveolus)
- Alveolus

The alveolus consists of three components (see fig. 18-2):

- The secretory alveolar epithelial cells, which synthesize the milk.
- The lumen into which the milk constituents are secreted by the alveolar cells and where they are temporarily stored.
- The myoepithelial cells. These surround the alveoli and contract when stimulated by the hormone oxytocin. The contraction of these cells causes milk letdown.

Definitions

Growth is increasing in size or weight.

Development is changing the anatomy of a tissue or organ (changing its biology) and potentially with new functions acquired.

 ## DEVELOPMENT OF THE MAMMARY GLAND AND ITS HORMONAL CONTROL

There are three stages in the development of the mammary gland or mammogenesis: before puberty, after puberty, and pregnancy.

Before Puberty

Before puberty, the mammary gland consists of a rudimentary simple duct system, in parenchymal tissue, leading to the teat or nipple together with mammary adipose tissue. The mammary gland grows at a rate proportionately with body growth.

After Puberty

Beginning at puberty, the duct system undergoes considerable development becoming very branched. There is growth of both the parenchymal and adipose tissues at a higher rate than the body is growing. This growth and development of the duct system are occurring under the influence of estrogens, particularly estradiol. Lobular-alveolar development is evoked by estrogen plus progesterone. In addition, the development of the mammary gland requires growth hormone (GH) (or somatotropin) and adrenal cortical hormones, such as cortisol and aldosterone.

Pregnancy

During pregnancy, the mammary gland completed its development. The duct system becomes even further branched, and at the ends of each duct will be an alveolus. The alveolus consists of glandular cells (the alveolar cells) surrounded by specialized smooth muscle cells, the myoepithelial cells. There is also further growth of both the parenchymal and adipose tissues.

As you might expect, hormones play a critical role in mammary gland development during pregnancy. Among the important hormones stimulating the development of the mammary gland are estrogens, progesterone, insulin, prolactin and placental lactogens, and GH.

Lactogenesis is the initiation of milk production. An animal needs to have milk available as soon as it gives birth. Lactogenesis requires a "cocktail" of hormones, including prolactin, insulin, cortisol, and GH, together with estrogens, progesterone, and thyroid hormones.

MILK PRODUCTION AND ITS HORMONAL CONTROL

Milk production and its maintenance at the optimal level (galactopoiesis) are critically important, first to the survival of the newborn animal, and then to it prospering and growing rapidly. The alveolar cells are exocrine cells playing the major role in the synthesis of milk. These cells synthesize the following:

- Lactose (a disaccharide of glucose and galactose) from glucose taken up from the blood.
- Casein and other proteins such as the major whey proteins β-lactoglobulin and α-lactalbumin from amino acids taken up from the blood. (Note that the phosphoprotein casein together with calcium ions forms a micelle. This is an insoluble dense protein granule that is in suspension in the milk.)
- Triglycerides and other lipids produced from precursors (such as acetate) taken up from the blood.

The need for glucose in ruminants creates a metabolic quandary. Ruminants absorb volatile fatty acids from the ruminal fermentation, not glucose as in other species. Therefore, ruminants must synthesize glucose from one volatile fatty acid, namely, propionic acid. This occurs in the liver by a process called *gluconeogenesis*.

Milk production has to meet the needs of the offspring. Therefore, milk production needs to increase as offspring get bigger and require more energy, protein, and other nutrients. Again, this is under hormonal control. The hormonal control of milk production is different in different animals. In rats and humans, prolactin is the major galactopoietic hormone with increases in prolactin secretion when there is suckling. Without prolactin, milk production ceases. In contrast, in ruminants such as cattle, the major hormone stimulating milk production is GH. This action of GH is used commercially in the dairy industry (see the section on approaches to improve milk production later in the chapter).

 ## MILK LETDOWN OR RELEASE AND ITS HORMONAL CONTROL

There are two mechanisms by which milk leaves the mammary gland:

1. Opening the streak canal such that milk in the teat cistern, udder cistern, and cistern ducts (galactophores) can exit passively. This is accomplished by negative pressure as with a milking machine, positive pressure as with hand milking, and a combination as with suckling.
2. The milk letdown response with milk being forcibly ejected from the mammary gland because of contractions of the myoepithelial cells that surround the alveoli (see figs. 18-3 and 18-4).

The milk letdown response is a neuron-endocrine reflex. When the mammary gland and, particularly, the teat are stimulated, nervous impulses pass to the brain. These lead to the release of a hormone, oxytocin, from the posterior pituitary gland. Oxytocin passes by way of the bloodstream to the mammary

FIGURE 18–3 The milk letdown reflex with stimulation of the teat leading to oxytocin release; this, in turn, stimulates the smooth muscles (myoepithelial cells) around the alveoli to contract, forcing the milk toward the teat.

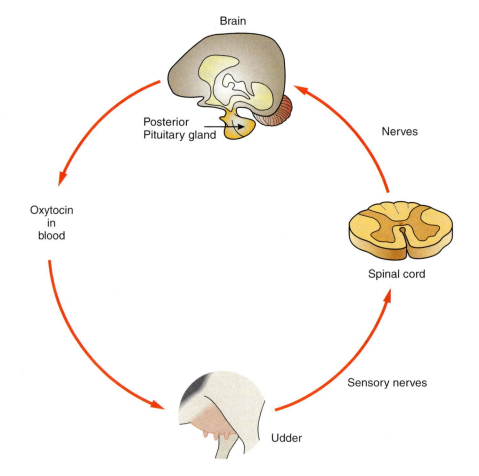

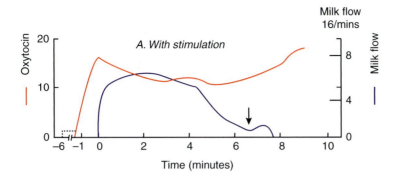

Milk flow
16/mins

A. With stimulation

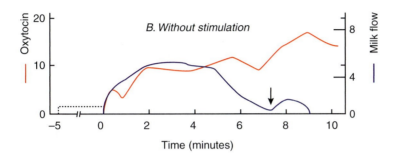

B. Without stimulation

FIGURE 18-4 Graph showing the effect of stimulation of the teat leading to oxytocin release (red line) and increased flow of milk (blue line). *A.* Shows milk flow and oxytocin concentrations when the teat has been stimulated. *B.* Shows milk flow and oxytocin concentrations when the teat has not been stimulated. Adapted from Mayer, Schams, Worstorff, and Prokopp (1984).

glands, where it stimulates the myoepithelial cells to contract. The reflex takes about 1 minute before the milk is under pressure in the mammary gland.

 ## APPROACHES TO IMPROVE MILK PRODUCTION

Management and Nutrition

Good management is essential for high milk production. It is important to have optimal health and nutrition. Moreover, stresses such as heat or cold should be avoided. In acute stress such as a barking dog where cows are milked, animals release epinephrine (adrenaline) from the adrenal gland (the adrenal medulla) and norepinephrine (noradrenaline) from the autonomic nerve terminals. Both epinephrine and norepinephrine interfere with oxytocin action in the milk letdown response.

One management approach is to terminate the lactation (or "dry the animal off") between lactations. During this time and naturally when an animal is neither pregnant nor lactating, the mammary gland undergoes involution. There is a breakdown of the lobular-alveolar system. This will need to be restored for a subsequent lactation.

Genetics

There has been a tremendous improvement in milk production per cow in the last 50 years. The available evidence supports that this is due to genetic selection for milk production with artificial insemination using semen from bulls with proven ability to sire superior heifer calves.

INTERESTING FACTOID

Neither new babies nor new mothers necessarily know what to do to ensure that milk is flowing from mother to newborn. If the mother becomes stressed, milk letdown will be impaired, and both the newborn and the mother will become frustrated and stressed.

FIGURE 18–5 Production of bST.

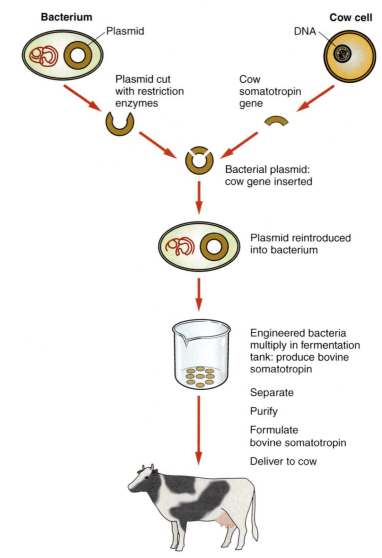

Bovine Somatotropin (bST)

In the United States, bST (GH) is widely used by dairy farmers to increase milk production and its efficiency. This is produced by bioengineering (see fig. 18-5). The advantages include more milk per cow, even in a very high-producing cow (see fig. 18-6); more milk per unit of feed; more milk per facility; and less animal waste per unit of milk produced.

There has been very strong opposition by activists to the use of bST. Some concerns have focused on animal health or human health. The U.S. Food and Drug Administration concluded, based on scientific evidence from multiple refereed papers, that there was no evidence supporting claims of adverse effects on either animal or human health.

Oxytocin

The synthetic hormone oxytocin can be used in cattle to cause milk letdown.

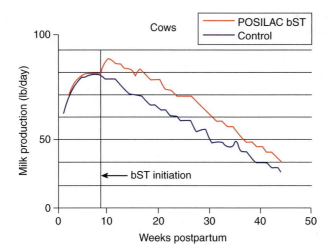

FIGURE 18-6 The effect of bST on milk production. Source: Delmar/Cengage Learning.

PROBLEMS WITH LACTATION

A major reason for reduced milk production or not saleable milk is mastitis. This disease is inflammation of one or more quarters of the mammary gland and is the most significant disease of dairy cattle. It causes losses because of reduced milk production and the presence of milk not saleable with a high concentration of somatic cells (leukocytes) together with long-term damage to the mammary gland, leading to culling. It is caused by the invasion of the mammary gland by pathogens, including various species of *Staphylococcus* such as *Staphylococcus aureus,* *Streptococcus* sp., and *Mycoplasma* sp.

REVIEW QUESTIONS

1. Are mature females of all species of mammals capable of producing milk?

2. What are the major components of milk?

3. Do the components of milk differ from species to species? If so, how?

4. What is crop milk?

5. What is colostrum?

6. How does colostrum incur passive immunity?

7. How many mammary glands do a cow, pig, mare, and dog have?

8. Where are the mammary glands of a cow, pig, mare, and dog?

9. What are the anatomic components of a mammary gland?

10. What is the role of the alveolar cells?

11. What is the role of the myoepithelial cells?

12. What are the stages of mammary development?

13. When do they occur?

14. What are the hormones that stimulate the mammary development?

15. What is mammogenesis?

16. What is lactogenesis?

17. What is galactopoiesis?

18. What is casein?

19. In what form is casein in milk?

20. What are the major whey proteins?

21. What is lactose?

22. Where is it produced?

23. Why does production of lactose produce a metabolic strain in ruminants?

24. How is the streak canal opened during milking or suckling?

25. What is milk letdown?

26. What is the milk letdown neuroendocrine reflex, and what role does oxytocin have in it?

27. Do nutrition, management, and genetics influence milk production?

28. Does stress affect milk production?

29. What is mastitis?

30. What is bST used for?

REFERENCES AND FURTHER READING

Bruckmaier, R. M., & Blum, J. W. (1998). Oxytocin release and milk removal in ruminants. *Journal of Dairy Science, 81*, 939–949.

Gorewit, R. C., & Gassman, K. B. (1985). Effects of udder stimulation on milking dynamics and oxytocin release. *Journal of Dairy Science, 68*, 1813–1818.

Mayer, H., Schams, D., Worstorff, H., & Prokopp, A. (1984). Secretion of oxytocin and milk removal as affected by milking cows with and without manual stimulation. *Journal of Endocrinology, 103*, 355–361.

Tucker, H. A. (2000). Hormones, mammary growth, and lactation: A 41-year perspective. *Journal of Dairy Science, 83*, 874–884.

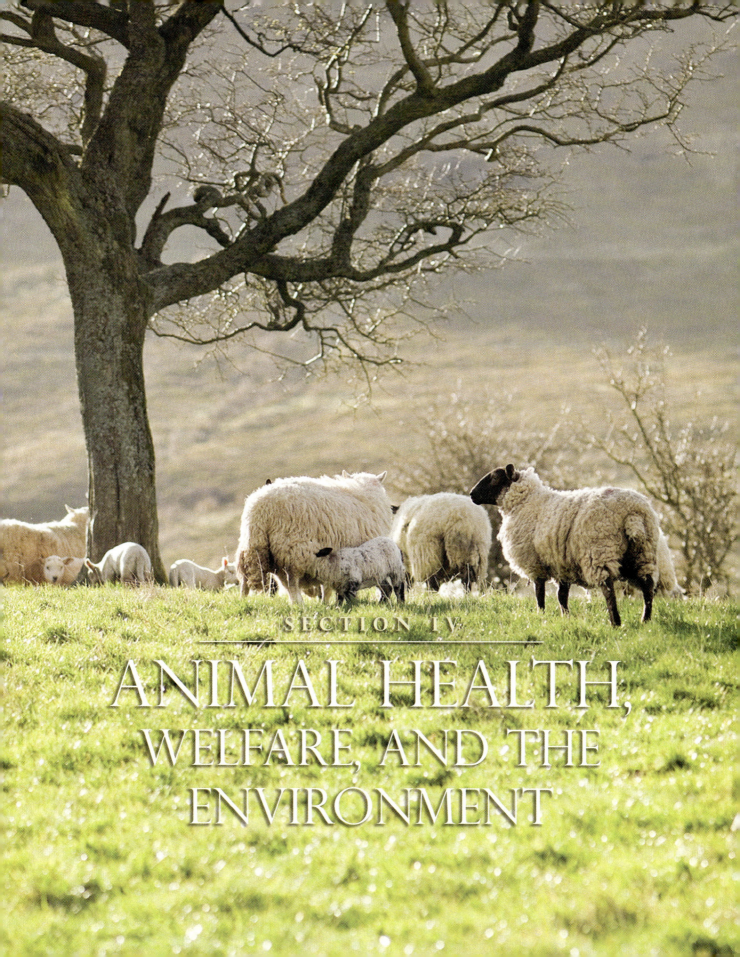

SECTION IV

ANIMAL HEALTH, WELFARE, AND THE ENVIRONMENT

CHAPTER 19

Animal Behavior, Animal Welfare, and the Environment

OBJECTIVES

This chapter will consider the following:

- Animal behavior, including aggression
- Animal stress
- Animal welfare
- Animal responses to high and low temperatures
- Animal housing

INTRODUCTION

Livestock and companion animals exhibit complex behaviors related to reproduction and care of offspring, and in response to external stresses, regardless of whether the stress is from the external environment or from people. In addition to behavioral responses to external stresses, there are physiologic responses. We will notice that if we are stressed, our mouths become dry, our hands are cold and sweaty, and our heart rate increases. This might be before a medical examination, a job interview, a date, or a scary movie. These are physiologic responses, and animals do something very similar.

There is increasing concern about the welfare of livestock and companion animals. Included in the responses to animals to stresses are extremes of temperature. Like us, livestock and companion animals also become stressed with very high or low temperatures. Adequate housing is important not only for our survival and prospering, but also for animals.

ANIMAL BEHAVIOR

There are many types of behaviors that are exhibited by animals, including aggression; reproductive behavior; grooming behavior; feeding, drinking, and elimination behaviors; behaviors to facilitate temperature homeostasis (cooling or heating); and communication.

Animal Aggression

There are two types of animal aggression: interspecies and intraspecies.

Interspecies Aggression

Interspecies aggression involves a series of behaviors between different species such as a predator attacking its prey, or antipredatory behaviors. The latter is aggression to protect against a predator. Examples with domestic animals are dogs and llamas as watch animals to prevent predation of sheep, or dogs to protect their owners.

Intraspecies Aggression

Intraspecies aggression is directed toward other members of the same species. There are different aggressive behaviors, including the following:

- Territoriality. This is where an animal, frequently the male, defends a territory against intruders. This may be a male defending a single female or a group, or "harem," of females such that the male genes are passed on, or both males and females defend an area to protect sources of feed for themselves and their offspring.
- Social hierarchy. The social hierarchy or *peck order* is important in social species and group-housed livestock to maintain stability and reduce the chances of injury to members of the group. The concept of a defined social hierarchy in animals is seen with the lead dog in a pack of dogs and was

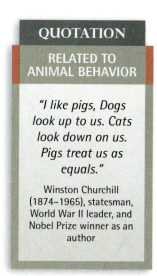

first described with chickens. The higher chicken in the hierarchy pecks at the next in the hierarchy and so on down the line. The lowest birds in the peck order will have a significant loss of feathers and appear the most bedraggled. A social hierarchy can be maintained by the "higher" animal(s) showing aggressive behaviors, whereas the subordinate shows submissive behavior. The social hierarchy can affect food intake if space at the feeder or feed bunk is limited because "lower-status" animals may not get sufficient time to eat.

- Other aggressive encounters include where another member of the same species is attacked and potentially killed or where there is forced mating.

Many aggressive behaviors are affected by the presence and amount of the male hormone testosterone acting on the brain, with castrated males not exhibiting aggressive behavior. There is evidence that the neurotransmitter serotonin and its amino acid precursor tryptophan are involved in brain control of aggression. Examples of intraspecies aggression will be included below for several domestic animals.

Aggression in Chickens

One of the earlier uses of chickens was cockfighting, which is an activity taking advantage of the natural proclivity of roosters to fight. Chickens were the first species in which researchers established the principle of pen orders, with the dominant animals showing a higher frequency of pecking toward subordinates. Interestingly, the frequency of another testosterone-related behavior, crowing, is higher at the top of the social hierarchy. There are both genetic links to aggression in chickens and to being a victim. Broiler breeder males are more aggressive than game chickens or layer-type white Leghorns or jungle fowl both toward other males and toward females. The aberrant behavior toward females includes increases in forced copulation, in the roughness during mating with females struggling, together with decreases in courtship behavior. This aberrant behavior is genetic and does not appear to be related to the feed restriction that broiler breeders experience during growth and development.

Feather pecking can lead to the death of the victim and to cannibalism. This is why beak trimming is performed in the poultry industry.

Aggression and Other Behavioral Problems in Dogs

Dogs show marked territoriality with barking if a strange dog or person enters their territory. Among the behavioral responses is an increase in barking. Interestingly, spayed dogs (i.e., with both the ovaries and uterus surgically removed) exhibit more offensive territorial aggression than other females, with both more barking and a lower pitch of vocalization.

Reports from Cornell University indicate that there are breed differences in the number of aggression problems reported by owners. There was a higher incidence than expected for Dalmatians, English springer spaniels, German shepherd dogs, and mixed-breed dogs, and a lower incidence with Labrador retrievers and golden retrievers. There were also differences in anxiety and phobias. In cats, there was a lower incidence of aggression and house soiling in domestic shorthair cats than other breeds but a higher incidence in males overall.

Reproductive Behavior

Animals have reproductive behaviors for several distinct reasons, such as the following:

- To increase the chances that the male will pass on its genes to as many offspring as possible.
- To increase the chances that the female gametes will be fertilized.
- To restrict mating to within the same species. Reproductive behaviors are often extremely species-specific, with major differences even in closely related species.
- To identify when a female is fertile (at estrus).
- Courtship behaviors before mating to ensure that the female is responsive and in the position to allow mounting (lordosis).
- Mounting, intromission, and insemination.
- Parental behaviors to feed and protect the offspring.

Examples of specific reproductive behaviors include the following:

- Female behaviors to show a male that she is in estrus. Examples include females winking her vulva (horses) or producing specific pheromones (cattle, dogs).
- Male behaviors to encourage an estrous female mammal (or a sexually mature female bird) to be receptive and adopt the position that will allow mating. The positioning of the female is called *lordosis*. The male behaviors range from circling movements by the rooster that are endearingly called a *waltz* in chickens, to the boar grunting and producing a pheromone in pigs, to licking and snorting around the genital area in cattle.

Feeding, Drinking, and Elimination Behaviors

Animals control the frequency of eating and drinking, and the volume of food and water consumed. This was discussed in Chapter 13.

Some species of animals also have specific behaviors related to urination and/or defecation. For instance, dogs mark their "territory" by urination, whereas pigs may defecate in specific areas of their housing.

Temperature-Related Behaviors

There are behaviors that help mammals and birds maintain their body temperatures. Examples of behaviors to promote cooling include panting, movement away from a heating source such as a brooder heater for young chicks, accessing shade for cattle, and wallowing in mud in pigs. Behaviors to assist heating include movement toward a heating source, groups of animals congregating together to share body heat, and making the body profile more compact to reduce heat loss.

> **INTERESTING FACTOID**
>
> *The Coolidge effect* is a behavior that increases the chances that the male will pass on its genes to as many offspring as possible. The *Coolidge* or *novelty effect* is the propensity of males to mate with different females. This was named after President Calvin Coolidge ("Silent Cal"), who observed a rooster mating with a series of different hens.

COMMUNICATION

Communication between members of the same species (intraspecies communication) can be accomplished by the following:

- Auditory detection (hearing) of vocalization such as sows responding by adopting the position of the body to allow mating (lordosis) in response to grunting from boars.
- Olfactory detection (smelling) of pheromones and other scents such as animals (e.g., dogs) marking their territories, or in the flehmen response of bulls to the urine of cows in heat or sows showing lordosis in response to the pheromone from a boar.
- Visual detection of signaling movements such as a submissive positioning in dogs, head movements (head bobbing) in poultry, or stallions responding to the "winking" of the vulva of an estrous mare.
- Various tactile interactions, particularly between offspring and parent, such as the neonate stimulating the mammary gland and then the milk letdown response (see Chapter 18).

 STRESS

Although there are multiple physiologic and behavioral responses to stresses, the classic response to stress is a large increase in the release of cortisol from the adrenal cortex (or release of corticosterone in poultry). This hormone acts to protect the animal by increasing the availability of glucose for metabolism and also suppresses the immune system. Cortisol is released in response to the pituitary hormone, adrenocorticotropic hormone, and this in turn is stimulated by the release of corticotrophin-releasing hormone from the hypothalamus, which is in the base of the brain.

What Is Stress?

Stress can be either physical or psychological. Physical stress includes extremes of temperature and abnormally low blood concentrations of glucose.

Psychologic stress includes transportation, a new environment, and being handled with or without pain, for example, branding in cattle, dehorning in cattle, tail docking in sheep, and castration.

Assessing Stress (Animal Welfare/Well-Being)

There are a variety of approaches to assess whether an animal is stressed, including the following:

- Behavioral indices (e.g., grooming in cats and preening in poultry, aggression, arousal, response to novel stimuli).
- Plasma concentrations of the adrenal stress hormones cortisol (mammals) and corticosterone (birds). A good indicator of stress is cortisol in the urine or saliva, and this has the advantage that it avoids the potentially stressful situation of taking a blood sample.

- Production metrics such as growth or milk production (dairy cattle) or egg production (laying hens).
- Health indices, including the frequency of injuries, disease, and mortality because stress impairs the immune response.
- Physiologic indices such as heart rate, which is increased by the two catecholamines: the adrenal hormone, epinephrine; and the sympathetic nerve neurotransmitter, norepinephrine.

By measuring the level of stress (or its reciprocal well-being) in animals using a spectrum of measures, it has been possible to demonstrate that in poultry, increased space in cages and the presence of perches improve the well-being of the birds. Examples of stress in domestic animals are discussed below.

Stress in Livestock

Cattle show distinct responses to stress. Painful procedures such as dehorning or branding are accompanied by an increased release of cortisol from the adrenal cortex and increased heart rate. Painful procedures are often associated with kicking behavior. The bedding used for dairy cows affects lesions on the legs, with compost packs being superior to sand beds or waterbeds, which were in turn superior to rubber-filled mattresses.

Cattle in commercial packing plants exhibit vocalization in response to pain/stress from the effects of such events as electric prods, slipping, sharp edges on equipment, or excessive pressure from a restraint device. Similarly, the load noise results in a short-term increase in cortisol in pigs.

PROFILE

Temple Grandin was educated at Franklin Pierce College (B.A.), Arizona State University (M.S. in animal science), and the University of Illinois (Ph.D. in animal science). She has written and spoken widely on the subject of autism, a subject about which she speaks from firsthand knowledge. Her research and knowledge of livestock behavior have led to improvements in animal-handling systems (curved chutes and race systems) in, for instance, processing plants with reductions, and even elimination, of fear and/or pain. Examples for her concepts, quoting her work, include the following:

- "Reducing stress during handling improves productivity."
- "In areas where animals are handled, illumination should be uniform and diffuse. Shadows and bright spots should be minimized."
- "Cattle and sheep are more sensitive than people to high frequency noises."
- "The sound of hanging metal can cause balking and agitation. Rubber stops on gates and squeeze chutes will help reduce noise."
- "An important concept of livestock handling is flight zone. The flight zone is the animal's 'personal space.' When a person enters the flight zone the animals will move away."
- "If an animal perceives a handling procedure or contact with a person as a threat, stress may increase."

Stress in Horses

In horses, the concentration of cortisol increases within minutes of a young or middle-aged horse being placed on a treadmill for an exercise test and stays high for at least an hour. No such response is observed in aged horses.

Stress in Dogs

When dogs are placed into animal shelters, they appear to be stressed based on the increase in release of cortisol from the adrenal cortex in the first few days. Interaction with people can reduce the stress response. There is also evidence not only that the novel environment is stressful but also that dogs perceive loud noises as both physical and psychologic stresses. It would seem reasonable to design kennels to reduce stress in dogs. Behavioral indicators of stress include auto-grooming, paw lifting, and vocalizing (barking), while those of well-being include tail wagging and higher posture.

Temperament

Temperament can be evaluated in domestic animals. In cattle, temperament is evaluated in a chute being scored between quiet and calm, and the opposite. Interestingly, cattle that were scored as more tranquil grew faster than those that were more agitated. This is consistent with the growth-impairing effects of the stress hormone cortisol not being as pronounced in more tranquil cattle.

 ## ANIMAL WELFARE

Producers are increasingly holding themselves accountable to consumers. Examples include the Pork Quality Assurance Plus assuring pork safety, swine welfare, and good production practices, and the United Egg Producer's poultry welfare standards. Third-party welfare assurance has been developed to audit production facilities to ensure animal welfare. This focus on animal welfare follows campaigns by activists and standards enunciated by major users of animal products, such as Burger King, Kentucky Fried Chicken Corporation, and McDonalds.

Burger King is increasing its purchase of cage-free eggs and stall-free pork. In addition, it is giving preference to processors that use controlled atmosphere stunning. Kentucky Fried Chicken Corporation, with more than 5,000 restaurants in the United States, has committed to humanely raised poultry, and it audits to ensure that the standards are being met. The new standards include education and training of poultry personnel, hatchery operation conditions, nutrition and feeding, health care, adequate space, routine inspections, and catching and transportation.

Kentucky Fried Chicken Corporation has requested that the U.S. Department of Agriculture and Department of Labor review the feasibility of gas killing chickens as an alternative to current methods.

TEMPERATURE CONTROL

Domestic animals, including livestock, poultry, and companion animals, are homeotherms. These animals maintain their deep or core body temperatures within narrow ranges or set points (see Table 19-1).

Domestic animals are homeotherms, but they can also be termed *endotherms* because they generate heat. In contrast, there are some animals that are exotherms because they gain heat from the environment. Homeotherms thermoregulate to maintain core body temperatures. The site in the body controlling temperature is in the hypothalamus, a region of the brain.

Homeotherms produce heat because of metabolism, that is, basal and metabolism related to production such as growth, reproduction, and lactation; activity; and digestion.

Homeotherms lose heat due to the latent heat required to form water vapor. This occurs during respiration and from the skin. It is known as *latent* or *insensible heat loss*. In addition, there is sensible heat loss because of conduction, convection, or radiation heat losses to the environment.

Heat Balance

The heat retained or lost (Δ heat) from an animal is the heat balance. This can be seen from the equation: Δ heat retained/lost or heat balance = heat production for maintenance, growth, reproduction, lactation, and activity − evaporative heat loss (respiration) ± heat loss to/gain from the environment from conduction, convection, and radiation.

In a steady-state situation, the heat produced by the body is balanced by the heat lost to the environment. At temperatures above the thermo-neutral zone, mammals or birds lose heat by the following:

- Increasing evaporative heat loss: evaporative heat loss = respiratory heat loss + evaporative heat losses from the skin via water permeability and sweating
- Increased blood flow to the skin (vasodilation)
- Behavioral responses such as wallowing in ponds

At temperatures below the thermo-neutral zone, mammals or birds respond in two overall manners:

1. They increase thermogenesis or the production of heat because of the following:
   - Increased metabolism (due to thyroid hormones) with both shivering and nonshivering thermogenesis

Definitions

Basal metabolic rate is heat production in a thermo-neutral environment at rest and at least 12 hours after the previous meal. Body temperature is measured as deep or core temperature because the temperature at the surface or periphery varies.

Circadian rhythm (*circa diēs* is about 1 day) is an endogenous rhythm (within an animal) with a periodicity of about 24 hours. It is entrained to exactly 24 hours by environmental factors such as the light–dark cycle.

Evaporative heat loss is heat lost because of the evaporation of water. It requires 539 cal to evaporate 1 g of water compared with 100 cal to heat the same amount of water from 0°C to 100°C.

Fever is a temporary increase in the body temperature of a person or animal about the set point range due usually to infectious disease.

Heat stress is when the temperature of an animal goes above the set point because of extremes of heat and the inability of the physiologic thermoregulatory processes to cope fully.

Homeotherms maintain their body temperatures within narrow ranges because of thermoregulation. Mammals and birds are homeotherms. In contrast, reptiles, amphibians, and fish are poikilotherms.

Hypothermia is when the core body temperature decreases below the normal range.

TABLE 19-1 Core body temperature ranges of selected domestic animals and a comparison to humans

SPECIES	CORE BODY TEMPERATURE °F	CORE BODY TEMPERATURE °C
Human	98–99	36–37
Sheep	102–103	39–39.5
Chicken	102–106	39–41.5
Cat or dog	100–102	38–39

Relative humidity is a percentage relative to the maximum humidity at a given temperature.

Thermogenesis is metabolism to increase the body temperature to within the set point range. It can be without shivering (nonshivering thermogenesis) because of increased metabolism, particularly in brown adipose tissue, and shivering thermogenesis by muscle contractions.

The *thermo-neutral zone* is an environmental temperature where a mammal or bird has to produce or dissipate little heat. Thermo-neutral zones in different species vary but may be as low as 15°C and as high as 35°C.

Units of heat include the following:

- A British thermal unit (BTU) is the amount of heat required to increase the temperature of 1 lb of water by 1°F; 1 BTU equals 1055.06 J.
- Calories (or cal) equal the amount of heat required to heat 1 g water by 1°C.

2. They reduce loss of heat by the following:
   - Vasoconstriction in the skin and extremities to greatly reduce blood flow and, therefore, lessen loss of heat
   - Increased insulation by laying down a layer of fat (poorly conductive of heat) in the skin or increase hair growth before the winter
   - Behavioral responses such as increasing the insulation from hair or feathers by ruffling or "puffing up"

Animals are less efficient at temperatures below and above the thermo-neutral zone, as can be seen in Figure 19-1.

Homeotherms show some changes in core body temperature. For example, there is a daily or circadian rhythm of core body temperatures—low at night, higher during the day. In the neonate, the thermoregulatory system is often not fully developed. In many infections, there is a fever with the body temperature increasing to levels above the set point range.

Sources of heat to an animal include sun (see fig. 19-2), litter or bedding, other animals, lights, and mechanical heat from automated auger-type feeders.

Responses to Very High Temperatures (Heat Stress)

High temperatures create significant problems to livestock and companion animals. Animals address the challenges of very high temperatures by reducing activity (virtually all), panting (dogs or poultry), sweating (horses; see "Interesting Factoid" sidebar on page 376), licking their surface (cats), or wallowing in mud (pigs).

Panting

In a number of different species, including poultry and dogs (see fig. 19-3), there is panting at times when there are high environmental temperatures and consequent heat stress. The panting promotes evaporative heat loss. Thermal panting involves increases not only in the rate of breathing (respiration rate) but also in the depth of breathing (tidal volume). The increased ventilation reduces the concentration of carbon dioxide/bicarbonate in the blood and increases the blood pH.

FIGURE 19-1 Effect of environmental temperature on energy usage in livestock and poultry. Note that energy consumed (feed intake) declines as temperatures increase, and the lowest energy for growth (or other production) or best-feed conversion efficiency occurs in the comfort or thermo-neutral zone. Source: Delmar/Cengage Learning.

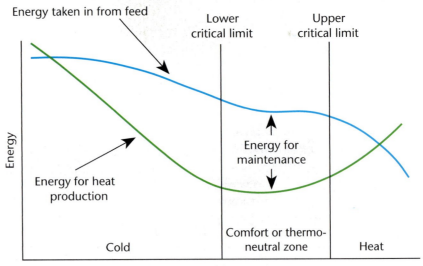

FIGURE 19-2 Heat stress is a major problem for livestock and companion animals. © Beneda Miroslav, 2010. Used under license from Shutterstock.com.

FIGURE 19-3 Dogs and also poultry use panting and, therefore, evaporative cooling to reduce body temperatures. Reproduced by permission from Sonya Etchison. © 2010 by Shutterstock.com.

- Kilocalorie (or kcal) equals 1,000 cal.
- kcal equals 4.187 kJ (kilojoules) (International System of Units).
- Rate of heating kcal/hour: 1 kcal/hour equals 1.16 W (watts).

Windchill is the combined effect of wind and low temperatures.

Respiratory evaporative heat loss = respiratory minute volume × latent heat of vaporization of water × (g water per liter of expired air − [relative humidity of inspired air × g water per liter of air saturated at ambient environmental temperature]).

As the temperature increases, more and more water can be carried in air. Therefore, as the temperature gets close to body temperature and/or humidity close to 100%, there is a problem because respiratory heat loss tends toward zero, and cooling does not occur. Without cooling, heat stress can occur. This can lead to loss of appetite, decrease in or even arrest of growth or reproduction, or even death.

Water

It is essential that domestic animals have access to clean and cool drinking water. This becomes progressively more important as the temperatures increase.

Shelter

It is important to provide shelter for domestic animals when maintained outside. This can include a sunshade to protect animals from intense sunlight and high temperatures. Temperatures in the shade are quite a bit lower than those in the sun.

Ventilation

Animals need ventilation. Without adequate ventilation, there is a build up of noxious chemicals such as ammonia in the air. Moreover, ventilation removes humidity from the air around the animal. At higher temperatures, animals reduce their temperature by the process of evaporation of water. This occurs regardless of whether the animal cools itself by panting as with dogs and poultry, sweating in horses or people, or through the skin, by wetting it by licking or wallowing in mud. At high humidities, evaporation decreases with there being no net evaporation at 100% humidity. Ventilation is important at all temperatures but critically important at high environmental temperatures.

Environmental Problems of Extremely Low Temperatures

Livestock on the range and outdoor companion animals can experience extremely low temperatures. This can lead to reduced performance in livestock, frostbite, hypothermia, and even death.

Windchill

The combined effect of wind and low temperatures is called *windchill*. The U.S. Government (National Oceanic and Atmospheric Administration)'s National Weather Service revised the calculation of windchill. A summary is shown in Table 19-2.

It is readily apparent that as temperatures decrease, windchill has a progressively larger effect. Windchill has two deleterious effects. It results in frostbite, which is the freezing of body tissues, and a tremendous wasteful loss of the dietary energy for the maintenance of body temperature. Providing windbreaks for range animals, shelter for dogs and horses, or complete confinement prevents some or all of the problems associated with low temperature and windchill.

TABLE 19-2 Windchill as calculated by the National Weather Service

WIND SPEED	EFFECTIVE TEMPERATURE IN °F (°C)					
0 (calm)[a]	40 (4)	30 (−1)	20 (−7)	10 (−12)	0 (−18)	−10 (−23)
10 mph (16 km/h)	34 (1)	21 (−6)	9 (−13)	−4 (−20)	−16 (−27)	−28 (−33)
20 mph (32 km/h)	30 (−1)	17 (−8)	4 (−16)	−9 (−23)	−22 (−30)	−35 (−37)
30 mph (48 km/h)	28 (−2)	15 (−9)	1 (−17)	−12 (−24)	−26 (−32)	−39 (−39)

Note: Temperatures shown in red result in frostbite in people with as little as 30 minutes of exposure.
[a]*Effective temperature and measured temperature in calm conditions are the same.*

Water

For animals maintained outdoors, provision must be made for a water supply that does not freeze.

Shelter

Shelter is equally important for animals exposed to cold weather. Windbreaks developed from stands of trees can reduce the problem of windchill. Placing animals in open feedlots such that they can congregate together reduces temperature loss.

Confinement

When animals are reared in confinement facilities with environmental control and ventilation (see fig. 19-4), they are not subjected to the extremes of temperature.

FIGURE 19-4 Ventilation on a modern hog farm. Courtesy of the U.S. Department of Agriculture. Photo by Ken Hammond.

REVIEW QUESTIONS

1. Give examples of aggressive behavior.

2. What is a social hierarchy or peck order?

3. Give examples of reproductive behavior.

4. How do animals communicate?

5. How can stress be measured or assessed?

6. What is the relationship between the hormone cortisol and stress?

7. Do dogs get stressed? If yes, how?

8. What is temperament?

9. Give examples of welfare requirements being sought by the food industry.

10. What is a kilocalorie?

11. Why do homeotherms generate heat?

12. How do homeotherms generate heat?

13. What is sensible and latent/insensible heat loss?

14. What is evaporative heat loss?

15. What is thermogenesis?

16. What are vasoconstriction and vasodilation, and why are they important to thermoregulation?

17. What is the circadian rhythm of body temperature?

18. Are there physiological problems with panting in animals?

19. Is the humidity of the air a problem for animals or people? If yes, why?

20. Why are extremes of temperature detrimental to animals?

21. What is windchill?

22. Why is provision of cool but not frozen water important?

23. Why is shelter important for animals exposed to cold weather?

24. What is the advantage of windbreaks?

25. What are the advantages of confinement facilities related to the extremes of temperature?

REFERENCES AND FURTHER READING

Bamberger, M., & Houpt, K. A. (2006). Signalment factors, comorbidity, and trends in behavior diagnoses in dogs: 1,644 cases (1991–2001). *Journal of the American Veterinary Medical Association*, *229*, 1591–1601.

Bamberger, M., & Houpt, K. A. (2006). Signalment factors, comorbidity, and trends in behavior diagnoses in cats: 736 cases (1991–2001). *Journal of the American Veterinary Medical Association*, *229*, 1602–1606.

Barnett, J. L., & Hemsworth, P. H. (2003). Science and its application in assessing the welfare of laying hens in the egg industry. *Australian Veterinary Journal*, *81*, 615–624.

Beerda, B., Schilder, M. B., van Hooff, J. A., de Vries, H. W., & Mol, J. A. (1999). Chronic stress in dogs subjected to social and spatial restriction. I. Behavioral responses. *Physiology & Behavior, 66*, 233–242.

Centers for Disease Control. *Cryptosporidiosis*. Retrieved October 29, 2009, from http://www.dpd.cdc.gov/dpdx/HTML/Cryptosporidiosis.htm

Dennis, R. L., Muir, W. M., & Cheng, H. W. (2006). Effects of raclopride on aggression and stress in diversely selected chicken lines. *Behavioural Brain Research*, *175*, 104–111.

Grandin, T. (2005). Well-being and handling. In W. G. Pond & A. W. Bell (Eds.), *Encyclopedia of animal science* (pp. 877–879). New York: Marcel Dekker.

Hennessy, M. B., Davis, H. N., Williams, M. T., Mellott, C., & Douglas, C. W. (1997). Plasma cortisol levels of dogs at a county animal shelter. *Physiology & Behavior, 62*, 485–490.

Houpt, K. A. (2005). Well-being assessment: Physiological criteria. In W. G. Pond & A. W. Bell (Eds.), *Encyclopedia of animal science* (pp. 887–889). New York: Marcel Dekker.

Jensen, P., Keeling, L., Schultz, K., Andersson, L., Mormede, P., Brandstrom, H., et al. (2005). Feather pecking in chickens is genetically related to behavioral and developmental traits. *Physiology & Behavior*, *86*, 52–60.

Keeling, L., Andersson, L., Schultz, K. E., Kerje, S., Fredriksson, R., Carlborg, O., et al. (2004). Chicken genomics: Feather-pecking and victim pigmentation. *Nature*, *431*, 645–646.

Kim, H. H., Yeon, S. C., Houpt, K. A., Lee, H. C., Chang, H. H., & Lee, H. J. (2005). Acoustic feature of barks of ovariohysterectomized and intact German Shepherd bitches. *The Journal of Veterinary Medical Science/the Japanese Society of Veterinary Science*, *67*, 281–285.

Kim, H. H., Yeon, S. C., Houpt, K. A., Lee, H. C., Chang, H. H., & Lee, H. (2006). Effects of ovariohysterectomy on reactivity in German Shepherd dogs. *Veterinary Journal (London, England: 1997)*, *172*, 15415–15419.

Lay, D. C., Jr., Friend, T. H., Bowers, C. L., Grissom, K. K., & Jenkins, O. C. (1992). A comparative physiological and behavioral study of freeze and hot-iron branding using dairy cows. *Journal of Animal Science*, *70*, 1121–1125.

Lay, D. C., Jr., Randel, R. D., Friend, T. H., Jenkins, O. C., Neuendorff, D. A., Bushong, D. M., et al. (1997). Effects of prenatal stress on suckling calves. *Journal of Animal Science*, *75*, 3143–3151.

Malinowski, K., Shock, E. J., Rochelle, P., Kearns, C. F., Guirnalda, P. D., & McKeever, K. H. (2006). Plasma beta-endorphin, cortisol and immune responses to acute exercise are altered by age and exercise training in horses. *Equine Veterinary Journal Supplement*, *36*, 267–273.

McClone, J. J. (2005). Well-being assessment: Concepts and definitions. In W. G. Pond & A. W. Bell (Eds.), *Encyclopedia of animal science* (pp. 883–886). New York: Marcel Dekker.

Millman, S. T., Duncan, I. J., & Widowski, T. M. (2000). Male broiler breeder fowl display high levels of aggression towards females. *Poultry Science*, *79*, 1233–1241.

Ratey, J. J., Grandin, T., & Miller, A. (1992). Defense behavior and coping in an autistic savant: The story of Temple Grandin, PhD. *Psychiatry*, *55*, 382–391.

Shea, M. M., Mench, J. A., & Thomas, O. P. (1990). The effect of dietary tryptophan on aggressive behavior in developing and mature broiler breeder males. *Poultry Science*, *69*, 1664–1669.

Swanson, J. C., & Rasette, M. (2005). Well-being assessment: Behavioral indicators. In W. G. Pond & A. W. Bell (Eds.), *Encyclopedia of animal science* (pp. 880–882). New York: Marcel Dekker.

Voisinet, B. D., Grandin, T., Tatum, J. D., O'Connor, S. F., & Struthers, J. J. (1997). Feedlot cattle with calm temperaments have higher average daily gains than cattle with excitable temperaments. *Journal of Animal Science, 75*, 892–896.

Whay, H. R., Main, D. C., Green, L. E., Heaven, G., Howell, H., Morgan, M., et al. (2007). Assessment of the behaviour and welfare of laying hens on free-range-units. *The Veterinary Record*, *161*, 119–128.

Wohlt, J. E., Allyn, M. E., Zajac, P. K., & Katz, L. S. (1994). Cortisol increases in plasma of Holstein heifer calves from handling and method of electrical dehorning. *Journal of Dairy Science*, *77*, 3725–3729.

Web Sites on Temple Grandin

American Registry of Professional Animal Scientists. (2002). Behavioral principles of livestock handling. (With 1999 and 2002 updates on vision, hearing, and handling methods in cattle and pigs.) Retrieved July 22, 2009, from http://lamar.colostate.edu/~grandin/references/new.corral.html

Grandin, T. (N.d.). *Recommended basic livestock handling. Safety tips for workers.* Retrieved July 22, 2009, from http://lamar.colostate.edu/~grandin/behaviour/principles/principles.html

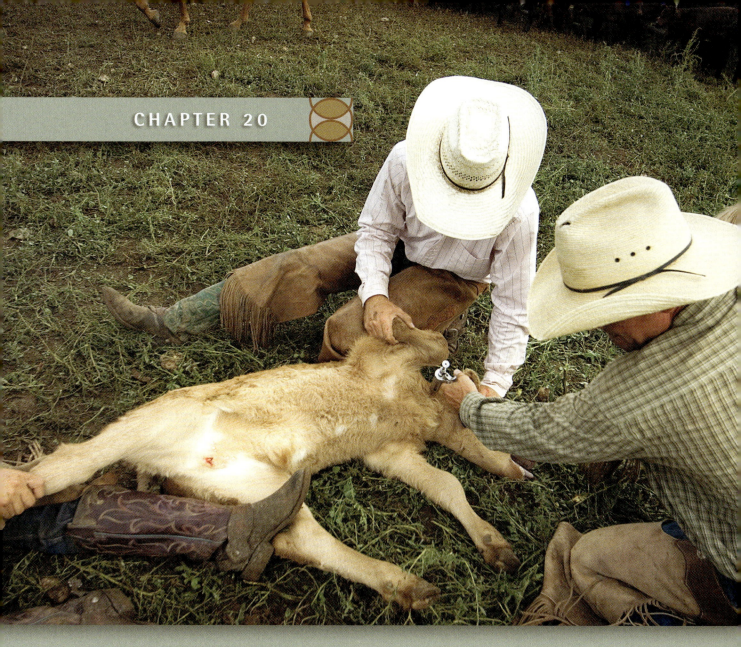

CHAPTER 20

Animal Diseases

OBJECTIVES

This chapter will consider the following:

- Animal disease, including discussion of animal diseases that are transmitted to people (zoonoses)
- Transmissible diseases caused by prions, and infectious diseases caused by viruses and bacteria
- Effects of parasites, including protozoa, flukes, roundworms, tapeworms, and external parasites
- Noninfectious diseases, including metabolic diseases

Definitions

Contagious diseases are easily spread from one animal/person to another.

Infectious diseases are those with an infectious agent or pathogen.

Morbidity is the number of sick animals (overt clinical infections).

Mortality is the number of animal deaths.

Parasites live in close proximity to an animal and receive their nutrition from the host.

Pathogens are organisms that cause infectious disease. They include viruses, bacteria, or protozoa.

Prophylactic treatments are administered to prevent a disease rather than to treat the disease. Examples include drugs and vaccines.

Subclinical infection is when an animal is infected, but there is no overt sickness/ symptom. Subclinical infections can reduce production efficiency.

Therapeutic drugs are used to fight an active disease.

Vector is an organism that facilitates movement of a pathogen or a parasite from one host to another. Examples include fleas with the dog tapeworm, pork with *Trichinella spiralis*, mosquitos with malaria, deer ticks with Lyme disease, blood-sucking bugs with *Trypanosoma cruzi* causing Chagas' disease, and tsetse flies with *Trypanosoma brucei* causing sleeping sickness in people and nagana in cattle.

Zoonosis (plural *zoonoses*) is an infectious disease spread from animals to people.

OVERVIEW OF ANIMAL DISEASE

Animal diseases are a threat to humans' well-being. Infectious animal diseases can cross species barriers and cause serious diseases in people. Such animal diseases are called *zoonotic diseases*. Animal diseases reduce productivity of livestock production by animal mortality (i.e., a dead animal does not produce a product) and by much reduced productivity, including depressed growth rates, lower reproductive efficiency, and increased spontaneous abortion. Animal diseases kill or incapacitate valuable and/or much-loved horses and companion animals.

Animal diseases are caused by infectious agents such as viruses and bacteria, or transmissible agents such as prions and/or may be noninfectious diseases. In addition, parasites such as protozoa, flukes, roundworms, tapeworms, and external parasites reduce productivity of livestock, impair performance of horses, and in general reduce the health of animals, rendering them susceptible to infections.

IMPACT OF ANIMAL DISEASE ON LIVESTOCK

There is still a tremendous impact of animal diseases in North America, Europe, Africa, Asia, and Latin America. This is despite an armory of mechanisms to combat animal diseases, including good management, sanitation, biosecurity, veterinary care, monitoring and depopulation in the event of an outbreak, vaccines, prophylactic drugs, and therapeutic drugs.

Examples of the economic impact of animal diseases based on government estimates include the following:

- Porcine reproductive and respiratory syndrome, costing U.S. pork production $560 million per year.
- The last major foot-and-mouth disease outbreaks costing the British economy about $17 billion and the outbreak in Taiwan costing $1.3 billion.
- Mastitis impacting dairy production by an average of over $100 per cow in the United States.
- Because of tsetse flies and trypanosomes, cattle production is not possible, or its efficiency is severely impacted, in Sub-Saharan Africa.
- Newcastle disease in nonvaccinated poultry has wiped out poultry in whole villages in parts of Africa.
- Animal diseases impact commerce and the price of animal products with the exporting of livestock or animal products restricted when there is a disease outbreak. For instance, the identification of bovine spongiform encephalopathy (BSE)–infected cattle in the United States led to the closing of major export markets.

ZOONOSES (IMPACT OF ANIMAL DISEASE ON PEOPLE)

A zoonosis is an infectious disease spread from animals to people directly through a bite, by aerosol, in food, or through a vector such as a mosquito. Examples of zoonoses are the following:

- Viral diseases:
  - Rabies

- Avian influenza
- Equine encephalitis virus
- Bacterial diseases:
  - Psittacosis (parrot fever) caused by *Chlamydophila psittaci* and spread from birds.
  - Bubonic plague caused by *Yersinia pestis* and spread from rats to people by fleas. It caused the Black Death around 1350, killing about a third of the population of Europe.
  - Salmonellosis caused by salmonella bacteria that infect many species of animals.
  - Anthrax caused by *Bacillus anthracis*, infecting, for instance, cattle.
- The prion-based disease variant Creutzfeldt-Jakob disease (vCJD), which is discussed in detail in the BSE section below.
- Trichinosis caused by the larval stage of the roundworm—*T. spiralis*.
- Cryptosporidiosis is an emerging zoonotic protozoan disease is caused by *Cryptosporidium parvum*. This chronic diarrheal disease is highly infectious. It can be transmitted from person to person, and from the feces of dogs and cats. It is a leading cause of death in AIDS patients. Cattle form a reservoir for *C. parvum*. The organism is resistant to inactivation in the environment, and there is no good pharmaceutical therapy.

INFECTIOUS AND TRANSMISSIBLE DISEASES

Infectious diseases will be considered below under the types of organisms that cause them from viruses to bacteria. Prion diseases or transmissible spongiform encephalopathies will also be considered in this section. Parasitic diseases will be covered in a separate section.

Prions

Prions are infectious agents made of protein. They cause the abnormal folding of native amyloid proteins. This often occurs in the nervous system to cause progressive degenerative brain disease. In 1997, the Nobel Prize was awarded to Stanley Prusiner for his discovery of the prions. Transmissible spongiform encephalopathies are prion-caused diseases. Examples include the following:

- BSE or mad cow disease in cattle
- Scrapie in sheep
- Chronic wasting disease in deer and elk
- Kuru in humans, which is found in New Guinea and thought to be associated with cannibalistic funeral rites
- Transmissible mink encephalopathy
- Transmissible spongiform encephalopathy in cats
- Creutzfeldt-Jakob disease (CJD) in humans
- Variant Creutzfeldt-Jacob Disease (vCJD) in humans

Bovine Spongiform Encephalopathy (BSE)

BSE or mad cow disease is a progressive fatal brain disease of cattle with neural degeneration. It was first observed in the United Kingdom, where 178,000 cattle were infected (in 35,000 herds) in the late 1990s and early 2000s. It spread to Western Europe and Canada. By 2007, there had been 11 cases of BSE in Canada and two cases in the United States. The first case in the United States was reported in December 2003 in Washington state, which was in a cow brought from Canada.

It is thought that the outbreak of BSE in Europe was spread by feeding ruminant meat and bonemeal to calves. This meat and bonemeal would contain BSE prions in nervous tissue, for example, brain and spinal cord from infected cattle were one of the ingredients of the meat and bonemeal. It is not known how BSE originated. It is speculated that a form of the infectious agent (the prions) for scrapie in sheep may have led to the formation of the BSE prions. BSE is not transmitted via the milk.

Creutzfeldt-Jacob Disease (CJD) is a rare but serious disease of people. There is progressive dementia, loss of memory, and balance and coordination together with speech impairment because of the death of brain cells. The incidence of CJD is one case per million of population. In Britain, a variant of CJD was observed with similar effects but a higher incidence. This variant Creutzfeldt-Jacob Disease (vCJD) has an incubation period of 5–20 years. It is thought that vCJD is caused by the same prions that cause BSE and through eating contaminated beef.

The federal Food and Drug Administration concluded that "there is strong epidemiologic and laboratory evidence for a causal association between vCJD and BSE." From 1995 to 2004, there were 147 cases of vCJD in the United Kingdom, seven in France, and one in each of Canada, Ireland, Italy, and the United States. The U.S. government has taken a series of approaches to prevent or reduce the chances of BSE becoming established in the United States, including the following:

- Prohibiting meat from suspect animals (downer cattle) getting in human food
- Banning animal (mammalian) products in ruminant feed
- The U.S. Department of Agriculture (USDA) sampling 40,000 animals per year and testing for BSE
- Stopping importing live animals or animal products from countries with BSE
- Prohibiting the use of the entire carcass of cattle in which there is a higher risk, such as cattle older than 30 months of age or in which the brain and spinal cord have not been effectively removed during processing from animal feed

There is a very low probability of BSE being established in the United States. The National Cattlemen's Beef Association states that the U.S. beef is safe from BSE because of the federal programs and the absence of tissues that potentially harbor BSE prions, such as brain and spinal cord from the meat. BSE prions have not been found in muscle or fat.

 # VIRUSES

Viruses can have either DNA or ribonucleic acid (RNA) as the genetic material. Under the Baltimore system of classification of viruses, they are assigned to one of seven groups based on the form of the DNA or RNA and not based on the diseases they cause: I, double-stranded DNA viruses; II, single-stranded DNA viruses; III, double-stranded RNA viruses; IV, positive-sense single-stranded RNA viruses, including foot-and-mouth virus; V, negative-sense single-stranded RNA viruses, including influenza and rabies; VI, reverse-transcribing diploid single-stranded RNA viruses; and VII, reverse-transcribing circular double-stranded DNA viruses.

Viroids are smaller than viruses and are known to be plant pathogens.

Viral Diseases

Viral diseases of dogs include distemper, rabies, parvovirus, foot-and-mouth disease, and avian influenza. The canine distemper virus causes distemper. In the absence of vaccination, distemper is a major cause in the death of puppies. Symptoms of the disease include fever, loss of appetite, vomiting, diarrhea, coughing, and discharge from the eyes and nose.

Rabies is a viral disease of many mammalian species, including bats, dogs, cats, foxes, raccoons, skunks, and people. It is spread chiefly by animal bites. Rabies is almost always fatal in people and most animals. Rabies vaccination is required legally for dogs and cats in all states in the United States. People who have a high risk of animal bites, such as veterinarians, receive prophylactic vaccination against rabies. If someone is bitten by an animal suspected of having rabies (being rabid), a series of vaccinations are used. Orally active vaccines at baiting stations are used to control rabies in wildlife.

Foot-and-Mouth Disease

Foot-and-mouth disease, also known as hoof-and-mouth, is a highly contagious viral disease of cattle, pigs, and sheep. The disease is caused by a RNA virus, the foot-and-mouth disease virus. North America has been foot-and-mouth disease free since 1954, with the last reported case in the United States in 1929. There was a significant outbreak in the United Kingdom in 2001, with 7 million animals culled. This spread to Western Europe. Vaccines against foot-and-mouth disease virus are only used in countries where foot-and-mouth disease is found because vaccination prevents exports and does not allow surveillance for the presence of the disease.

Avian Influenza

Avian influenza is caused by one of a series of viruses. The disease not only affects poultry but also can infect both game and wild birds. The latter is particularly a problem because they fly around and can spread the disease locally or, for those that migrate, spread the disease into different regions and countries.

Avian influenza viruses are classified on two bases:

1. The degree of pathogenicity, that is, the ability of the virus to cause disease. The virus can have low pathogenicity or high pathogenicity. Infection with a low-pathogenicity avian influenza virus will cause little overt signs of disease. However, the low-pathogenicity avian influenza virus can mutate into a high-pathogenicity virus, with serious consequences.
2. The presence of two glycoproteins on the surface of the virus:
   a. Hemagglutinin proteins (H). There are 16 (H1–H16).
   b. Neuraminidase proteins (N). There are nine (N1–N9).

A particularly serious form of the avian influenza virus is the H5N1 virus because it is highly contagious among birds, and the disease is frequently fatal to them. Moreover, there are reports of infection by people with the H5N1 virus. At the moment, the infection is bird (either live or during food preparation, or from contaminated surfaces in a poultry facility) to person, and not person to person.

Influenza in People and Domestic Animals

In people, pathogenic viruses cause influenza. It is thought that avian influenza viruses mutate into forms that are pathogenic to people and/or domestic animals. Avian influenza type A viruses may be transmitted from animals to humans in two main ways: directly from birds (live or dressed) from avian virus–contaminated environments to people, or through an intermediate host, such as a pig. Transfer of the virus to pigs and horses may be from people to domestic animals.

The human influenza viruses can be categorized as A, B, and C types. The type A viruses are further classified by subtype based on surface proteins—hemagglutinin (H) and neuraminidase (N). Subtypes of the virus that have caused influenza in mammals include H3N2, H2N2, H1N1, and H1N2 in humans; H1N1 and H3N2 in pigs; H7N7 and H3N8 in horses; and H3N8 in dogs.

PANDEMIC: 675,000 DEAD IN THE UNITED STATES

The influenza pandemic of 1918 killed between 25 and 50 million people in the world, and 675,000 in the United States alone. The high virulence of the virus among the infected led to mortality rates in the United States of over 2.5%. Other influenza epidemics had death rates of less than 0.1%. Moreover, unlike other influenza epidemics that disproportionately affect the very young and the elderly, the 1918 pandemic had a high death rate in young adults (15–34 years old).

This influenza pandemic was called the Spanish flu because the first reports were from Spain. The reason for the first reports was that there was not the censorship in Spain in contrast to the countries fighting World War I.

What caused the pandemic? The infectious agent was an unusually deadly influenza A virus strain of subtype H1N1. What do we know about the virus? The RNA of the 1918 influenza A virus was sequenced both from formalin-fixed lung autopsy materials and unfixed lung tissues from an Alaskan influenza victim buried in the permafrost. The 1918 virus gene sequences are much more similar to avian H1N1 viruses than other mammalian H1N1 viruses. Therefore, it seems that the 1918 influenza pandemic was a form of avian influenza.

The virus has been reconstituted in a high-containment biosafety level 3 enhanced laboratory (see figs. 20-1 and 20-2). The 1918 virus genes conferred the unique high-virulence phenotype in three models: live mice with the end-point death; embryonated chicken eggs with the end point being death of the embryo (see fig. 20-3); and human bronchial epithelial cells with the end-point viral growth.

There are already a number of deaths of people infected with the avian H1N1 virus due to contact with the infected birds. As yet, no virus with the ability to be easily passed from person to person and cause an epidemic has been found. If mutations in an avian H1N1 virus led to a high-virulence virus affecting people with the virus passed from person to person in 1918, could it happen again?

Should we worry? The United Nations World Health Organization is following avian influenza outbreaks and sequencing the RNA of the viruses. With national governments and the Food and Agriculture, there are surveillance programs to monitor avian influenza in poultry, wild birds, and in people working with poultry. Avian influenza is spread by wild birds and by movement of poultry. A critically important prevention of avian influenza is biosecurity of poultry facilities.

Control measures include culling of infected poultry, sanitation of poultry facilities, and preventing the poultry products from being consumed by the public. As yet, poultry are not vaccinated against avian influenza. There are stockpiles of vaccines against avian H1N1 viruses and antiviral therapeutics to be released in the event of a human outbreak. In the United States, the Centers for Disease Control and the USDA are working together, and with state public health and agriculture agencies with programs of surveillance and planned responses in the event of an outbreak.

Novel 2009 H1N1 Influenza (So-Called Swine Flu)

The first case of a new influenza in people was identified in the United States in April 2009. According to the Centers for Disease Control (CDC), "the virus was originally referred to as 'swine flu' because laboratory testing showed that many of the genes in this new virus were very similar to influenza viruses that normally occur in pigs (swine) in North America." It is thought that the disease originated in pigs. The virus evolved around September 2008, and the first cases in people were seen in Mexico City early in 2009. Since then, the disease has spread throughout the world and was named a pandemic by the World Health Organization (WHO). The new influenza A virus is distinct from seasonal flu and is referred to as novel H1N1, pandemic H1N1, or 2009 H1N1:

- 2009 refers to the year of its first discovery of the subtype of the virus.
- H is for the type of a specific viral protein, haemagglutinin, that allows the virus to bind to the cell that it will then attempt to enter.
- N is for the type of the enzyme, neuraminidase, that breaks down mucus to allow the virus to get to the host cell.

Definitions

Antiviral drugs such as oseltamivir (Tamiflu) and zanamivir (Relenza) inhibit the viral enzyme neuraminidase.

Epidemic (or *epidemic disease*) is an infectious disease that spreads rapidly causing many people (or animals) to become sick or ill; showing obvious clinical symptoms of the disease (such as fever).

Epidemiology is the science of the causes, spread and control of diseases.

A *pandemic* is an epidemic effecting a very large number (or proportion) of people and occurring across the world.

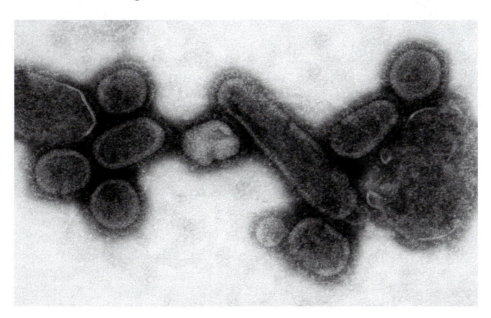

FIGURE 20-1 Transmission electron micrograph showing a recreated 1918 influenza virus. Courtesy of Cynthia Goldsmith and the Centers for Disease Control.

FIGURE 20-2 Colorized transmission electron micrograph of avian influenza A H5N1 viruses (*gold*) grown in Madin-Darby canine kidney cells (*green*). Courtesy of the Centers for Disease Control/ Cynthia Goldsmith, Jacqueline Katz, and Sherif R. Zaki.

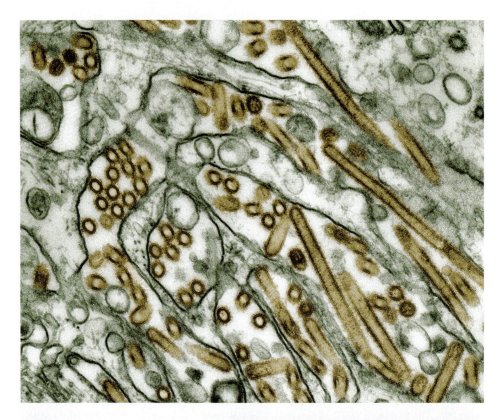

FIGURE 20-3 Photograph of Dr. Taronna Maines inoculating 10-day-old embryonated hens' eggs with an H5N1 avian influenza virus. Courtesy of the Centers for Disease Control/ Greg Knobloch.

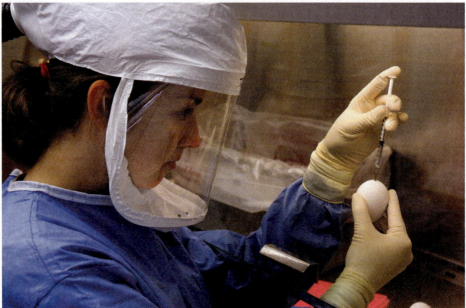

The effects of the 2009 H1N1 virus range from mild to severe. Most people recover without medical treatment but hospitalizations and deaths occur in the minority of cases. This is predominantly observed in people at particular risk with seasonal flu such as pregnancy, diabetes, heart disease, asthma, and kidney disease. However, unlike seasonal flu, there does not appear to be any greater vulnerability to people over 65 years of age and children younger than 5 years of age. Interestingly, people exposed to the 1918 influenza virus have antibodies to the novel H1N1 virus.

According to the World Health Organization, the new influenza A (subtype H1N1) virus "has never before circulated among humans and is very different from what normally circulates in North American pigs." When two viruses infect the same cell simultaneously, there can be exchange of genes during replication with North American and Eurasian swine lineages.

It is thought that the 2009 H1N1 virus is derived by reassortment of the genomes from viruses endemic in pigs (two separate lineages), birds, and people; at least one of them a derivative of the 1918 human virus. The genomes of the 1918 human and swine H1N1 are descended from avian influenza A virus.

BACTERIA

Bacteria are a simple microscopic form of life. They can produce all the proteins needed for life. They use DNA as their genetic material. They are called *prokaryotes* because they do not contain a nucleus and membrane-bound organelles. In contrast, eukaryotes (animals, plants, fungi, and protozoa) do contain a nucleus or cellular organelles.

Bacterial Diseases

Bacterial diseases include mastitis, the secondary infection of shipping fever in cattle, *Borrelia burgdorferi*, causing Lyme disease in dogs, cats, and people, and bubonic plague. Many food-borne diseases are caused by bacteria such as *Campylobacter jejuni* and *Campylobacter coli*, *Escherichia coli* O157:H7 (*E. coli* O157:H7), salmonellosis caused by *Salmonella* bacteria (see fig. 20-4), and *Listeria*.

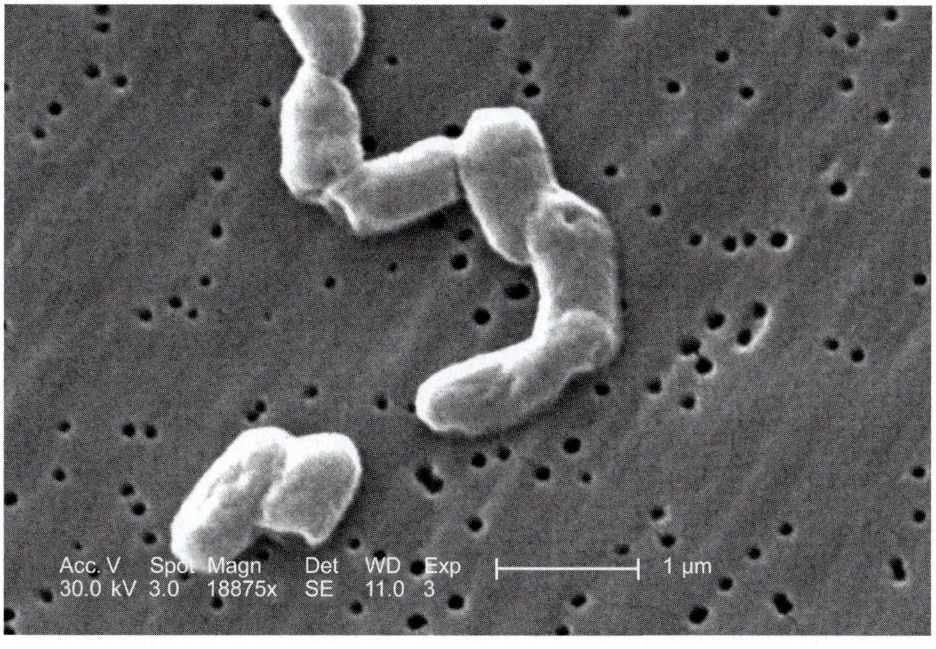

CLASSIFICATION

The classification of bacteria is domain, Bacteria; and other domains, Eukarya (containing the Animalia, Fungi, Plantae, and Protista kingdoms) and Archaea.

FIGURE 20-4 Scanning electron micrograph of rod-shaped Gram-negative *Salmonella* bacteria. *Salmonella* is an important disease in domestic animals and is a food-borne disease for people (original magnification × 18875). Courtesy of the Centers for Disease Control/Janice Carr.

Acc.V Spot Magn   Det WD Exp
30.0 kV 3.0 18875x   SE   11.0 3   1 μm

TABLE 20-1 Incidence of significant health problems/diseases in U.S. dairy cattle in 2001

HEALTH PROBLEMS/DISEASES	CATTLE AFFECTED (%)
Mastitis	14.7
Lameness	11.6
Retained fetal membranes	7.8
Milk fever	5.2
Abomasal displacement	3.5
Heifers dying before weaning	8.7

Source: Based on the National Animal Health Monitoring System.

Mastitis

Mastitis is the number one disease of dairy cattle (see Table 20-1). It results in much-reduced milk production, the loss of milk not saleable with a high concentration of somatic cells (leukocytes), and long-term damage to the mammary gland. It is caused by the invasion of the mammary gland by pathogens, including various species of *Streptococcus*, *Staphylococcus*, and *Mycoplasma* bacteria.

The innate immune response (discussed in Chapter 21) plays a significant role in how the mammary gland deals with pathogenic bacteria. The teat canal is a barrier preventing pathogens from entering the mammary gland. Between milking and during the dry period, the teat canal is sealed by keratin, which is a plug. This is derived from the stratified epithelial lining of the canal. Inflammation is part of the innate immune, and mastitis is inflammation of the mammary gland.

Mastitis can be either of the following: clinical or overt (readily seen by observation), or subclinical.

PARASITES AND DISEASE

A parasite is an organism that lives in close proximity to an animal and completely depends on it. Parasites receive their nutrition from the host, using the host's blood or absorbing nutrients in the host's intestine. Although the parasites rarely kill the host, they do inflict significant harm. Parasites adversely affect the quality of life for companion animals and livestock; they also reduce production efficiency of livestock and may result in the death of an animal.

Internal Parasites (Endoparasites)

Internal parasites include protozoa, roundworms or nematodes, flatworms or trematodes, cestodes or tapeworms, and some flies such as a botfly.

PROTOZOA

Protozoa are single-celled eukaryotes (see the "Classification" sidebar at left) with a nucleus and intracellular organelles. They impact the livestock and companion animals as parasites, as zoonotic diseases, and by symbiotic ciliates participating in the fermentation in the rumen.

CLASSIFICATION

The classification of protozoa is domain, Eukarya (which is in contrast to the other domains Archaea and Bacteria); and kingdoms, Protista (Protozoa) together with Animalia, Plantae, and Fungi.

Trypanosomes

Trypanosomes are protozoa with flagella that can cause serious diseases in animals and people. Examples include the following:

- The trypanosome *T. brucei* causes African trypanosomiasis (or sleeping sickness) in humans and nagana in cattle in Africa. The vector is the tsetse fly. The presence of the tsetse belt effectively closes some 10 million km^2 in Africa to efficient cattle production.
- *T. cruzi* causes Chagas' disease in Central and South America. Its vector is the blood-sucking bug *Rhodnius prolixus*.
- *Trypanosoma equiperdum* are sexually transmitted in horses.

Coccidia

Coccidia are intracellular parasites of the intestinal cells. The following are examples of coccidian and coccidial diseases

- *Cryptosporidium* is a widespread zoonosis (see fig. 20-5).
- Coccidiosis in dogs and cats is due to coccidia of the genus *Isospora*.
- Toxoplasmosis is due to ingestion of protozoan parasites, *Toxoplasma gondii*, in raw or undercooked meat, or through contact with an infected cat.
- Bovine coccidiosis.
- Coccidiosis in poultry due to *Eimeria tenella* (see fig. 20-6). Without coccidiostats, production of poultry would be severely impacted with significant mortality (deaths), morbidity (sickness), and reduced growth and poorer feed conversion efficiency. A number of coccidiostats are available.

Other Protozoa

Giardia intestinalis is a common parasite of cats. It also infects dogs, cattle, sheep, and people. Protozoan parasites *Theileria annulata* and *Theileria parva* infect cattle

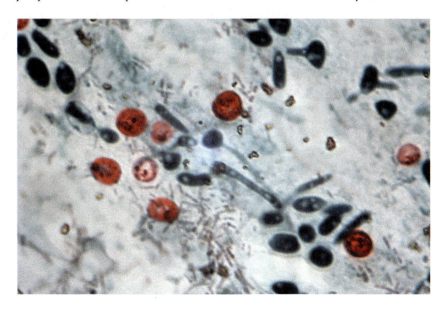

FIGURE 20-5
Cryptosporidium is a zoonotic disease. It can be shed in the feces from young cattle and pass to people when the drinking water is inadequately treated. It causes severe diarrhea, particularly in children. The image shows oocytes, each containing four sporozoites with four stained nuclei. Courtesy of the Centers for Disease Control.

FIGURE 20-6 Coccidiosis is a major disease in poultry. The organism that causes coccidiosis is the protozoan *Eimeria*, and is illustrated in this figure. Courtesy of the USDA.

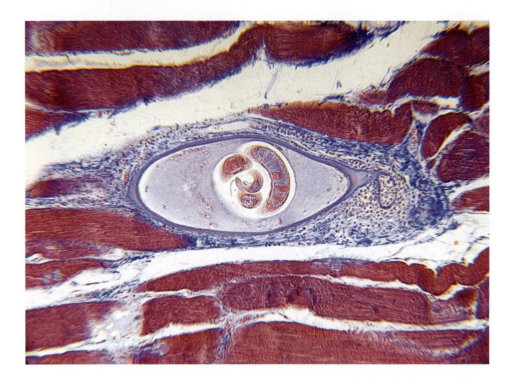

FIGURE 20-6 Coccidiosis is a major disease in poultry. The organism that causes coccidiosis is the protozoan *Eimeria*, and is illustrated in this figure. Courtesy of the USDA.

in sub-Saharan Africa. This greatly impacts the potential development of the industry. Ticks are the vector. Another protozoan parasite is *Histomonas meleagridis*, which causes a disease called *blackhead* in chickens and turkeys.

 ROUNDWORMS OR NEMATODES

CLASSIFICATION
Roundworms are in the Nematoda phylum.

Roundworms are from the phylum Nematoda, and include both parasitic and free-living species.

Examples of roundworm parasites are the following:

- Intestinal roundworms (ascarids):
  - Dogs, with two species of roundworms (*Toxocara canis* and *Toxascaris leonina*)
  - Cats, with two species of roundworms (*Toxocara cati* and *T. leonina*)
  - Pigs, with the large roundworm (*Ascaris suum*)
  - Horses, with the large roundworm in horses (*Parascaris equorum*)
  - Chickens and turkeys (*Ascaridia galli*)
- Strongyles; for example, in horses, there are large strongyles (also known as red worms or bloodworms) and small strongyles.
- Pinworms found in, for instance, horses and people.
- Heartworm (*Dirofilaria immitis*), which is found in dogs and cats.
- *T. spiralis*. Although this is rare in the United States, larval *T. spiralis* present in pig muscle causes trichinosis in people who eat pork. The USDA has a trichinae certification program. See Figure 20-7 for images on muscle with *Trichinella* and an image of organization.

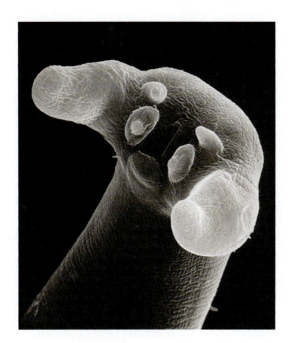

FIGURE 20-7 If a person or pet eats food contaminated with *Trichinella*, the organism can develop in either the muscles or brain. The image shows *T. spiralis* in the muscle of a dog. Courtesy of USDA Agricultural Research Service Biosystematics and U.S. National Parasite Collection.

- Hookworms, which are small nematodes that attach to the gastric mucosa. There are hookworm species that infect cats, dogs, and people, with the potential for anemia with severe infections.

Intestinal Roundworms (Ascarids)

The intestinal roundworms in dogs (*T. canis* and *T. leonina*) are about 3 in (10 cm) long. Large numbers of these give a puppy a "pot-bellied appearance" and a failure to thrive. They may cause diarrhea or vomiting, and, in extreme circumstances, blockage of the intestine. A similar situation exists as the intestinal roundworm of cats (*T. cati* and *T. leonina*). The intestinal large roundworm in horses is *P. equorum*. This can be 10 in long (25 cm) and about a third of 1 in wide, and can be a problem in foals and young horses. Growth of the infected horse is greatly reduced, and there is the potential for intestinal blockage. The female large roundworm produces 200,000 eggs per day that can then infect other horses on the same pasture. It is easy to see how easily roundworms are spread.

Strongyles

The large strongyles in horses are the bloodworms or red worms (*Strongylus vulgaris*, *Strongylus equinus*, and *Strongylus edentatus*). As the name implies, these live in the blood vessels of horses. The adults are red-brown in color due to their consumption of the host's blood. *S. vulgaris* is ¾ in (2 cm) long, whereas *S. equinus* is 2 in (5 cm) long. The life cycle of the bloodworm *S. vulgaris* involves the horse ingesting the larval stage 3 from pasture. This migrates through the intestinal wall to the blood vessels, which develops into larval stage 4, and then throughout the body of the horse. They then migrate to the colon and cecum, where they develop into sexually mature adults (see the life cycle diagram in Figure 20–9). These feed

---

CLASSIFICATION

The classification of ascarids or parasitic nematodes is phylum, Nematoda; class, Secernentea; and order, Ascaridida.

The classification of strongyles or parasitic roundworms is phylum, Nematoda; class, Secernentea; and order, Strongylida (Strongyles).

on "plugs" of intestinal mucosa and the host's blood from the resulting bleeding. The difference in the life cycle for *Strongylus equines* and that of *S. edentatus* is the stage 4 *S. edentatus* larva inhabiting the liver. The bloodworms cause anemia in the infected horse because of damage to the intestinal blood vessels and bleeding, cause loss of weight due to damage to the intestine, and may cause colic.

In horses, there are about 30 species of small strongyles or small bloodworms in the United States from four genera: *Cyathostomum*, *Cylicocyclus*, *Cylicodontophorus*, and *Cylicostephanus*. They are red in color, and between 0.25 and 2.0 in (1.2–5 cm) long. Small strongyles are less harmful than the large strongyles and do not migrate through the body, living in the intestinal wall. The stage 3 larvae are consumed when the horse grazes. Development of the larva occurs in the intestine, with migration toward the large intestine such that the adults are found in the cecum and colon. The small strongyles can cause mild colic and intestinal ulcers (see figs. 20-8 and 20-9).

Heartworm (*D. immitis*)

Adult heartworms are about 7 to 10 in long, with the females being somewhat larger than the males. They are found in the heart and blood vessels in the lung. The females produce several thousand eggs per day. After fertilization, there is the formation of the first larval stage: microfilariae. These can live in the bloodstream of the host for up to 2 years but do not develop further unless they are present in blood ingested by a mosquito. In the mosquito, the microfilariae develop, undergoing molts to produce the third larval stage. This migrates to the head of the mosquito, ready to infect a new host. If this occurs, they develop under the host's skin into a fourth-stage

---

CLASSIFICATION

The classification of heartworms is phylum, Nematoda; class, Secernentea; and order, Spirurida.

FIGURE 20–8 Laval stage of a strongylus stained with Oil Red O to show large amounts of lipid being stored. Kindly provided by Stacey Lettini and Michael Sukdheo at Rutgers, The State University of New Jersey.

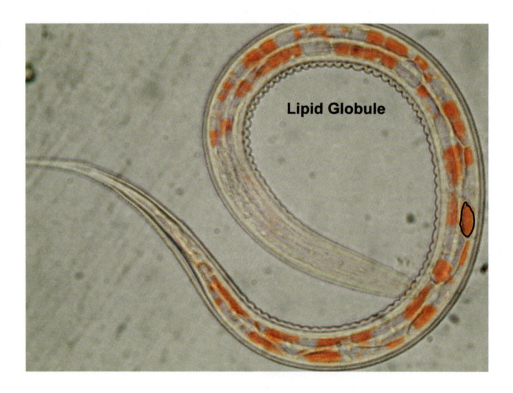

Lipid Globule

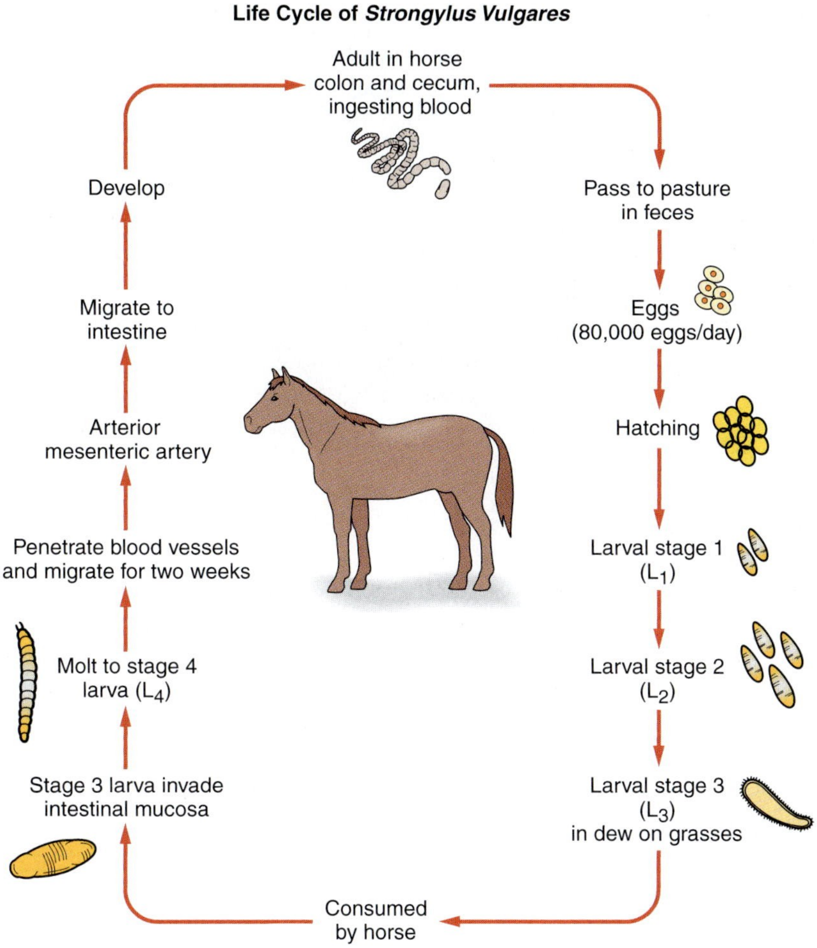

Life Cycle of *Strongylus Vulgares*

Adult in horse colon and cecum, ingesting blood

Develop

Pass to pasture in feces

Migrate to intestine

Eggs (80,000 eggs/day)

Arterior mesenteric artery

Hatching

Penetrate blood vessels and migrate for two weeks

Larval stage 1 (L$_1$)

Molt to stage 4 larva (L$_4$)

Larval stage 2 (L$_2$)

Stage 3 larva invade intestinal mucosa

Larval stage 3 (L$_3$) in dew on grasses

Consumed by horse

FIGURE 20-9 Life cycle of *S. vulgaris.*

larva. These then migrate to muscles and develop/molt into a fifth stage or immature adult. These in turn are carried by the bloodstream to the pulmonary arteries where they grow into the final stage of the life of heartworms: the sexually mature adult. The life cycle of the heartworm is summarized in Figure 20-10.

Heartworms inflict damage on the dog or cat, reducing its ability to exercise, causing loss of weight, and ultimately can kill the host by causing congestive heart failure. The effects are dependent on the number of heartworms, with overt symptoms with 25 heartworms, circulatory problems with over 50, and a life-threatening situation with over 100.

PINWORMS

Pinworms (*Oxyuris equi*) reside in the cecum and colon of horses. They are very thin, but female adults can be up to 6 in (15 cm), whereas the males are only one-half in (1 cm) long. They have a simple life cycle, with horses ingesting the eggs that have been shed in the feces. Eggs are released by the pinworms in a sticky fluid. This causes irritation around the anus, obvious discomfort, and, therefore, a need for rubbing or biting in that area. The latter behavior leads to reinfection.

CLASSIFICATION

The classification of pinworms is phylum, Nematoda; class, Secernentea; and order, Oxyurida.

FIGURE 20-10 Schematic diagram of the life cycle of the heartworm (*D. immitis*). Adult heartworms in the right ventricle and pulmonary artery of the dog or cat (primary host) release large numbers of microfilariae into the bloodstream. These are taken into a mosquito when feeding on the dog's blood. The microfilariae molt to form second-stage and then third-stage larva in the mosquito (secondary host or vector), migrating to the mouth parts of the mosquito. When the mosquito feeds on dog or cat blood, the infective larvae infect the dog or cat. These molt to fourth-stage larva in about 7 days in the tissue and then the adult heartworm (stage 5) in 2–3 months. These then migrate via the bloodstream to the heart, where females grow to about 14 in long and males about 10 in long. After 3–4 months, reproduction occurs, with the male heartworms fertilizing the ova from the females, producing microfilariae.

Infection of dogs or cats with multiple heartworms leads to coughing and reduced ability to/interest in exercise in mild cases, and when severe, shortness of breath, fainting, weight loss, fever, abdominal swelling, and even death. It is critically important to test for heartworm and to administer prophylactics such as ivermectin plus pyrantel.

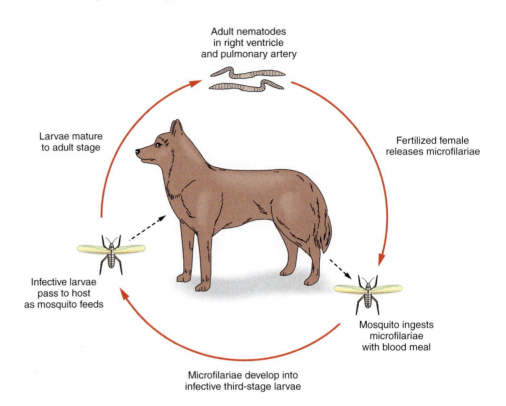

Adult nematodes in right ventricle and pulmonary artery

Larvae mature to adult stage

Fertilized female releases microfilariae

Infective larvae pass to host as mosquito feeds

Mosquito ingests microfilariae with blood meal

Microfilariae develop into infective third-stage larvae

FLATWORMS

An example of a flatworm is the sheep liver fluke *Fasciola hepatica*.

Cestodes (tapeworms)

Tapeworms live in the intestines of their primary host. The scolex or head of the tapeworm embeds itself in the upper small intestine using hooks and suckers. From the head emerges a chain of proglottids or independent segments, each about the size of a grain of rice. Additional proglottids are being continually produced, and, therefore, with time, they move down the gastrointestinal tract to be eliminated with the feces. The tapeworm has no digestive system. The proglottids just absorb nutrients that otherwise would go to the host. When mature, each proglottid will have both a male and female reproductive system, and when a terminal proglottid(s) breaks off the chain, each will contain about 20 eggs.

The common tapeworm of dogs and cats (*Dipylidium caninum*) is up to 25 in (62 cm) long. The intermediary host is the flea, with dogs and cats eating the fleas during grooming. The eggs hatch in the fleas and develop into infectious larvae. Signs that a dog or cat has a tapeworm infestation include the presence of proglottids, white large seed-like structures with a squirming motion, in the feces or around the anus. In addition, a dog may exhibit a scooting behavior if the anus is irritated. The most important pathogenic tapeworms to humans are the pork tapeworm (*Tenia solium*) and beef tapeworm (*Tenia saginata*). In these cases, people are the primary host, and pigs and cattle are, respectively, the intermediary hosts (see fig. 20-11). The beef tapeworm causes cysticercosis in cattle.

CLASSIFICATION

The classification of flatworms is phylum, Platyhelminthes; and class, Trematoda.

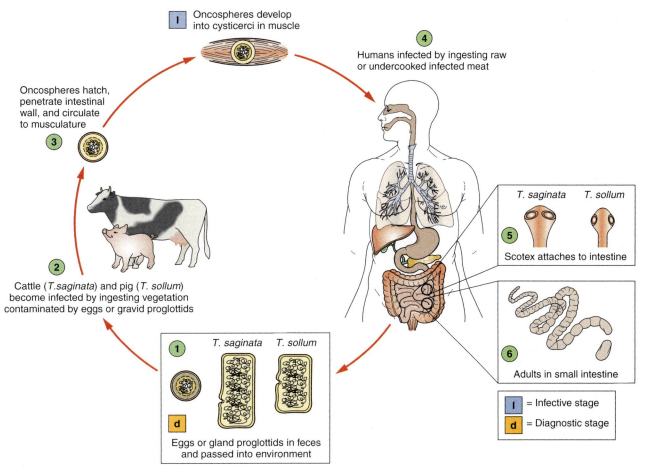

FIGURE 20–11 Life cycle of the beef and pork tapeworms.

Three species of tapeworms infect the horse: *Anoplocephala perfoliata*, which is the most common tapeworm; *Anoplocephala magna*; and *Paranoplocephala mamillana*. Horses with infections are not at their healthiest. Diagnosis is difficult, with subclinical infections causing damage to the gastrointestinal tract and reducing nutrient absorption, but not causing visible disease.

 FLIES

Botflies are true flies with a parasitic stage in their development. They lay their eggs on the skin of a horse. These can then be transferred to the mouth and, therefore, intestines. The larva migrates from the intestines to just under the skin. The bot will then emerge through the skin. The eggs are also infectious to people.

 EXTERNAL PARASITES (ECTOPARASITES)

Ectoparasites can be either insects or arachnids (ticks and mites, bugs, fleas, flies, lice, or mosquitos).

CLASSIFICATION

The classification of tapeworms is phylum, Platyhelminthes; and class, Cestoda.

The classification of flies is phylum, Arthropoda; class, Insecta; and order, Diptera (true flies).

FIGURE 20–12 Stages in the life of cattle ticks. *A,* Young nymph. *B,* Nymph partially engorged with blood. *C,* Nymph fully engorged with blood. *D,* Adult female. *E,* Adult male.

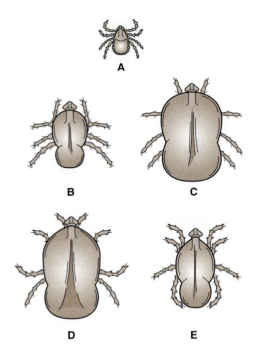

Ticks and Mites

Ticks and mites live on the skin of animals and exist by sucking blood, but some use dead skin. Like spiders, these are arachnids. See Figure 20-12 for examples of the life stages of ticks in cattle.

Other examples of ticks and mites include the following:

- American dog tick (*Dermacentor variabilis*).
- Deer tick (*Ixodes scapularis*), which is the vector for Lyme disease caused by a spirochetal bacteria.
- Lone Star tick (*Amblyomma americanum*).
- The mite, *Sarcoptes scabiei,* infecting dogs, cats, pigs, and people. In dogs, this sarcoptic mite causes mange, also known as sarcoptic mange or canine scabies. Another mite that infects dogs is *Demodex canis.* This may result in skin disease (see fig. 20-13).
- Poultry mites – the Northern Fowl Mite (*Ornithonyssus sylviarum*) and the Chicken Mite (or Red Roost Mite).

INSECTS

Examples of parasitic insects include the following (see fig. 20-14):

- Lice (order, Phthiraptera) (nomenclature, singular louse; a louse egg is a nit). Lice can be either blood sucking or chewers on the skin. An example is poultry lice, Menacanthus stramineus.
- Fleas (order, Siphonaptera), such as the cat flea (*Ctenocephalides felis*) and dog flea (*Ctenocephalides canis*).
- Blood-sucking bugs (order, Hemiptera), e.g., *R. prolixus.*
- Blood-sucking true flies, e.g., mosquitoes and tsetse flies.

FIGURE 20–13 The mite, *S. scabiei*, causes skin problems and irritation in dogs. Reproduced by permission from Sebastian Kaulitzki. © 2010 by Shutterstock.com.

A

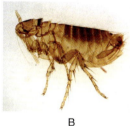

B

C

FIGURE 20–14 Examples of insects that are ectoparasites are (*A*) lice (courtesy of the Centers for Disease Control/ Frank Collins) and (*B*) fleas (courtesy of the Centers for Disease Control, and photo by John Montenieri), or are blood-sucking true flies such as (*C*) mosquitos (courtesy of Getty Images) and tsetse flies. These insects are frequently also vectors or secondary hosts for pathogenic bacteria (e.g., the plague-causing bacteria causing bubonic plague: *Yersinia pestis*), or are parasites such as protozoa (trypanosomes causing sleeping sickness, plasmodium causing malaria), tapeworms, and heartworms.

 ## NONINFECTIOUS DISEASES

Noninfectious diseases include diseases in which there are nutritional, metabolic, stress-related, and genetic causes. Examples of such diseases are milk fever; founder or laminitis (inflammation and edema of the laminae of the hoof) in horses, which is associated with dietary carbohydrate; ascites (accumulation of fluid in the body cavity) in meat-type chickens; and leg weakness/tibial dyschondroplasia in meat-type poultry. Stress diseases include heat stress. In cats and dogs, obesity, leading for instance to diabetes mellitus, is a growing problem. In addition to noninfectious diseases, there are also bone fractures and other injuries.

Milk Fever

Milk fever is a metabolic disease in cattle occurring at the transition between late pregnancy and early lactation. In milk fever, cattle have hypocalcemia or reduced concentrations of calcium in the blood. This can be either subclinical, in which there is hypocalcemia (calcium <2 mmol/L) but with no obvious clinical signs of the disease (but reduced muscle function and depressed milk production), or clinical, in which the hypocalcemia is so severe (calcium <1.4 mmol/L) that nerve and muscle functioning is greatly perturbed and the cow collapses (a "downer" cow). The mortality with milk fever is about 5%, and there is a 3.5-year reduction in the life of affected cattle. This is not a new disease; it was first reported in Germany in 1793.

The incidence of milk fever in the United States is between 5% and 10%. It has been reduced by improved nutrition to increase the capability of the cow to absorb dietary calcium and mobilize bone calcium. These measures include magnesium supplementation, shifting the anion to cation ratio in the diet, and the use of vitamin D or its metabolites. Cows with early signs of milk fever can be treated with a drench of either calcium chloride or calcium propionate. Where there are overt symptoms, calcium is administered intravenously.

Ascites

Ascites is a metabolic disease in poultry, with fluid accumulation in the body cavity, an inability to supply tissues with sufficient oxygen, and a flaccid but enlarged heart. It is estimated that 8% of the mortality of meat-type chickens is due to ascites.

Noninfectious Diseases in Dogs and Cats

Not only do companion animals suffer from infectious disease, but also there are diseases related to metabolism, aging, or mental health such as cancer, osteoarthritis, heart disease, endocrine diseases, anxiety or confusion, some skin diseases, and obesity.

Obesity in Companion Animals

The problem of obesity in dogs and cats is being increasingly recognized by veterinarians and pet owners. Obesity predisposes the animals to serious health consequences.

An obese dog is defined as one in which the body weight is 15% above ideal. The incidence of obesity in dogs is between 20% and 40% in the United States. Dogs can be placed on weight-reduction programs with daily caloric intake to achieve a 1% loss of body weight per week. As with people, the dogs may be stressed from the hunger. Recently, the Food and Drug Administration approved a drug, dirlotapide, specifically for canine weight reduction.

It is estimated that about one third of all domestic cats in the United States are overweight or obese, as indicated by body mass index, percent body fat, and girth. This can lead to diabetes mellitus, lower urinary tract disease, and lameness in cats. One factor influencing obesity is neutering or removal of the gonads. In both male and female cats, there are increases in the amount of adipose tissue after gonadectomy if the cats have free access to food (are fed ad libitum). What seems to be happening is that food intake increases after removal of the gonads, but there is no change in the basal rate of metabolism. However, this is also likely to be less expenditure of energy associated with exercise, with the decreased fighting and roaming behaviors observed in male cats after castration. At present, reduction in food or reducing the fat/caloric content of the feed is the only approach available to address the problem.

1. What is a zoonotic disease (zoonosis)?
2. Give examples of zoonotic diseases.
3. How do diseases affect animals?
4. What are the major classes of pathogens?
5. What are the major classes of parasites?
6. What are contagious and infectious diseases?
7. What is the difference between mortality and morbidity?
8. What is the difference between a clinical infection or disease and a subclinical disease?
9. What is the difference between a prophylactic and a therapeutic treatment?
10. What is the significance of a vector in disease transmission?
11. What are examples of vectors?
12. On an economic basis, why is animal disease important?
13. What is a prion?
14. What is the significance of prions in the diseases of livestock?
15. What is bovine spongiform encephalopathy (BSE)?
16. What measures has the U.S. government taken to reduce the chances of BSE becoming established in the United States?
17. What is bovine variant Creutzfeldt-Jacob Disease?
18. What types of viruses exist?
19. Give examples of viral diseases in companion animals and livestock.
20. What is rabies?
21. Why is there currently a major concern about avian influenza?
22. How are bacteria classified by biologists?
23. Give examples of bacteria that cause food-borne diseases. Is there an issue for food safety?
24. What is mastitis?
25. Give examples of bacteria that can cause mastitis.
26. Give an example of a species of bacteria that can cause disease in dogs. What is the disease?
27. What is a parasite?
28. What are the major groups of parasites?
29. Give examples of diseases caused by trypanosomes.
30. What are coccidia?
31. Why are they a problem?

32. What diseases do they cause?

33. Where in the body do roundworms inhabit?

34. What is the impact of roundworms in dogs, horses, and pigs?

35. What is the impact of *T. spiralis* parasitism?

36. What are the effects of hookworm infestation?

37. What is a strongyle?

38. What species can be infected with strongyles?

39. What is a secondary host? How does this apply to strongyles?

40. Why are heartworm infestations serious for dogs?

41. To which class of parasites does the liver fluke belong?

42. What is a cestode?

43. What are the segments of a tapeworm called? What is present in each?

44. How does a tapeworm attach to the wall of the intestine?

45. How many eggs are present in each tapeworm segment or proglottid when it is shed?

46. What is the difference between an endoparasite and an ectoparasite?

47. Can flies ever be endoparasites? If yes, give an example.

48. Are ticks insects?

49. Why are we concerned about ticks?

50. What are the major types of insect ectoparasites?

51. Are all diseases caused by pathogens? If not, give examples of noninfectious diseases in cattle, dogs, cats, and poultry.

52. What causes milk fever? What is the impact of this problem? Explain your answer.

53. What prophylactic and therapeutic approaches can be used with a cow with subclinical and clinical milk fever?

54. Why should pet owners be concerned about obesity in their dogs and cats?

55. What causes obesity in dogs and cats?

56. What are the remedies for obesity in dogs and cats?

REFERENCES AND FURTHER READING

Bovine Spongiform Encephalopathy (BSE)

U.S. Department of Health and Human Services. *Federal agencies take special precautions to keep "Mad Cow Disease" out of the United States.* Retrieved July 23, 2009, from http://www.hhs.gov/news/press/2001pres/01fsbse.html

Avian Influenza

Department of Health and Human Services. Centers for Disease Control and Prevention. *Influenza viruses.* Retrieved July 23, 2009, from http://www.cdc.gov/flu/avian/gen-info/flu-viruses.htm

Tumpey, T. M. Basler, C. F., Aguilar, P. V., Zeng, H., Solórzano, A., Swayne, D. E., et al. (2005). Characterization of the reconstructed 1918 Spanish influenza pandemic virus. *Science, 310,* 77–80.

Novel 2009 H1N1 Influenza

British Medical Journal. *A/H1N1 influenza virus: The basics.* Retrieved September 6, 2009, from http://www.bmj.com/cgi/content/full/339/jul24_2/b3046?view=long&pmid=19633037

Centers for Disease Control. *2009 H1N1 flu (swine flu) and you.* Retrieved September 6, 2009, from http://www.cdc.gov/h1n1flu/qa.htm

Gallaher, W. R. (2009). Towards a sane and rational approach to management of influenza H1N1 2009. *Virology Journal, 6,* 51.

Itoh, Y. et al. (2009). In vitro and in vivo characterization of new swine-origin H1N1 influenza viruses. Retrieved September 6, 2009, from http://www.nature.com/nature/journal/vnfv/ncurrent/abs/nature08260.html

Morens, D. M., Taubenberger, J. K., & Fauci A. S. 2009. The persistent legacy of the 1918 influenza virus. *New England Journal of Medicine, 361,* 225–229.

World Health Organization. *Pandemic (H1N1) 2009.* Retrieved September 6, 2009, from http://www.who.int/csr/disease/swineflu/en/

Milk Fever

Bethard, G., Verbeck, R., & Smith, J. F. *Controlling milk fever and hypocalcemia in dairy cattle: Use of dietary cation-anion difference (DCAD) in formulating dry cow rations.* New Mexico Agricultural Experiment Station Technical Report 31. Retrieved July 23, 2009, from http://cahe.nmsu.edu/pubs/research/dairy/TR31.pdf

Horst, R. L., Goff, J. P., Reinhardt, T. A., & Buxton, D. R. (1997). Strategies for preventing milk fever in dairy cattle. *Journal of Dairy Science,* 80, 1269–1280.

Obesity in Cats

Backus, R. C., Cave, N. J., & Keisler, D. H. (2007). Gonadectomy and high dietary fat but not high dietary carbohydrate induce gains in body weight and fat of domestic cats. *The British Journal of Nutrition*, *98*, 641–650.

Kanchuk, M. L., Backus, R. C., Calvert, C. C., Morris, J. G., & Rogers, Q. R. (2003). Weight gain in gonadectomized normal and lipoprotein lipase-deficient male domestic cats results from increased food intake and not decreased energy expenditure. *The Journal of Nutrition*, *133*, 1866–1874.

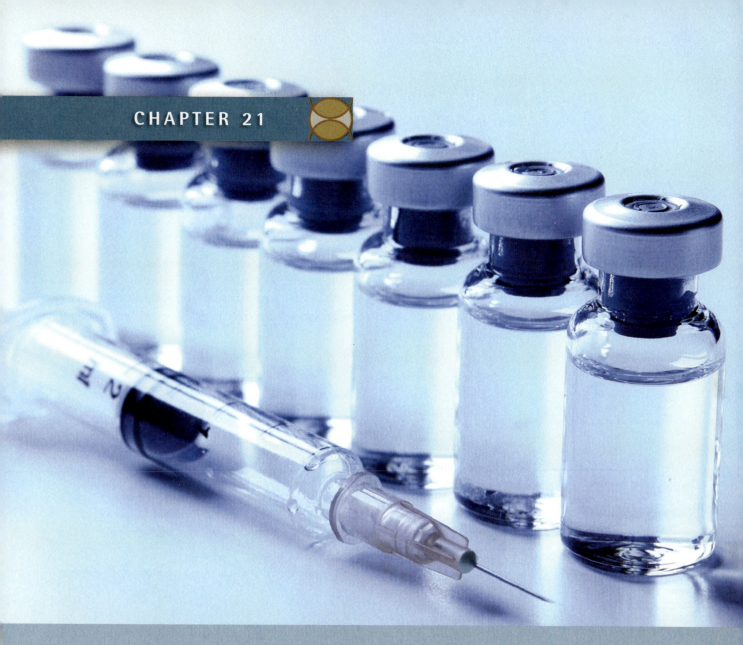

CHAPTER 21

Animal Health, Therapeutics, Immunology, and Vaccines

OBJECTIVES

This chapter will consider the following:

- An overview of animal health
- The defense mechanism of animals, including immunology
- Prophylactic and therapeutic drugs
- Vaccines

AN OVERVIEW OF ANIMAL HEALTH

There has been a shift in our thinking from treating clinical disease to disease prevention. There are a series of steps to promote animal health regardless of whether considering livestock species, horses, or companion animals. These include biosecurity and, therefore, reducing the opportunity for a pathogen to come in contact with the animal.

BIOSECURITY

Biosecurity is a series of measures attempting to prevent pathogens from reaching a herd or flock of animals. A biosecurity plan involves first assessing where the risks are. There then needs to be the development and implementation of policies with the training of employees and those who visit the facilities. Measures might include the following:

- A change of clothing.
- Shower in–shower out.
- Plastic overshoes.
- Disposable overalls.
- Footbaths with disinfectant.
- High-pressure hosing of vehicles before they can visit the facility.
- Prohibiting employees from having the same species of animal at home.
- Limiting visitor access (see fig. 21-1).
- Sanitation and, therefore, killing or washing away large numbers of the pathogenic organisms. A rule of thumb is that sanitation should remove 99.99% (or four orders of magnitude) of the pathogens. High-pressure

FIGURE 21–1 The importance of biosecurity as seen from the entrance to a hog farm. Courtesy of the U.S. Department of Agriculture. Photo by Ken Hammond.

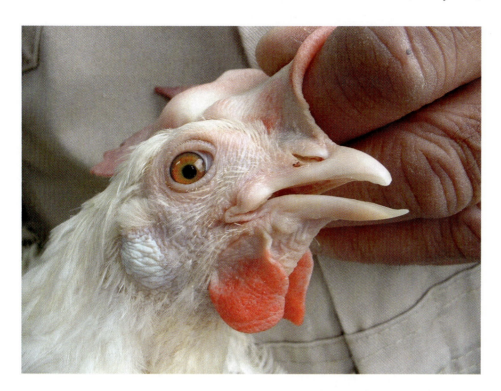

FIGURE 21-2 Visual health inspection of meat-type chicken. Courtesy of Ralph Stonerock.

hosing removes pathogens using the principle of diluting the concentration of the organism. In addition, soaps, detergents, and sanitizers such as sodium hypochlorite kill or render harmless pathogenic organisms.

- Monitoring health status/veterinary care (see fig. 21-2).
- Optimizing nutrition.
- Reducing stress, for example, by providing clean shelter from the extremes of weather and reducing parasite infestation.
- Prophylaxis.
  - Vaccination
  - Prophylactic drugs
- Therapeutic drugs.

Animal health is important to livestock producers, horse owners and trainers, and the owners of companion animals. Animal health companies or animal pharmaceutical companies produce drugs, vaccines, and diagnostic tests for companion animals, horses, and livestock.

An indicator of the magnitude of the animal health industry comes from the sales figures, which are summarized in Table 21-1.

 ## THE DEFENSE MECHANISM OF ANIMALS, INCLUDING IMMUNOLOGY

How do animals protect themselves against invading microorganisms? There are two systems protecting the animal and mounting a defense to the pathogens: innate immunity and adaptive immunity. These systems exist throughout the

TABLE 21-1 Animal health companies' 2006 sales

ANIMAL HEALTH COMPANY	2006 SALES (BILLION $)
Pfizer Animal Health	2.3
Merial Ltd.	2.2
Intervet	1.4[a]
DSM Animal Nutrition	1.2
BASF	1.0
Novartis Animal Health	0.9
Fort Dodge Animal Health	0.9
Schering-Plough Animal Health	0.9
Elanco Animal Health	0.9[b]
Idexx Laboratories	0.8

Source: Feedstuffs March 19, 2007.
[a]Intervet acquired by Schering-Plough in 2007.
[b]Elanco acquired Ivy Vet Life in 2007.

vertebrates, including aquaculture finfish, poultry, livestock, horses, dogs, cats, and people.

 INNATE IMMUNITY

Innate immunity is the first line of defense against invading pathogens. The vast majority of pathogens either are prevented from entering the animal because of physical and chemical barriers, or are destroyed within minutes or hours by innate defenses. The innate immune response is nonspecific.

Barriers

Skin

This provides both a physical and chemical barrier with the keratinized epidermis, which is impermeable to microorganisms. There is also sebum—an oily secretion—which contains fatty acids and a low pH. These provide an environment not conducive for the pathogens.

Mucous Membranes

Mucous membranes consist of the gastrointestinal tract, respiratory tract, mammary glands, and reproductive tracks. These would appear to be a ready site for invasion, but they usually are not. For instance, enteric pathogens are continually attempting to invade or colonize the gastrointestinal tract. The intestinal mucosa gut-associated lymphoid tissues of the intestine are part of the defense mechanism.

Protections against pathogenic invasion include the following:

- Enzymes such as pepsin (digestive enzyme that hydrolyzes proteins).
- Mucus films such as those of mucin glycoproteins in the intestine that are secreted by goblet cells moved by cilia.
- Low pH (particularly in the stomach). In addition, elements of the innate immune system provide further defenses.

- Antimicrobial chemicals such as bile salts, which have a detergent effect on the cell wall of microorganisms, and cryptidins produced in the intestinal crypts of the small intestine and defensins.
- Surfactant proteins in the lungs that facilitate phagocytosis.

Antimicrobial Peptides and Proteins

Antimicrobial peptides and proteins are an integral part of the epithelial defense barrier. In the innate immune response, there is an immediate response with no memory of prior exposure. There are genome-coded pattern recognition receptors or pattern recognition molecules. These recognize molecules found on pathogens, such as lipopolysaccharide on the outer membrane of Gram-negative bacteria, carbohydrates, N-formyl peptides, and teichoic acid on the cell membrane of Gram-positive bacteria, but not those found in the animal. This allows the phagocytes to engulf invading pathogens.

The innate host defense system also includes the production of defensins. These proteins are antimicrobial. They are produced by phagocytic cells, lymphocytes, and the epithelial cell linings of the gut, genitourinary tract, or the trachea, bronchus, and bronchioles. Defensins also signal adaptive immune responses.

Phagocytes

Phagocytes, which are resident macrophages and recruited neutrophils, not only engulf invading organisms (see fig. 21-3) but also respond with a respiratory burst. This is the release of highly reactive oxygen, including the superoxide radical ($O_3^{\cdot}$) and hydrogen peroxide. The respiratory burst helps kill internalized bacteria and acts as a second messenger activating signaling pathways within the cell and to other immune cells.

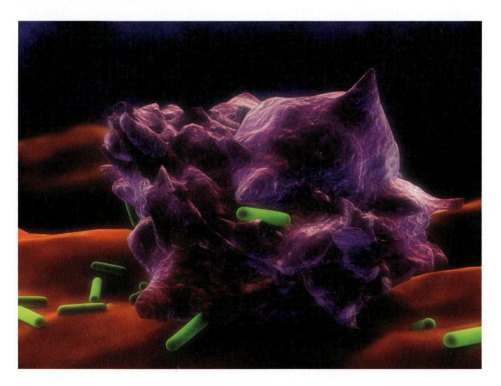

FIGURE 21-3 Phagocyte engulfing pathogenic (disease-causing) bacteria. Reproduced by permission from Sebastian Kaulitzki. © 2010 by Shutterstock.com.

Definitions

Active immunity is when the immune system of an animal responds to foreign protein such as pathogenic microorganisms with the production of antibodies and a cell-mediated response.

An *antibody* is a protein produced by the immune system (plasma cells). The antibody binds to foreign proteins or glycoproteins (e.g., the surface proteins of bacteria or viruses) specifically and facilitates their destruction.

An *antigen* is a substance, usually a protein, that is foreign to the body (such as viruses and bacteria) and that stimulates the immune system to produce antibodies that bind specifically to the antigen.

Immunogen is a substance, usually a protein that is foreign to the body (such as viruses and bacteria), that stimulates the immune system to produce antibodies that bind specifically to antigenic sites on the immunogen.

Immunoglobulin (Ig) is the globulin fraction of serum containing the antibodies.

Passive immunity is where antibodies are transferred from one animal to another. such as through the placenta, in colostrum, or when an antivenom is given after a snakebite.

Prophylactic treatments are those that prevent a disease rather than treat the disease. Examples include drugs and vaccines. Therapeutic drugs are used to "fight" an active disease.

Inflammation

Inflammation with increased blood flow because of vasodilation, redness, and swelling is one of the first immune responses to infection. Inflammation is produced by histamine produced by mast cells, eicosanoids (e.g., prostaglandins), and cytokines, released by injured or infected cells.

Inflammation is followed by phagocytosis, with microorganisms engulfed by phagocytic cells, including monocytes, macrophages, neutrophils, and dendritic cells. Other cells can become phagocytic. Once engulfed by the phagocytic cell, the microorganisms are in a phagosome. These will then fuse with lysosomes, and the ingested pathogen is digested. There is also the release of reactive oxygen intermediates, which are toxic to the pathogen.

Other Aspects of the Innate Immune System

Other aspects of the innate immune system include interferons produced after viral infection and inhibiting viral replication, complement-serum proteins that can destroy pathogens, and cytokines or chemical messengers.

 ## ADAPTIVE IMMUNITY

There are two types of adaptive immunity: humoral (antibodies), and cell-mediated. Tissues of the adaptive immune system include the primary immune organs. The thymus is where the T cells of cell-mediated immunity originate. The bursa Fabricius, which is close to the cloaca, in birds or bone marrow in many mammals is where the B cells of humoral immunity originate. In a few species such as sheep and cattle, the Peyer's patch is the primary lymphoid organ where the B cells develop. In sheep and cattle, this is 3–6 ft (1–2 m) in length along the end of the small intestine. B and T cells migrate to the secondary lymphoid organs, and once there, establish humoral immunity.

There are also secondary lymphoid organs:

- Spleen.
- Lymph nodes. However, birds lack well-defined lymph nodes.
- Gut-associated lymphoid tissues including the Peyer's patches. These are secondary immune organs in many species of mammals. The gut-associated lymphoid tissues contain both cell-mediated and humoral immune-functioning lymphocytes, effecting acquired immunity to enteric pathogens. In birds, there is Meckel's diverticulum and cecal tonsils. The former is a remnant of the yolk sac and is found in the middle of the small intestine.
- Harderian gland.
- Bone marrow in some species is a secondary lymphoid organ.

Humoral Immunity or Antibody Mediated Immunity

Antibodies are very effective in eliminating extracellular antigens. There are five classes of antibodies or immunoglobulins (Igs) in mammals:

- IgA, which is a dimer with two of the basic Ig units joined together by a connecting J chain together with the secretory domain. It is in the mucus secretions on mucosal surfaces and provides an important component to the first line of defense against pathogens.

- IgD, which is a monomer with a single basic Ig unit. It is the antigen receptor on B cells.
- IgE, which is a monomer with a single basic Ig unit. This induces the release of histamine from mast cells in inflammation and allergy.
- IgG, which is a monomer with a single basic Ig unit. It is the major humoral agent combating disease. There are subtypes of IgG: IgG1, IgG2, and IgG3.
- IgM, which is a pentamer with five of the basic Ig units joined together. It is the first humoral agent combating disease.

Birds have IgY, with many of the characteristics of IgG but do not have either IgD or IgE. IgD is thought to be absent in pigs and some ruminants.

Antibodies can be considered conceptually as having a Y shape. We might consider the IgG class of antibodies as the basic type for simplicity. They are composed of four protein chains: two identical light chains and two identical heavy chains (see fig. 21-4). There are two antigen-binding sites comprising elements of both a light chain and a heavy chain. As might be expected in view of the tremendous number of different antigens, there need to be antibodies that can bind to the huge variety of antigens at the antigen-binding site. The polypeptide sequence adjacent to the antigen-binding sites shows much variation. This is known as the *variable region* or *domain* for each of the light and heavy chains. Interestingly, the antibodies of camels have the unique characteristic of having only two heavy chains and no light chains.

During development, there is a molecular rearrangement of the multiple genes of both the heavy and light chains in each precursor immune cell. This results in each cell having the ability to produce a single unique antibody. Some of the antibodies would bind to proteins in the animal itself and, if produced, could do damage to the animal. To prevent this problem, during development cells that are capable of producing antibodies to an animal's own proteins are destroyed. It is important for the immune system to "recognize self."

If a B cell is exposed to an antigen from a foreign organism, it then responds by producing antibody, by dividing so there are multiple daughter cells producing antibody, and by generating memory cells ready to respond the next time the foreign protein is detected. There are multiple genes for the light and heavy chains. In cattle, there is largely a single family of variable heavy genes.

Antibodies provide protection by preventing pathogens from entering the body through mucosal surfaces (e.g., secretory IgA), by killing or helping to kill pathogens in the blood and other extracellular environments in the body (IgG and IgM), or by initiating the process of inflammation with increased phagocytosis (IgE).

Antibodies (IgG and IgM) act by binding to the foreign organism. This then leads to the demise of the pathogen by a number of mechanisms:

- The complement pathway. The pathogen is lysed (broken open and killed) by the presence of complement plus antibodies.
- Opsonization (or coating) of the antigen so it can be readily phagocytosed; effector mechanisms include activation directly on the antigen.

Antibodies (IgG and IgM) also bind to toxins produced by pathogens. The antibody toxin complex is less toxic and much more easily removed from the body.

FIGURE 21–4 Structure of IgG-type antibody. *A* is reproduced by permission from Blamb. © 2010 by Shutterstock.com. *B* is reproduced by permission from ynse. © 2010 by Shutterstock.com.

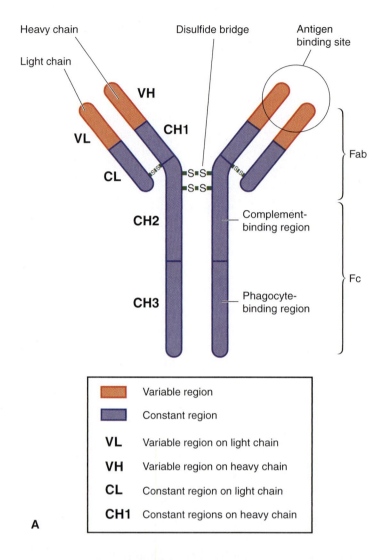

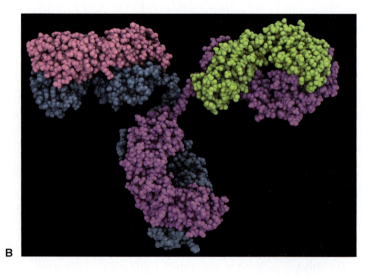

Antibodies help the natural killer cells eliminate infected or neoplastic (cancer) cells. The antibody binds to the antigen on the cell surface, and the killer cell binds to the antibody.

Passive Immunity

Animals are born or, in the case of birds, hatched before their immune systems are fully functional. To protect newborn birds and mammals against pathogens in the environment, the mothers provide antibodies. There is maternal transfer of antibodies, particularly IgG, to the progeny via three mechanisms:

1. Transport across the placenta, as is the case for people, rodents, and dogs.
2. Transfer in the first milk or colostrum, with the antibodies passing through the intestinal wall of the newborn before this becomes a barrier. This is the case for cattle, sheep, pigs, cats, and dogs.
3. Transfer to the chick embryo through ingestion of the yolk, which contains appreciable amounts of antibodies.

Antibodies from the mother also provide protection after the neonate's intestine has become impermeable to proteins. Antibodies continue to be present in milk, and these can bind to pathogens in the gut lumen. The importance of this is apparent from the observation that early weaning of pigs can be accompanied by diarrhea.

Cell-Mediated Immunity

Cell-mediated immunity is crucial when the pathogen has entered a cell. The cells that provide cell-mediated immunity are the T-immune cells. Humoral immunity is much less useful when the pathogen is in a cell because the antibodies do not cross the cell membrane. The different types of T cells have different cell markers and different functions. Cells include the following:

- Cytotoxic cells
- T-helper cells (Th)
  - Type-1 Th cells (Th1), which direct the cell-mediated response
  - Type-2 Th cells (Th2), which facilitate humoral responses
- T-suppressor cells

Nutrition and Disease Resistance

Deficiency of a required nutrient impairs the ability of an animal to fight invasion by pathogens. The obvious remedy is to restore the required nutrient to the diet. There is growing evidence that the published requirement for nutrients, although suitable for optimizing growth and other performance indicators, may not optimize immune function. Higher levels of some amino acids improve immune function, whereas vitamin E or β-carotene (vitamin A) may improve antioxidant status.

Genetics and Disease Resistance

There is considerable variation in the susceptibility of animals to diseases. Frequently, in livestock, selection for improved performance (such as growth and lactation) has reduced disease resistance. It is now possible to select for disease resistance.

Definition

Colostrum is the first milk produced in a lactation. It is rich in proteins, particularly antibodies. These antibodies provide passive immune protection to the neonate against pathogens before the newborn has a fully developed immune system, that is, before it is fully immune competent.

VACCINES AND DRUGS

Vaccines and drugs are as critical to the health of animals as they are to people.

Vaccines

Vaccination is a method to stimulate the immune system such that it shows a larger and quicker response to a specific disease microorganism or pathogen (see fig. 21-5). Vaccines are either prophylactic, which stimulate the immune system such that it is prepared to respond to a future infection, or therapeutic, which stimulate the immune system to combat an existing infection, as is the case when someone is bitten by a rabid animal and receives rabies vaccination.

A vaccine is a mixture of the antigen and an adjuvant. A vaccine can be of four types:

* Killed organism.
* Live attenuated or weakened organism in the modified live vaccine.
* Protein from an organism, usually recombinant derived.
* Toxoid, for example, a tetanus-causing organism, *Clostridium tetani*, produces a toxin or toxic protein. The toxicity of this can be greatly reduced without affecting its immunogenicity in the toxoid, or its ability to provoke an antibody response.

An antigen is a protein foreign to the body, such as viruses and bacteria, which stimulates the immune system to produce antibodies that bind specifically to the antigen. The efficacy of a vaccine depends on two key factors: the antigen(s) included, and the adjuvant. An adjuvant is a mixture of chemicals that acts to stimulate both cell-mediated and humoral immune responses, and to deliver the antigen to the body's antigen-presenting cells.

FIGURE 21-5 Examples of vaccines used to prevent diseases (in a prophylactic manner) in cattle, pigs, poultry, and companion animals. Reproduced by permission from Steven Mann. © 2010 by Shutterstock.com.

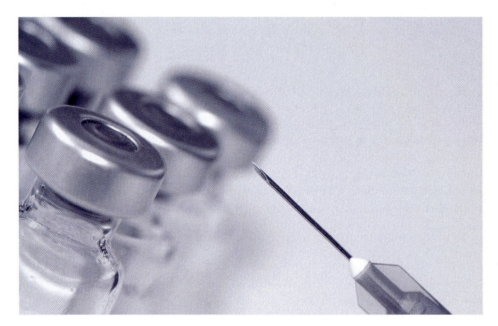

Vaccines for Specific Species

Cattle

Examples of vaccines used for cattle include infectious bovine rhinotracheitis; parainfluenza 3; bovine virus diarrhea (BVD), which protects against shipping fever; bovine respiratory syncytial virus; and *Leptospira canicola*, *Leptospira grippotyphosa*, *Leptospira hardjo*, *Leptospira icterohaemorrhagiae*, and *Leptospira pomona*.

Cattle are immunized against BVD as a protection against the virus and, therefore, shipping fever. The BVD virus, with its two genotypes types 1 and 2, is a widespread problem for both beef and dairy cattle. Shipping fever is a secondary infection associated with infection with the BVD virus.

Pigs

Examples of pig diseases for which vaccines are recommended include atrophic rhinitis, erysipelas, mycoplasma pneumonia, *Actinobacillus* pleuropneumonia, parvovirus, mycoplasmal pneumonia, transmissible gastroenteritis, *Escherichia coli*, leptospirosis, and porcine reproductive and respiratory syndrome virus. Porcine reproductive and respiratory syndrome virus costs the U.S. swine industry $560 million per year. Commercial vaccines are available and effective against the porcine reproductive and respiratory syndrome virus using either killed viruses or modified live viruses.

Poultry

Examples of poultry diseases for which vaccines are recommended include Marek's disease, Newcastle disease virus, infectious bursal disease virus or Gumboro disease, avian pneumovirus, infectious bronchitis, laryngotracheitis, fowl pox, fowl cholera, and avian encephalomyelitis. A number of countries have adopted the policy of vaccinating against avian influenza. At the time of this writing, the United States was not one of them.

Horses

According to a report for the American Association of Equine Practitioners, guidelines for vaccination of horses recommend vaccination against the following: West Nile virus, tetanus toxoid, encephalomyelitis, influenza, rhinopneumonitis (equine herpesvirus types 1 and 4), strangles, rabies, Potomac horse fever, and equine viral arteritis. The vaccine against equine herpesvirus types 1 and 4 reduces the respiratory diseases caused by these viruses.

Coggins Test

Equine infectious anemia is a serious viral disease of horses. Infection can be acute (with about 30% of infected horses dying), chronic, or subclinical, with no symptoms. There is no vaccine available. Horses with a subclinical infection are carriers of equine infectious anemia. Horses are tested by state laboratories for the presence of antibodies to the equine infectious anemia virus using the Coggins test. A negative Coggins test indicates the lack of antibodies to equine infectious anemia virus and, therefore, that the horse is safe to be around other horses and is required to move a horse across state lines and for horse shows.

INTERESTING FACTOID

The first vaccine was the administration of cowpox to people in 1796 by Edward Jenner to protect against smallpox. In the 1860s, Louis Pasteur developed vaccines for fowl cholera and anthrax using dead organisms. Vaccines against rabies and smallpox using attenuated or weakened organisms were developed in the 1870s.

DOG AND CAT VACCINATION

According to the University of California, Davis's recommendation for vaccination of dogs, the core vaccination is a modified live vaccine against canine parvovirus, canine distemper virus, and canine adenovirus type 2, with booster shots after 1 year and then every 3 years. Dogs should receive a single vaccination against rabies. There are optional vaccinations depending on locality and likely exposure for canine parainfluenza virus and *Bordetella bronchiseptica*, preventing kennel cough; canine distemper virus, preventing measles; *Canine leptospira*; and *Borrelia burgdorferi*, preventing Lyme disease.

The American Veterinary Medical Association and the American Animal Hospital Association canine vaccine task force list vaccines to canine coronavirus, *Giardia* spp., and canine adenovirus-1 as not generally recommended. There is insufficient information on the efficacy of canine rattlesnake vaccines.

For the core vaccination of cats, the University of California, Davis recommends a modified live vaccine against feline herpesvirus 1, *Feline calicivirus*, and feline panleukopenia virus, with booster shots after 1 year and then every 3 years. Cats should have a single vaccination against rabies. There are optional vaccinations depending on locality and likely exposure for feline leukemia virus; feline immunodeficiency virus; feline infectious peritonitis; *Chlamydophila felis*; B. *burgdorferi*, preventing Lyme disease; and *Giardia spp.*

Dogs and Cats

The recommended vaccines against infectious diseases in dog and cats are summarized in the boxed text above.

Vaccine Against Cancer

There is now a vaccine for the treatment of canine melanoma. This is the first therapeutic vaccine for the treatment of either animal or human cancer. Previously, treatment of oral melanoma in dogs was by surgery, radiation, or chemotherapy, but none of these was very effective.

 ## ANIMAL HEALTH DRUGS

Animal health drugs are used for the following:

- Prophylactic uses, such a preventing heartworm
- Growth promotants and production efficiency–enhancing agents, such as antibiotics in poultry and pigs
- Antibiotics to resolve a bacterial disease and/or protozoal diseases
- Dewormers
- Drugs to improve reproductive efficiency and resolve reproductive problems, which are considered in Chapter 16
- Anesthetics and analgesics
- Companion animal drugs

Prophylactic Drugs

Prophylactic treatments are those to prevent the disease, not to treat it. There are numerous examples of such drugs. Owners of dogs and cats routinely give their animals long-lasting drugs against fleas, ticks, heartworm, hookworm, and other parasites. These are frequently flavored, for instance with liver, so the animal perceives them as a "treat." Prophylactic approaches using drugs are used for livestock and horses to reduce to the impact of both internal and external parasites.

Ivermectin, which is produced by Merial (a major animal health company), is one of the most widely used drugs to improve the health of animals. It is effective against many ectoparasites and endoparasites in livestock, horses, and companion animals. A related compound, eprinomectin, has even broader activity. An alternative drug against many ectoparasites and endoparasites in livestock, horses, and companion animals is selamectin, which is produced by Pfizer. This is used, for instance, for dogs and cats, where it is applied to the skin. In cats, Revolution (selamectin) is used to kill parasites, including adult fleas (*Ctenocephalides felis*), while also preventing the eggs from hatching; heartworm (*Dirofilaria immitis*); ear mites (*Otodectes cynotis*); roundworms (*Toxocara cati*); and hookworms (*Ancylostoma tubaeforme*).

There are a variety of other drugs effective against roundworms, hookworms, and heartworms. These other dewormers include pyrantel, moxidectin, and fenbendazole. Other dewormers used with horses include moxidectin (Quest), fenbendazole (Safe-Guard), pyrantels (Pyrantel Pamoate and Strongid), and quasi-quantals.

IVERMECTIN

Ivermectin has broad-spectrum antiparasite effects against intestinal parasites, including roundworms, most mites, and some lice. Ivermectin (22,23- dihydroavermectin) is derived from compounds produced by a specific species of bacteria from the soil, *Streptomyces avermitilis*. It was introduced in the early 1980s after a series of U.S. Food and Drug Administration approvals. It is rapidly absorbed through the intestine and acts by interfering with the parasite's nervous system, binding to glutamate-gated chloride channels.

Ivermectin alone is effective against larval heartworms or microfilariae in the circulation. It is not effective against adult heartworms, tapeworms, flukes, fleas, ticks, or flies. The most common uses in small animal practice for ivermectin would include monthly prevention of heartworm infection; treatment of ear mite situations; clearing heartworm larvae in active heartworm infection; and treatment of sarcoptic, notoedric, or demodectic mange.

Examples of products containing ivermectin are Ivomec (for cattle), Ivomec Plus (for cattle), Heartgard 30 (for dogs), Heartgard Plus (for dogs), Iverhart Plus (for dogs), and Heartgard (for cats). Some dogs, particularly collies, show toxic signs in response to ivermectin. It is also used in people to treat onchocerciasis (river blindness). Ivomec Plus also contains clorsulon to be effective against liver flukes.

In horses, ivermectin combined with praziquantel is marketed as Equimax, which is "an all-in-one dewormer" that combines ivermectin and praziquantel to safely eliminate all major internal parasites, including tapeworms. These tapeworms include ascarids *Strongylus vulgaris* and other large strongyles (*S. vulgaris*, *Strongylus edentatus*); and small strongyles and pinworms.

Praziquantel (EquiMax, Zimectrin Gold, and Quest Plus) is used in combination with ivermectin or moxidectin.

Antibiotics

Antibiotics are used therapeutically to treat specific diseases such as the use of Micotil (tilmicosin injection) to treat bovine respiratory disease. They are also used to improve the efficiency of livestock production as growth promotants.

Production Efficiency in Livestock

Drugs and hormones to increase the efficiency of livestock production include the following:

- Antibiotics increasing the growth of poultry and pigs
- Estrogen and other sex hormones enhancing growth in cattle
- GH or Somatotropin (POSILAC, Lactotropina), improving the efficiency of milk production in dairy cattle
- Adrenergic agonists, increasing the growth and feed efficiency of pigs and cattle
- Ionophores, such as Rumensin, improving the efficiency of fermentation in the rumen of cattle

Anesthetics and Analgesics

Examples include the following:

- Sedatives/analgesics used, for instance, for minor surgical and diagnostic procedures, such as Dormosedan (detomidine hydrochloride) for horses.
- Arthritis pain reduction using nonsteroidal anti-inflammatory drugs. One nonsteroidal anti-inflammatory drug used for pain relief in horses is phenylbutazone (Bute). This drug is also abused.

Companion Animal Drugs

There are multiple drugs used for dogs and cats, including drugs for heart disease, osteoarthritis, canine weight reduction, skin disease, and endocrine disorders such as thyroid insufficiency. To make it easier to give dogs medication, some of the drugs are produced in chewable liver-flavored tablets.

Osteoarthritis

Nonsteroidal anti-inflammatory drugs are used in dogs and other companion animals. These inhibit the enzyme cyclooxygenase and, therefore, reduce prostaglandin production and the pain associated with osteoarthritis.

Heart Disease

There are drugs available to treat heart failure in dogs. They act by lowering blood pressure and reducing fluid in the lungs.

Endocrine Diseases

A drug is available to treat Addison's disease, which is the insufficient release of cortisol from the adrenal cortex or hypoadrenocorticism in dogs.

Skin Disease

There are drugs available to treat skin diseases. One contains an antifungal agent, an antibiotic, and the anti-inflammatory compound, dexamethasone, to address three likely pathogenic scenarios. In addition, companion animals may develop allergies. There are, for instance, drugs available to treat skin allergies in dogs.

Mental Health for Dogs

The tricyclic drug clomipramine acts by inhibiting the reuptake of norepinephrine and serotonin at the synapse. It is used as an antidepressant in people for whom it also alleviates the anxiety component of depression. This drug is also used to reduce separation anxiety in dogs. The selective serotonin reuptake inhibitor fluoretine (Prozac) is used as an antidepressant in people. It has been approved by the FDA for the treatment of separation anxiety in cats and dogs. In addition, there are research findings that fluoretine reduces aggression and compulsive behavior in dogs. There is also a drug available for cognitive dysfunction syndrome, in which older dogs show signs of confusion and other behavioral changes.

Other drugs used in dogs and cats include antibiotics, pregnancy termination, which is described in Chapter 16, and endocrine diseases.

 ## NOVEL APPROACHES TO COMBAT DISEASES

Novel approaches being researched to combat infectious diseases in animals include the following:

- Genetic selection, including the use of candidate genes to increase the resistance of livestock to diseases.
- Probiotics are benign microorganisms that improve the environment in the gastrointestinal tract.
- Bacteriophage is a virus that infects a specific bacterium. The bacteriophages multiply within the bacteria and ultimately lead to their death.
- Bacteriocins are chemicals produced by bacteria that kill other bacteria, either a related species or different strain.

SYMPTOMS OF COGNITIVE DYSFUNCTION SYNDROME (CDS)

Based on the Web site CDSinDogs.com, symptoms of cognitive dysfunction syndrome shown in a dog include disorientation/confusion. The dog appears lost or confused in the house or yard. It fails to recognize familiar people or respond to verbal cues or name. It has difficulty finding the door or stands on the hinge side of the door. It appears to forget the reason for going outdoors.

Regarding interaction with family members, the dog seeks attention less often and walks away when being petted. It shows less enthusiasm upon greeting you and no longer greets family members.

The dog sleeps more during the day and sleeps less during the night. It wanders or paces more.

Regarding house-training, the dog urinates indoors and has accidents indoors soon after being outside. It forgets to ask to go outside.

Source: Pfizer (2008).

REVIEW QUESTIONS

1. How has there been a change in thinking from treating clinical disease, to disease prevention?

2. What is biosecurity?

3. Give examples of measures that might be taken to ensure biosecurity.

4. Why is sanitation important for an animal facility or farm?

5. How is sanitation achieved?

6. Give examples of prophylactic measures taken by animal owners.

7. What are four of the major animal health companies? What types of products do they produce?

8. What types of barriers are there to prevent pathogens invading the body?

9. Give examples of chemicals that the body produces that kill bacteria in a nonspecific manner.

10. What is the difference between innate and adaptive immunity?

11. How does the innate immune system protect the body against pathogens?

12. How does the innate immune system recognize pathogens?

13. What is phagocytosis?

14. What type of cells phagocytose pathogens?

15. What is an oxidative burst? How is it relevant to protection against invading pathogens?

16. What is inflammation? How is it relevant to protection against invading pathogens?

17. What is an antibody?

18. What is an antigen?

19. What is passive immunity?

20. What is colostrum? How does it enhance the chances of a newborn calf surviving?

21. What is the difference between humoral and cell-mediated immunity?

22. What are the primary lymphoid organs? Give examples of secondary lymphoid organs.

23. What are the cell types that lead to humoral and cell-mediated immunity?

24. What are the major types of immunoglobulin? What is their overall chemical structure? Where are they found, and what do they do?

25. What is the basic structure of an immunoglobulin? What is the antigen-binding site? How do animals produce antibodies to so many antigens?

26. How does the immune system differentiate between self and foreign proteins?

27. Why are memory cells so important to immunity?

28. How do antibodies fight an infection?

29. Give examples of T-immune cells. What does each do?

30. How do genetics and nutrition affect an animal's ability to be protected against a disease?

31. What is a vaccine?

32. Who were the first to develop vaccines?

33. What is the principle of a vaccine? How is it produced?

34. Give examples of vaccines for each of the following species: cattle, pigs, poultry, horses, dogs, and cats.

35. Are there vaccines for any animal cancer? If yes, which?

36. Give examples of drugs used for livestock, horses, dogs, and cats.

37. What is ivermectin used for?

38. What are antibiotics used for?

39. What are steroid drugs used for?

40. Are there drugs for a dog's mental health? If yes, what are they used for?

REFERENCES AND FURTHER READING

Ibanez, M., & Anzola, B. (2009). Use of fluoxetin, diazepam, and behavior modification as therapy for treatment of anxiety-related disorders in dogs. *Journal of Veterinary Behavior: Clinical Applications and Research*, *4*, 223–229.

Irimajiri, M., Luescher, A. U., Douglass, G., Robertson-Plouch, C., Zimmerman, A., & Hozak, R. (2009). Randomized, controlled clinical trial of the efficacy of fluoxetine for treatment of compulsive disorders in dogs. *Journal of the American Veterinary Medical Association*, *235*, 705–709.

Pfizer. (2008). CDSinDogs.com. Retrieved October 30, 2009, from http://www.cdsindogs.com/cds_checklist.asp

Pfizer Animal Health. *CattleMaster® GOLD™ Protection. Like no other.™ Redefining killed BVD protection.* Retrieved July 23, 2009, from http://www.cattlemastergold.com/display.asp?country=US&lang=EN&species=OO&drug=C1&sec=300

Dog and Cat Vaccinations

University of California, Davis. *Recommendations on companion animal care.* Retrieved August 2, 2009, from http://www.vetmed.ucdavis.edu/CCAB/veteri~2.htm#vaccinations

Parasites

American Veterinary Medical Association. Retrieved August 2, 2009, from http://www.avma.org/communications/brochures/parasites/parasites_brochure.pdf

Horses

Wilson, J. *Equine vaccinations and deworming*. Retrieved July 23, 2009, from http://www.extension.umn.edu/distribution/livestocksystems/DI8540.pdf

Immunity

Caldwell, D. J., Danforth, H. D., Morris, B. C., Ameiss, K. A., & McElroy, A. P. (2004). Participation of the intestinal epithelium and mast cells in local mucosal immune responses in commercial poultry. *Poultry Science*, *83*, 591–599.

Erf, G. F. (2004). Cell-mediated immunity in poultry. *Poultry Science*, *83*, 580–590.

Goff, J. P. (2006). Major advances in our understanding of nutritional influences on bovine health. *Journal of Dairy Science*, *89*, 1292–1301.

Le Blanc, S. J., Lissemore, K. D., Kelton, D. F., Duffield, T. F., & Leslie, K. E. (2006). Major advances in disease prevention in dairy cattle. *Journal of Dairy Science*, *89*, 1267–1279.

Lillehoj, H. S., Min, W., & Dalloul, R. A. (2004). Recent progress on the cytokine regulation of intestinal immune responses to *Eimeria*. *Poultry Science*, *83*, 611–623.

Scott, T. R. (2004). Our current understanding of humoral immunity of poultry. *Poultry Science*, *83*, 574–579.

Yasuda, M., Jenne, C. N., Kennedy, L. J., & Reynolds, J. D. (2006). The sheep and cattle Peyer's patch as a site of B-cell development. *Veterinary Research*, *37*, 401–415.

CHAPTER 22

Animal Agriculture and the Environment

OBJECTIVES

This chapter will consider the following:

- Animal waste
- The impact of animals on greenhouse gases

ANIMAL WASTE

Animal waste comprises both urine and feces. Urine consists of water together with nitrogenous waste from the animal (urea in mammals and uric acid in birds) and other soluble compounds that pass through the kidney. Also present may be antibiotics and other drugs such as growth stimulants. Feces are the undigested and/or unabsorbed materials that are voided by the animal. In livestock, the feces are called *manure*. The composition of manure depends on the animal species and the composition of animal feed (e.g., protein content). Feces also contain the following:

- Microorganisms, including pathogens, that live in the gut.
- Any unabsorbed minerals, such as phosphorus as phytate phosphorus and metals such as calcium chelated by the phytate or minerals such as copper and zinc present in the diet of pigs at levels that exceed the capacity of the body to absorb.
- Products of bacterial fermentation in the intestine. These are frequently odorous/volatile (smelly) and can create problems for a producer with neighbors. An example of a chemical found in animal waste with an extremely bad smell is skatole. This gives feces its characteristic smell.

Animal waste can be used as a source of plant nutrients, particularly nitrogen and phosphorus. This has been used traditionally in agriculture, with the manure being applied to the land. Manure is used by organic producers often as the principal source of plant nutrients to fertilize (see fig. 22-1 and Table 22-1).

Definitions

The following are U.S. Environmental Protection Agency (EPA) definitions.

An *animal feeding operation* (AFO) is an agricultural operation in which animals are raised in confined situations.

A *concentrated animal feeding operation* (CAFO) is an agricultural operation in which animals are raised in confined situations. CAFOs tend to be larger than AFOs. About 15% of AFOs are CAFOs. CAFOs must meet both of the following criteria:

- Animals are confined for at least 45 days in a 12-month period.
- There is no grass or other vegetation in the confinement area during the normal growing season.
- CAFOs are regulated by the U.S. EPA as point sources of pollution.

AN OUTSTANDING RESEARCHER ON ANIMAL WASTE

Growing up in Troy, New York, Wendy Powers dreamed of working with exotic animals as a zoo veterinarian. She planned on turning that dream into reality when she entered the animal science program at Cornell University. But after a few courses, she found her curiosities shifting from monkeys and giraffes to cows, specifically to the relationship between dairy production and the environment (see fig. 22-2).

Wendy found herself wrapped up in the Cornell Dairy Fellows program, where her interest in dairy management flourished. After graduation, she took a position as a herdsman at a Florida dairy farm. At the same time, she pursued a master's of science degree from the University of Florida in dairy nutrition, with a focus on distillers' grains in cow diets. That led her to a Ph.D. research project that took a careful look at anaerobic digestion as an odor control strategy for dairy farms. Along the way, she beefed up her knowledge by taking classes in agricultural engineering, soil sciences, environmental engineering, and microbiology, in addition to the required courses in animal science and nutrition and human nutrition. After completion of her Ph.D., Wendy accepted a faculty position at Iowa State University, where she conducted research and extension programming for 10 years.

Now, Dr. Powers is the director of environmental stewardship for animal agriculture at Michigan State University (see fig. 22-2). As a nutritionist, a major focus of her current research program is to study how diets can be modified to reduce nutrient excretions and air emissions. She led the development of a state-of-the-art research facility to conduct this work. The facility is the largest in the world and is fully automated to collect air emission data. Dr. Powers has both a research and an extension appointment at Michigan State University, thereby moving her scientific findings into practice in the field. She works across disciplines, engaging both agronomists and engineers in helping to solve problems and implement solutions. Dr. Powers has been recognized as a leader in her field.

There is the potential for the constituents of the animal waste to run off the land, for instance during heavy rains or with the thawing of snow cover. Moreover, manure is usually applied either before planting or after harvesting in, respectively, the spring or fall. Thus, the animal waste needs to be stored

A

B

FIGURE 22-1 *A*, Animal waste or manure. Courtesy of the U.S. Department of Agriculture. Photo by Jeff Vanuga. *B*, Land application of animal waste provides valuable plant nutrients such as nitrogen and phosphate to the soil. Courtesy of the U.S. Department of Agriculture. Photo by Tim McCabe.

TABLE 22-1 Composition of animal waste (percentage)

ANIMAL	NITROGEN	PHOSPHATE	POTASSIUM
Cattle (dried)	1.3	0.9	0.8
Horse (fresh)	0.6	0.3	0.5
Pig (fresh)	0.6	0.5	0.4
Poultry with litter	2.8	2.8	1.5

Source: Kansas State University Extension.

before application, with the potential, again, for losses into the watershed. Approaches have been developed whereby manure can be applied most of the year (see fig. 22-1).

Animal waste also contains pathogens, including the following six human pathogens: *Campylobacter* spp. (bacteria), *Salmonella* spp. (nontyphoid, protozoan) (see fig. 22-3), *Listeria monocytogenes* (protozoan), *Escherichia coli* O157:H7 (protozoan), *Cryptosporidium parvum* (protozoan), and *Giardia lamblia* (protozoan). These organisms account for over 90% of food borne and water-borne diseases in people, and frequently also affect dogs and cats. They are discussed in some detail in Chapter 24. The use of manure to fertilize fruits and vegetables (such as salad) that are not cooked can be a problem for human health.

Animal waste can be considered as "non-point pollution" with animals on pasture or ranch, or when there is land application of the animal waste. The animal waste can wash into rivers, creeks, and streams, for example, during a rainstorm. A similar situation exists for synthetic fertilizer when land applied. In 2003, the U.S. EPA finalized a set of new rules for CAFOs because these are defined by the U.S.

FIGURE 22-2 Dr. Wendy Powers with a Holstein cow. Photo by Kurt Stepnitz, Michigan State University–University Relations.

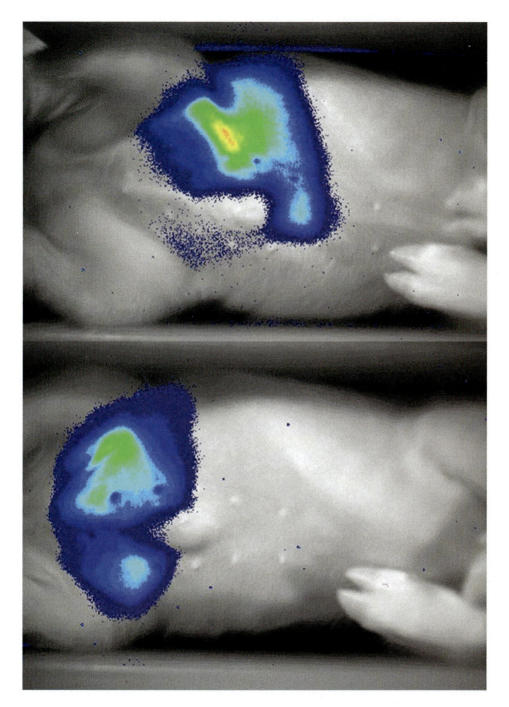

FIGURE 22-3 Animal feces contribute pathogens to the environment. *Upper* (time, 24 hours) and *lower* panels (48 hours) show biophotonic images of pigs challenged with salmonella organisms bioengineered using a luciferase to emit light. Note how the salmonella is being moved to the posterior of the pigs and will ultimately be voided to the environment. Image courtesy of Scott T. Willard and Peter L. Ryan, Mississippi State University.

EPA as point sources of pollution. The National Pollutant Discharge Elimination System program regulates the discharge of pollutants from point sources to waters of the United States. The confined animals or their waste can come into contact with surface water. There are specific regulations for the different sizes of CAFOs (see Table 22-2). The goal of the regulations is to reduce pollution of America's waterways.

TABLE 22-2 The U.S. Environmental Protection Agency (EPA) definitions of different sizes of Concentrated Animal Feeding Operations (CAFOs)

ANIMAL	SIZE THRESHOLDS (NUMBER OF ANIMALS)		
	LARGE CAFOs	MEDIUM CAFOs	SMALL CAFOs
cattle or cow/calf pairs	1,000 or more	300–999	less than 300
mature dairy cattle	700 or more	200–699	less than 200
veal calves	1,000 or more	300–999	less than 300
swine (weighing more than 55 pounds)	2,500 or more	750–2,499	less than 750
swine (weighing less than 55 pounds)	10,000 or more	3,000–9,999	less than 3,000
horses	500 or more	150–499	less than 150
sheep or lambs	10,000 or more	3,000–9,999	less than 3,000
turkeys	55,000 or more	16,500–54,999	less than 16,500
laying hens or broilers (liquid manure–handling systems)	30,000 or more	9,000–29,999	less than 9,000
chickens other than laying hens (other than liquid manure–handling systems)	125,000 or more	37,500–124,999	less than 37,500
laying hens (other than liquid manure handling–systems)	82,000 or more	25,000–81,999	less than 25,000
ducks (other than liquid manure handling–systems)	30,000 or more	10,000–29,999	less than 10,000
ducks (liquid manure–handling systems)	5,000 or more	1,500–4,999	less than 1,500

Source: http://www.epa.gov/guide/cafo/.

ISSUES *for discussion*

ANIMAL WASTE

1. What are the pros and cons of intensive animal agriculture?
2. How should communities, states, and the federal government ensure that agriculture does not negatively impact air and water quality?
3. What, if any, should be the responsibility of livestock producers?
4. What, if any, should be the responsibility of livestock producers with odors from animal waste?
5. What, if any, should be the responsibility of the neighbors of livestock producers?

ATMOSPHERE

Agriculture contributes to the shifts in the composition of Earth's atmosphere, including ammonia (NH_3); greenhouse gases, which are carbon dioxide (CO_2), nitrous oxide (N_2O), and methane (CH_4); and odiferous compounds affecting the neighbors of animal units.

Ammonia

It is estimated that over half of the global emissions of ammonia (NH_3) come from agriculture and about a third of the total from livestock (see Table 22-3). Ammonia

TABLE 22-3 Global emissions of atmospheric ammonia (NH3) in million metric tons NH3-N per year (FAO data)

Livestock	21
Synthetic fertilizers	9
Undisturbed ecosystems	10
Croplands	4
Biomass burning	4
Human excrement	3
Oceans	8
Biofuel combustion	2
Total emission	61

Source: http://www.fao.org/docrep/W5146E/w5146e04.htm.

in the atmosphere is frequently found in aerosols. High concentrations of ammonia in animal facilities can adversely affect the health of both the workers and the animals.

Of the 21 million metric tons of ammonia being added to the atmosphere every year, the vast majority are returned to the surface, predominantly in rain. This ammonia comes from cattle (12.9 million metric tons), sheep and goats (1.5 million metric tons), pigs (3.4 million metric tons), poultry (1.9 million metric tons), and other animals such as buffalo (1.9 million metric tons).

Soil Effects on Ammonia

Ammonia is oxidized to nitric oxide and, therefore, nitrate or to N_2O by nitrification in the soil as part of the nitrogen cycle. Nitric oxide and nitrates can be converted in the soil to N_2O and nitrogen gas by nitrification in the soil.

Greenhouse Gases

With increasing human and livestock populations and industrialization, there are changes in the composition of Earth's atmosphere (see Table 22-4).

There is a scientific consensus that changes in the composition of the atmosphere are the result of human activities, as is shown in Table 22-5. Some scientists and policy makers disagree with the prevailing viewpoint.

In the past 100 years, average surface temperatures in the world have increased by about 0.8°C. Figure 22-4 shows the real (measured) and projected changes in the average temperature in different geographic regions of the world.

TABLE 22-4 Increases in the concentration of "greenhouse" gases in the world's atmosphere over history since before the Industrial Revolution

YEAR	CARBON DIOXIDE (CO_2) (Ppmv)	METHANE (CH_4) (Ppbv)	NITROUS OXIDE (N_2O) (Ppbv)
1750	276	700	270
2000	380	1745	314
2008	386	1788	322

Source: World Meteorological Organization, retrieved August 2, 2009, from http://www.wmo.int/pages/prog/arep/gaw/ghg/ghgbull06_en.html; and National Oceanographic and Atmospheric Administration, The NOAA Annual Greenhouse Gas Index (Aggi), retrieved August 2, 2009, from http://www.esrl.noaa.gov/gmd/aggi/.

Definitions

Units of concentration in the air:

ppmv = parts per million by volume

ppbv = parts per billion by volume

UNITS OF WEIGHT

- 1 kg = 1,000 (10^3) g (or equivalent to 2.2 lb)
- 1 mega-gram or metric ton = 1 million (10^6) g
- 1 Gg (or giga-gram) = 1 billion (10^9) g
- 1 Gg (giga-gram) = 1 million (10^6) kg
- 1 Gg (giga-gram) = 1,000 (10^3) metric tons
- 1 tera-gram (Tg) = 1 trillion (10^{12}) g
- 1 tera-gram (Tg) = 1 billion (10^9) kg
- 1 tera-gram (Tg) = 1 million (10^6) metric tons
- Next units: peta- (10^{15}), exa- (10^{18}), zetta- (10^{21}), yotta- (10^{24}), and xona (10^{27})

TABLE 22–5 Estimates of global "greenhouse" gas emissions in billion metric tons (or thousand Tg) CO_2 equivalent[1] in 2000

SECTOR	CARBON DIOXIDE (CO_2)	METHANE (CH_4)	NITROUS OXIDE (N_2O)
Energy	23.4	1.6	0.2
Agriculture	7.6	3.1	2.6
Industry	0.8	0	0.2
Waste	0	1.3	0.1
Global total	31.9	6.0	3.1

[1]*Based on the different ability of different gases to cause climate changes or global warming potential (GWP): $CO_2 = 1$, $CH_4 = 21$, and $N_2O = 310$.*
Source: U.S. Environmental Protection Agency.

FIGURE 22–4 Global warming. Real (measured) changes in average global surface temperatures. Image courtesy of National Aeronautics and Space Administration.

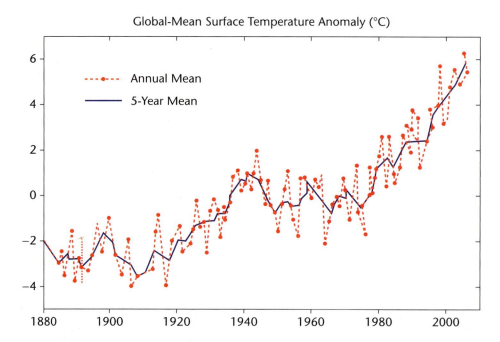

The changes in greenhouse gases in the atmosphere are thought by a scientific consensus to be playing a significant role in global warming, as is exemplified by a United Nations scientific review in 2007, the Kyoto treaty, and the U.S. EPA.

There are many scientists and political leaders who consider that the link between man's activities and climate change has been fully demonstrated. Some large agribusinesses (e.g., Cargill and Smithfield) have announced that they are to reduce greenhouse gas emissions by 6% before 2010.

Different gases that are released into the atmosphere have a different size of effect on global warming: CO_2 has a global warming potential of one, CH_4 of 21, and N_2O of 310.

The U.S. EPA has estimated the contribution that various sectors of the U.S. economy are making to the emission of greenhouse gases.

Carbon Dioxide (CO_2)

The United States produced 1,580 million metric tons of CO_2 in 2003, coming from the use of petroleum (42%) and coal (36%).

Conservation tillage reduces emission of CO_2 from soil. CO_2 can also be sequestered in the soil.

Methane (CH_4) and Nitrous Oxide (N_2O)

Agriculture is a net emitter of greenhouse gases globally and in the United States. Conclusions, data, and estimates from the U.S. EPA state the following:

- Agriculture is responsible for 63% of human activity–related non-CO_2 emissions globally.
- Agriculture is responsible for 52% of human activity–related CH_4 released into the atmosphere globally.
- Agriculture is responsible for 84% of human activity–related N_2O released into the atmosphere globally.
- It is projected that agriculture-related CH_4 and N_2O emissions will increase by over 20% by 2020.
- In the United States, agriculture-related emissions of CH_4 and N_2O contribute 13% of non-CO_2 and less than 1% of all greenhouse gas emissions.

These figures do not include the generation of CO_2 from fossil fuel to run farm equipment or to produce fertilizers.

Where Do the CH_4 and N_2O Come From?

CH_4 is released into the atmosphere as the result of the fermentation in the rumen of cattle and other ruminants. This is called *enteric CH_4* and is a major contributor to CH_4 release globally and in the United States. Globally, 86 Tg of CH_4 is of enteric origin, and this represents 32% of global non-CO_2 greenhouse gas emissions. In the United States, CH_4 from livestock enteric origin is 27% of agricultural non-CO_2 emission and 2% of all greenhouse gas emissions (see Table 22-6).

Manure is responsible for the emission of both CH_4 and N_2O. In 2000, the U.S. EPA estimated that livestock manure was responsible for the release of 10.7 Tg CH_4 and 0.63 Tg N_2O into the atmosphere. This represents just less than 8% of global agricultural non-CO_2 greenhouse gas emissions and is projected to increase by 21% by 2020. Another major source of CH_4 is rice production.

TABLE 22-6 Estimates of the U.S. emissions of methane in teragram (Tg) CO_2 equivalents (1 Tg = 2.2 billion lb)

SOURCE	1990	2003
Landfills	172	131
Natural gas systems	128	126
Livestock—enteric	118	115
Coal mining	82	54
Livestock—manure	31	39
Waste water	25	37
Petroleum systems	20	17
Other	29	26

Source: U.S. Environmental Protection Agency.

Mitigation of the Effects of Agriculture on Greenhouse Gases

There are numerous approaches whereby agriculture can reduce its effect on greenhouse gas emissions:

- Changes in agronomic practice, such as shifts to reduce tillage and increase carbon sequestration.
- Improvements in the efficiency of animal production. This entails reducing the amount of greenhouse gases produced per unit of meat, milk, or eggs. For instance, the use of bovine somatotropin to increase the efficiency of milk production reduces CH_4 emission per unit of milk produced.
- Manure management to reduce CH_4 and N_2O.

It should be noted that ethanol production does not greatly affect the agricultural "footprint" on greenhouse gas emissions but will reduce that of petroleum somewhat.

ISSUES *for discussion*

GREENHOUSE GASES AND GLOBAL WARMING

1. Are you concerned with global warming?
2. Do you accept the link between climate change and human activity?
3. What are the likely impacts, if any, of global warming in your lifetime?
4. What should be policy changes, if any, to combat global warming?
5. Are the Kyoto treaty provisions fair to the United States?
6. What changes in agriculture should be considered to mitigate against global warming?
7. How does increasing the efficiency of animal production influence global warming?

REVIEW QUESTIONS

1. What is animal waste?

2. What are the major constituents of animal waste?

3. What can animal waste be used for?

4. What problems are associated with animal waste?

5. Are there problems with land application of animal waste when fresh fruits and vegetables for salads are being produced?

6. What is an animal feeding operation (AFO)?

7. What is a concentrated animal feeding operation (CAFO)?

8. Who defines an AFO and a CAFO?

9. Is animal waste considered as a point source of pollution?

10. Are CAFOs regulated by the U.S. EPA as point sources of pollution?

11. What is the rationale behind U.S. EPA regulation?

12. Does livestock influence the atmosphere?

13. What are the negative effects of odors from livestock?

14. How much do agriculture and livestock contribute to the release of ammonia into the atmosphere?

15. Does ammonia contribute to greenhouse gases?

16. What is ppmv?

17. What is ppbv?

18. Is there is a scientific consensus that changes in the composition of the atmosphere are the result of human activities?

19. What is a giga-gram (Gg)?

20. What is a tera-gram (Tg)?

21. What is the average increase in global temperatures over the last 100 years?

22. What is the global warming potential of greenhouse gases?

23. Can agriculture reduce emission of the greenhouse gas carbon dioxide (CO_2)?

24. What is the contribution of agriculture to methane release into the atmosphere?

25. Where does methane (CH_4) come from?

26. What is the contribution of agriculture to nitrous oxide (N_2O) release into the atmosphere?

27. Can agriculture mitigate greenhouse gases?

REFERENCES AND FURTHER READING

Asman, W. A. H. (2007). *Global emission inventory for ammonia, with emphasis on livestock and poultry*. Retrieved May 3, 2007, from www.adsa.org/discover/intersummaries/asman.doc

Kansas State University Extension. *Approximate composition of some natural fertilizer materials. SolidWaste Management Fact Sheet No. 8*. Retrieved July 24, 2009, from http://www.oznet.ksu.edu/library/solw2/EP8.PDF

Marland, G., Boden, T. A., & Andres R. J. (2006). Global, regional, and national CO_2 emissions. In trends: A compendium of data on global change. Oak Ridge, TN: Carbon Dioxide Information Analysis Center, Oak Ridge National Laboratory, U.S. Department of Energy. Retrieved August 5, 2007, from http://cdiac.ornl.gov/trends/emis/tre_usa.htm

U.S. Environmental Protection Agency. *Agriculture. Global mitigation of non-CO_2 greenhouse gases*. Retrieved July 24, 2009, from http://www.epa.gov/nonco2/econ-inv/downloads/GM_SectionV_Agriculture.pdf

U.S. Environmental Protection Agency. *Technical summary. Global mitigation of non-CO_2 greenhouse gases*. Retrieved July 24, 2009, from http://www.epa.gov/nonco2/econ-inv/downloads/GM_SectionI_TechnicalSummary.pdf

U.S. Environmental Protection Agency. *U.S. Emissions Inventory 2005: Inventory of U.S. greenhouse gas emissions and sinks: 1990–2003*. Retrieved August 3, 2007, from http://www.epa.gov/methane/sources.htm

Intensive Animal Agriculture

American Veterinary Medical Association. (2009). *Response to* The Final Report of the Pew Commission on Industrial Farm Animal Production. Retrieved November 17, 2009, from http://www.avma.org/advocacy/PEWresponse/PEW_report_response.pdf Accessed November 17, 2009.

Pew Commission on Industrial Farm Animal Production. (2009). *Putting meat on the table: Industrial farm animal production in America*. Retrieved November 17, 2009, from http://www.ncifap.org

SECTION V
ANIMAL PRODUCTS

CHAPTER 23

Introduction to Animal Products

OBJECTIVES

This chapter will consider the following:

- Where food is purchased, with food sales as an introduction
- The health effects of animal products
  - Essential fatty acids
  - Functional foods/zoonutrients such as conjugated linoleic acid (CLA)
  - The saturated fat versus trans fat issue

ANIMAL PRODUCTS IN FOODS PURCHASED AND CONSUMED

Most Americans consume animal products, often at every meal (see fig. 23-1). This may include the following:

- Breakfast: milk with cereal or eggs, bacon, and sausage
- Lunch: a hamburger or submarine sandwich with deli meats
- Dinner: meat with a starch and vegetables or pizza, followed by ice cream
- Party food such as cheese and crackers

Meat and dairy products are considered in more detail in chapters 25 and 26, respectively.

The Economic Research Service of the U.S. Department of Agriculture estimated that Americans in 2007 spent the following on food:

- Food at home, $583 billion: This figure includes food stamps; the federal Women, Infants, and Children program; and donated food, and represents about 5.7% of the household disposable income.
- Food away from home, $1,139 billion: This figure includes expense account and other business-related meals together with meals to inmates and patients. Excluding these meals, the food away from home decreases to $416 billion, which is 4.1% of the household disposable income.

We purchase most of the food we consume at home from large supermarkets. The Food Marketing Institute has data on the number of supermarkets (defined as >$2 million in sales per year) in the United States. In 2004, there were 34,052, and these employed 3.4 million people. The largest supermarket chains are listed in Table 23-1.

We spend over 40% of the disposable income used for food outside the home. This includes fast-food/quick-service restaurants, such as McDonald's (with

FIGURE 23-1 Children enjoying a nutritious meal, including a hamburger, at a day care facility. Courtesy of the U.S. Department of Agriculture. Photo by Ken Hammond.

TABLE 23-1 Largest supermarket chains in the United States by grocery sales in 2006

RANKING	COMPANY	SALES (BILLION $)	MARKET SHARE (%)
1	Walmart	92	18
2	Kroger	61	12
3	Safeway	38	8
4	SUPERVALU	35	7
5	Costco	32	6
6	Sam's Club	24	5
7	Ahold	24	5
8	Publix	22	4
9	Delhaize	17	4
10	H.E. Butt Grocery	12	3
11	Albertsons	10	2
12	A&P	9	2
13	Winn-Dixie	7	1
14	Meijer	7	1
15	Giant Eagle	6	1
16	Whole Foods	6	1
17	BJ's Wholesale Club	5	1
18	Hy-Vee	4	1
19	Wegmans	4	1
20	Target	4	1

Source: Directory of Supermarket, Grocery and Convenience Store Chains.

global sales of $22 billion per year), Burger King, Pizza Hut, Wendy's, Subway, Sonic, and Quiznos; national chains and local family restaurants, such as Applebee's, Chili's, and T.G.I. Fridays; and upmarket restaurants, including the steakhouses Outback Steakhouse and Ruth's Chris Steakhouse.

If we think of the signature product from virtually any of the national chains, it is "built" around animal products, for example, a hamburger, cheeseburger, Big Mac, Whopper, Meat Lover's Pizza, Egg McMuffin, sausage sandwich, deli sub, beef fajita, chicken enchilada, and large steaks.

ANIMAL PRODUCTS AND HUMAN HEALTH

Animal products have both positive and negative effects on the consumer. It is well established that animal products are a very important part of the human diet (see Table 23-2). Animal products provide an excellent source of protein, and both vitamins and minerals. Some of the important nutrients in animal products are the following:

- Protein, which when digested, provides an excellent balance of essential amino acids for growth, muscles, and replacing lost cells.
- Vitamins, particularly vitamin B_{12}, which is only found naturally in animal products, and vitamins A and D in milk. Milk in the United States is supplemented with these vitamins.

TABLE 23-2 Daily requirements of some critically important nutrients in the human diet and that can be provided predominantly by animal products

	PROTEIN (G/D)	CALCIUM (MG/D)	IRON (MG/D)
Children 4–8 y	15	800	4.1
Girls 9–13 y	28	1,300	5.7
Boys 9–13 y	27	1,300	5.9
Adolescent girls	38	1,300	7.9
Adolescent boys	44	1,300	7.7
Men	46	1,000[a]	6.6
Women	38	1,000[a]	8.1
Pregnant women	50	1,000	23
Lactating women	60	1,000	7.0

Source: National Academy of Science Food and Nutrition Daily Reference Requirements.
[a]Requirements for men and women over 50 years old increase to 1,300 mg/d.

- Vitamins in eggs such as choline (125 mg per egg), B vitamins, folate (essential for development), and lutein, which is important to functioning of the eye and particularly the macula.
- Key minerals:
  - Calcium, which is essential for bone development and potassium, lowering blood pressure; it is available from milk and dairy products.
  - Iron essential for hemoglobin, oxygen transportation, and, therefore, red blood cell formation together with mortar skill development; it is available from meat.
  - Zinc is a critical cofactor, with meat and eggs as excellent sources.
  - Copper is another critical cofactor, with meat as an excellent source.
  - Iodine is essential for thyroid hormone products and, therefore, normal metabolism, growth, and mental development, and is available from seafood.

Milk provides an excellent source of calcium. For instance, in the United States, women consume 64% of their calcium intake from milk and dairy products. Despite this, many women in the United States are not consuming the daily requirement of calcium as established by the National Academy of Science. There are also issues of iron deficiency. The Centers for Disease Control estimates the following levels of iron deficiency in the United States: 5% of American children, 16% of adolescent girls, 2% of men, and 12% of women.

There are marked racial disparities in iron deficiency in U.S. women, with iron deficiency found in 10% of white women, 19% of black women, and 22% of Latinas. This is linked to both the incidence of poverty and cultural issues.

Animal products from cattle may be considered as functional foods because they contain conjugated linoleic acid (CLA). This component has health-promoting effects. Moreover, consumption of fish and fish oils provides an excellent source of two of the essential fatty acids.

Animal products also contain saturated fat. The federal Food and Drug Administration (FDA) has concluded that saturated fat increases the blood concentration of low-density lipoprotein (LDL), which is the so-called bad cholesterol, and, therefore, elevates the risk of coronary heart disease. The major negative aspect of virtually any food is that when consumed in excess, it leads to obesity and consequent adverse effects of longevity and overall health.

ESSENTIAL FATTY ACIDS

There are three major types of omega-3 fatty acids that are ingested in foods and used by the body: a-linolenic acid, eicosapentaenoic acid, and docosahexaenoic acid. Once eaten, the body converts a-linolenic acid to eicosapentaenoic acid and docosahexaenoic acid, the two types of omega-3 fatty acids more readily used by the body. Extensive research indicates that omega-3 fatty acids reduce inflammation and help prevent risk factors associated with chronic diseases such as heart disease, cancer, and arthritis.

Crucial to health are omega-3 and omega-6 essential fatty acids. Consumption of fish and fish oils provides an excellent source of two of the essential fatty acids. The omega-3 and omega-6 essential fatty acids are polyunsaturated fatty acids and include the following:

- α-linolenic acid, which is referred to as 18:3;9,12,15 because there are 18 carbons in the hydrocarbon chain, with three double bonds at carbon numbers 9, 12, and 15. It is found in some plant oils and also in animal fats.
- Eicosapentaenoic acid (20:5[n-3]) is found in fish oils (see fig. 23-2). It is a precursor for biologic messenger compounds such as some prostaglandins, thromboxanes, and leukotrienes. Because of this, eicosapentaenoic acid reduces inflammation.
- Docosahexaenoic acid (22:6[ω-3]) is also from fish oils. Docosahexaenoic acid is a major fatty acid in phospholipids in the retina, in spermatozoa, and in the brain.

FIGURE 23-2 *A*, Structure of eicosapentaenoic acid. Source: Delmar/Cengage Learning. *B*, It can be found in omega-3 fish oil capsules. Reproduced by permission from vgstudio. © 2010 by Shutterstock.com.

- Linoleic acid (18:2:9,12) is the shortest chain omega-6 essential fatty acid, and is found in plant oils and also animal fats.

 ## FUNCTIONAL FOODS

Functional foods is a term used to describe the beneficial effects of the food that go beyond its traditional nutritive value. Functional foods can come from plants where they contain phytonutrients or from animal products where they contain zoonutrients. An example of a zoonutrient is conjugated linoleic acid (CLA).

Conjugated Linoleic Acids (CLAs)

CLAs, which are associated with improved health, are specific isomers of linoleic acid, namely, *cis*-9, *trans*-11 CLA isomer and the trans10, cis 12 CLA. Although these are trans-fatty acids, they are reported to have health benefits, including prevention of cancer, immune enhancement, and weight reduction, together with effects on the cyclooxygenase enzyme. The National Research Council of the U.S. National Academy of Science concluded that "CLA is the only fatty acid shown unequivocally to inhibit carcinogenesis in experimental animals."

The CLAs are produced by ruminants and the rumen microflora. Therefore, they are found in meat and dairy products.

 ## TRANS FATS

A trans fat is a vegetable fat, with unsaturated fatty acids, that has undergone hydrogenation such that the fatty acids are partially unsaturated. The fatty acids are trans isomers. Trans fats in foods are produced in industrial plants by partially hydrogenating plant oils such that they are solid at room temperature, are important for such products as margarine and cooking fats, and do not undergo spoilage or turn rancid (e.g., in items such as cookies or snack foods). According to the FDA, consumption of trans fats by people has the following detrimental effects: increasing circulating concentrations of LDL ("bad cholesterol"), decreasing circulating concentrations of high-density lipoprotein ("good cholesterol"), and increasing the risk of coronary heart disease.

Since 2006, the FDA has required that foods be labeled with their trans fat content (see fig. 23-3). Advocacy groups are pushing for trans fats to be banned. California together with some U.S. cities such as New York, Boston, and Philadelphia, and countries such as Switzerland and Denmark, have banned trans fats.

 ## SATURATED FATS AND CHOLESTEROL

Animal products are the primary source of saturated fat. Saturated fats are a major form of fat in animals. Some animal products contain a very high concentration such as butter, lard, and ice cream; others such as beef have a moderate amount depending on the cut, whereas other meats have low amounts of saturated fat such as skinned chicken breast. The FDA requires that processed foods be labeled with the amount of saturated fat (see fig. 23-3).

INTERESTING FACTOID

Butter has a very high content of saturated fatty acids, whereas margarine has both unsaturated fatty acids and hydrogenated fatty acids. So, is butter better for you than margarine? The medical consensus seems to be that it depends on the type of margarine. The more it is solid because of hydrogenation, the more trans fat and saturated fat it will contain.

FIGURE 23-3 The FDA requires that the nutrient content labels of processed foods include the amount of trans fats and saturated fats. Reproduced by permission from Stephen VanHorn. © 2010 by Shutterstock.com.

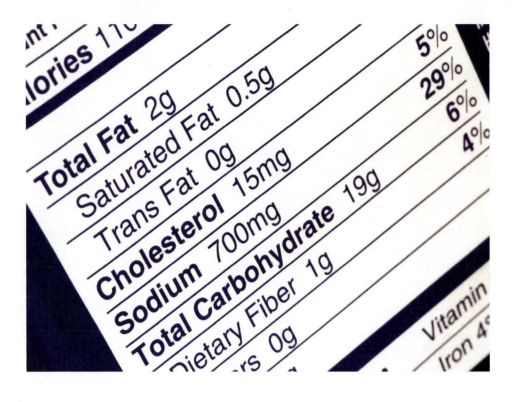

Saturated fat and cholesterol increase the circulating concentrations of LDL in human populations and, thereby, increase the risk of coronary heart disease. The Centers for Disease Control considers that "diets high in saturated fat have been linked to chronic disease, specifically, coronary heart disease." The National Institutes of Health estimates that more than 12.5 million Americans have coronary heart disease, with more than 500,000 deaths each year.

REVIEW QUESTIONS

1. What proportion of American's food dollar is for food consumed at home?

2. What are the top five supermarket chains in the United States?

3. What is a quick-service restaurant?

4. Give five examples of a quick-service restaurant.

5. Give examples of how animal products have a positive effect on human health.

6. What is an essential fatty acid?

7. What are the negative effects of animal products on human health?

8. What are high-density lipoprotein and low-density lipoprotein, and why should we worry about the level of low-density lipoprotein?

9. What is a trans fat?

REFERENCES AND FURTHER READING

National Research Council. (1996). *Carcinogens and anticarcinogens in the human diet*. Washington, DC: National Academy Press.

CLA and Zoonutrients

Bauman, D. E., Baumgard, L. H., Corl, B. A., & Griinari, J. M. (1999). Biosynthesis of conjugated linoleic acid in ruminants. In *Proceedings of the American Association of Animal Science*. Retrieved July 27, 2009, from http://www.asas.org/jas/symposia/proceedings/0937.pdf

Fat in the Diet

Centers for Disease Control and Prevention. *Dietary fat*. Retrieved July 27, 2009, from http://www.cdc.gov/nccdphp/dnpa/nutrition/nutrition_for_everyone/basics/fat.htm#saturated%20fat

U.S. Department of Agriculture. *Dietary guidelines for Americans 2005*. *Chapter 6 fats*. Retrieved July 27, 2009, from http://www.health.gov/DIETARY GUIDELINES/dga2005/document/html/chapter6.htm

Food Distribution

Food Marketing Institute. *Supermarket facts. Industry overview 2006*. Retrieved July 27, 2009, from http://www.fmi.org/facts_figs/superfact.htm

Food Expenditures

U.S. Department of Agriculture. Economic Research Service. *Food CPI and expenditures: Food expenditure tables*. Retrieved July 27, 2009, from http://www.ers.usda.gov/Briefing/CPIFoodAndExpenditures/Data/

Importance of Animal Products in Human Diet

Weaver, C. M. (2000). Calcium requirements of physically active people. *The American Journal of Clinical Nutrition, 72,* 579S–584S.

Iron Deficiency

Centers for Disease Control. *Iron deficiency—United States, 1999-2000. Table 1. Prevalence of iron deficiency—United States, national health and nutrition examination surveys, 1988-1994 and 1999-2000.* Retrieved July 27, 2009, from http://www.cdc.gov/mmwr/preview/mmwrhtml/mm5140a1.htm#tab1

Nutrition Labeling of Food

U.S. Department of Health & Human Services. U.S. FDA. *Food labeling guide.* Retrieved July 27, 2009, from http://www.cfsan.fda.gov/~dms/2lg-toc.html

Saturated and Trans Fat and Human Health

Advanced notice of proposed rulemaking—food labeling: Trans fatty acids in nutrition labeling. Retrieved August 2, 2009, from www.fda.gov/ohrms/dockets/dailys/03/oct03/100903/03n-0076-c000002-vol1.pdf

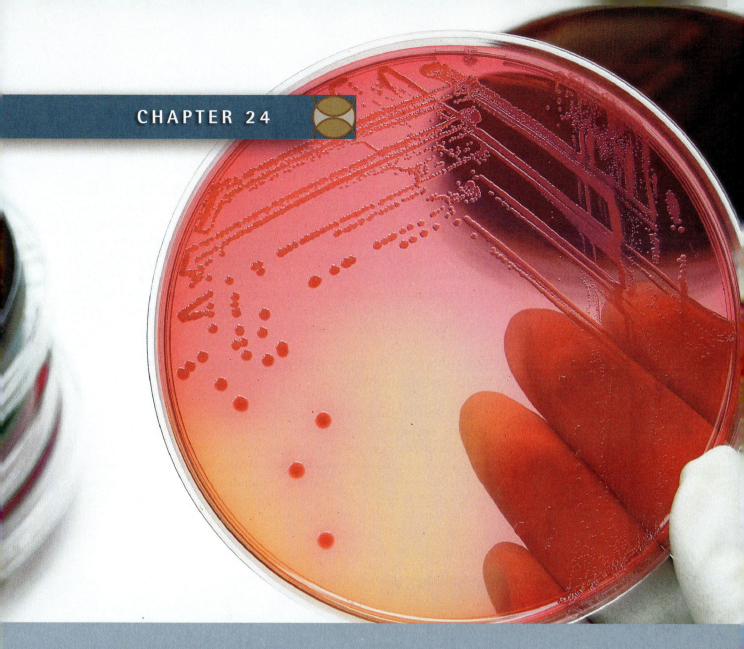

CHAPTER 24

Food Safety of Animal Products

OBJECTIVES

This chapter will consider the following:

- Overview of food safety
- Incidence of food-borne pathogens and their impact
- Food safety and the livestock industry before and after harvest, including irradiation
- Hazard analysis and critical control point (HACCP)
- Food safety in the home and restaurants
- Alternative approaches to food safety

Definition

Hazard analysis and critical control point (HACCP) is the approach developed to ensure the safety of food for astronauts more than 30 years ago. It is used by the Food and Drug Administration (FDA) and U.S. Department of Agriculture (USDA) to improve the safety of food in the United States. The approach is to focus on critical control points (CCPs) in the preparation of food before purchase. There are CCPs after purchase by the consumer, with food storage, preparation, and cooking being examples.

OVERVIEW OF FOOD SAFETY

Some people view pesticides together with food preservatives as a major threat to health despite these having gone through governmental approval processes. The reality based on scientific research is that food-borne pathogens (bacteria and viruses) are the real risks of unsafe food (see fig. 24-1).

There have been significant advances in food safety using the approaches of Hazard analysis and critical control points (HACCP) (see the sidebar). The U.S. Department of Agriculture's Food Safety Inspection Service is responsible for inspection of meat at packing plants. The Food Safety Inspection Service is shifting to a new system for inspection, namely, risk-based inspection (see fig. 24-2).

Internationally, the Codex Alimentarius Commission sets standards for food safety. The World Trade Organization recognizes only one body for sanitary and phytosanitary issues of food safety: the Codex Alimentarius Commission. This body uses a science-based risk assessment.

INCIDENCE OF FOOD-BORNE PATHOGENS AND THEIR IMPACT

Food-borne bacterial diseases are a significant health problem in the United States. In 2007, the USDA estimated that infection of people in the United States by the food-borne pathogen Shiga toxin producing *Escherichia coli* 0157 cost the nation $430 million because of hospitalization and loss of productivity. Moreover, the cost of another food-borne organism, *Salmonella*, is $2.4 billion per year. A published paper by Meade et al. at the Centers for Disease Control (CDC) and Prevention in 2005 estimated the severity of food-borne organisms in the United States to be 76 million illnesses per year, 325,000 hospitalizations per year, 5,000 deaths per year, and costs to the economy of $7 billion per year.

FIGURE 24-1 Researchers are making progress in developing techniques to improve food safety. Image courtesy of Agricultural Research Service, USDA. Photo by Scott Bauer.

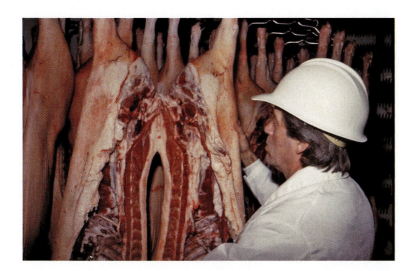

FIGURE 24–2 USDA inspection of meat. Courtesy of the USDA.

The CDC and Prevention estimated in 2005 that the top five food-borne bacterial diseases were the following:

- *Campylobacter* (see fig. 24-3), with about 2 million cases per year
- *Salmonella* nontyphoidal, with 1.3 million cases per year
- *E. coli* O157:H7, with 63,000 cases per year
- *E. coli* non-O157:H7, with 31,000 cases per year
- *Listeria monocytogenes* (see fig. 24-4), with 2,500 cases per year

The severity of the illness when infected with *Listeria* is particularly acute, with death rates close to 50%. Other diseases caused by bacteria that contaminate food or water are cholera and typhoid. Typhoid fever is a life-threatening illness caused by the bacterium *Salmonella typhi*. These pathogens only live in people. In the United States, there are 400 cases per year, with most acquired while traveling internationally. In the developing world, 21.5 million people per year are infected with typhoid. Cholera is an intestinal infection caused by food or water contaminated with the bacteria *Vibrio cholerae*. The bacteria produce an enterotoxin causing watery diarrhea. This can lead to severe dehydration and death. It is one of three diseases that require notification to the World Health Organization. There were 131,943 cases reported to the World Health Organization in 2005, with only 12 cases in the United States. With underreporting, the World Health Organization estimates that worldwide, there are between 1.3 and 2.6 million cases annually. There is 2% mortality with the disease, but this can reach over 40% in some countries.

There are also food-borne viral diseases. Viral diseases can be contracted from some animal products. For instance, hepatitis A can be carried by shellfish from water polluted by raw sewage. The hepatitis A virus causes hepatitis, which is a disease of the liver. Similarly, contaminated raw or partially cooked shellfish can be the source of noroviruses causing viral gastroenteritis. The infection can also be spread by food preparers infected with the virus.

Food-borne diseases are a particular problem with young children, the elderly, and others with impaired immune functioning, such as people infected with HIV

INTERESTING FACTOID

The safety of astronauts has improved our lives. We all know how unpleasant food-borne diseases are with diarrhea and vomiting. To have this in space would be worse for both the person with the problem and his or her compatriots. The National Aeronautics and Space Administration developed HACCP to greatly reduce the possibility of food-borne disease. They set a zero tolerance for *Salmonella* and a low acceptable level for coliforms.

FIGURE 24-3 Transmission electron micrograph of *Campylobacter*, which is a major food-borne pathogen. Courtesy of Irene Wesley, the National Animal Disease Center, Agricultural Research Service, and USDA.

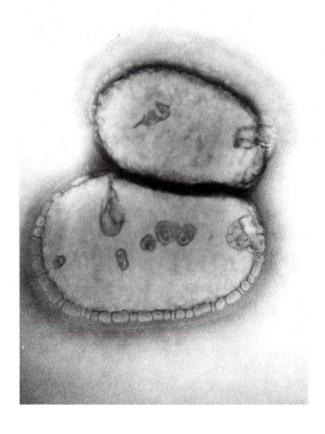

FIGURE 24-4 Transmission electron micrograph of *Listeria*, which is a major food-borne pathogen. Courtesy of Irene Wesley, the National Animal Disease Center, Agricultural Research Service, and USDA.

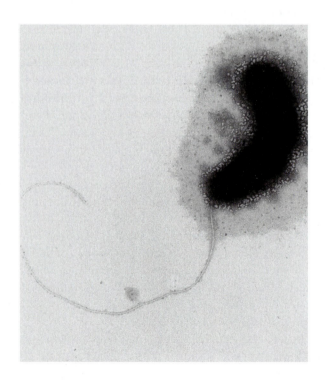

FIGURE 24-5 A pregnant woman using a kitchen knife to cut raw meat before cooking. Pregnant women, young children, the elderly, or people with immunosuppression can be at greater risk for acquiring food-borne illness, and need to take additional precautions when handling raw food products. Courtesy of the CDC. Photo by James Gathany.

and those receiving chemotherapy (see fig. 24-5). Risk factors for food-borne pathogens are the following:

- Age: young children (<4 or 5 years) and older people (>60 or 65 years)
- Pregnancy
- Immune system compromised because of HIV or chemotherapy
- Reduced liver or kidney function

 ## FOOD SAFETY AND THE ANIMAL PRODUCTS

It is essential to ensure the safety of the products of livestock and poultry production. Animal products can be a cause of concern for food safety because of the following:

- Chicken or pork can be contaminated with *Campylobacter* (see fig. 24-3), *L. monocytogenes* (see fig. 24-4), *Salmonella*, and *E. coli* non-O157:H7.
- Beef can be contaminated with *E. coli* O157:H7 and other food-borne bacteria.

It should be emphasized that food safety issues are not restricted to animal products because there are a considerable number of cases of food-borne diseases after consumption of vegetables contaminated with pathogens.

One of the horror stories in food safety was the case of the Jack in the Box restaurant chain in the 1990s. Four people died, and hundreds became ill from eating hamburger made with beef contaminated with *E. coli* O157:H7. This involved not only a problem with the beef, but also inadequate cooking. Heat will kill the bacteria. When HACCP started, the baseline presence of *Salmonella* was for chickens 20%, hogs 8.7%, cattle 2%, ground beef 7.5%, ground chicken 45%, and ground turkey 50%.

Preharvest Food Safety

Preharvest food safety is to have the healthiest animals or their products being harvested. Examples of successful preharvest food safety is the elimination of

INTERESTING FACTOID

Food-borne pathogens are not the only cause of diarrhea. There are numerous other causes. According to the CDC, the most common cause of severe childhood diarrhea is a rotavirus. The symptoms are watery diarrhea and vomiting. About 55,000 children are hospitalized annually in the United States because of this rotavirus infection. Over half a million children die annually from the disease worldwide. This virus passes from person to person due to fecal contamination of the hands and the like. There is now a vaccine for rotavirus.

Salmonella from flocks of laying chickens and the tremendous reduction of trichina in pork.

Postharvest Food Safety

There are a series of approaches to improve the safety of animal products after harvest, including HACCP, treatment of carcasses to reduce pathogens, pasteurization (addressed in Chapter 26), and irradiation.

Treatment of Carcasses to Reduce Pathogens

The USDA has approved the use of an acid rinse to be sprayed on meat carcasses. Contamination occurs when the hide or intestines are being removed. Acetic, lactic, and citric acid solutions are all effective in reducing bacteria contamination of the carcasses. Trisodium phosphate is used to reduce pathogens in broiler-processing plants.

Irradiation (or Cold Pasteurization)

Food irradiation is the treatment of food with either radiation from radioactive isotopes (cobalt-60 or cesium-137) or electrons from a linear accelerator to kill pathogens and spoilage microorganisms. The food is moved through the irradiation chamber on a conveyor. Irradiation does not increase the temperature of the food. The energy waves are not retained by the food; the food does not become radioactive.

Irradiation of food as a method of food preservation began in 1950. Currently, 40 countries have approved irradiation for use on a variety of foods. Because irradiation is classified as a food additive, the FDA must approve its use in the United States. All irradiated foods must carry the international symbol called a "Radura," along with a statement that they have been treated by irradiation (see fig. 24-6).

Foods currently approved for irradiation by the FDA include spices, herbs, and seasonings since 1983; pork since 1985; fruits and vegetables since 1986; poultry since 1990; and ground beef since the early 2000s. Hospitals irradiate food for patients who are immunocompromised.

Irradiation represents a significant advance in food safety, but it is only being implemented slowly because of organized opposition. The FDA has proposed easing labeling requirements for irradiated foods. There are organizations such as the Organic Consumers Association and Public Citizen that oppose irradiation of food. These cite concerns such as the following:

FIGURE 24-6 The international Radura symbol used to show that food has been irradiated to destroy food-borne pathogens. Courtesy of the USDA.

- Irradiation does not help the consumer or the farmer.
- Irradiation masks "filthy conditions in slaughterhouses."
- Irradiation leads to the formation of toxic compounds in the meat.
- Irradiation destroys nutrients such as vitamins.

These claims have been addressed by the U.S. CDC and Prevention and others. The U.S. CDC and Prevention have concluded:

> Treating raw meat and poultry with irradiation at the slaughter plant could eliminate bacteria commonly found raw meat and raw poultry, such as *E. coli* O157:H7, *Salmonella*, and *Campylobacter*. These organisms currently cause millions

of infections and thousands of hospitalizations in the United States every year. Raw meat irradiation could also eliminate Toxoplasma organisms, which can be responsible for severe eye and congenital infections. Irradiating prepared ready-to-eat meats, like hot dogs and deli meats, could eliminate the risk of *Listeria* from such foods. Irradiation could also eliminate bacteria like *Shigella* and *Salmonella* from fresh produce.

The technique certainly aids the consumer by protecting against food-borne disease. The argument that the irradiation only masks poor conditions in packing plants is belied by the federal inspection of these plants, and by the notion that both irradiation and cooking kill pathogens. Should cooking be banned? Vitamins sensitive to heat are also sensitive to irradiation. It is sometimes claimed that irradiation leads to the formation of volatile toxic chemicals in meat. Although this may be associated with off-flavors, the CDC and Prevention have concluded that "an overwhelming body of scientific evidence demonstrates that irradiation does not harm the nutritional value of food, nor does it make the food unsafe to eat."

INTERESTING FACTOID
Organizations that support food irradiation are the American Medical Association, American Veterinary Medical Association, Institute of Food Technologists, Scientific Committee of the European Union, United Nations Food and Agriculture Organization, World Health Organization, National Food Processors Association, and U.S. CDC and Prevention.

ISSUES *for discussion*

FOOD SAFETY

1. What level of risk is acceptable for your food?
2. Should we use/misuse science as the basis of public policy?
3. Why do you think there is opposition to irradiation?
4. How many people have become sick or died because of slowness to implement irradiation widely?
5. Who are the winners and losers with irradiation?
6. Who gains by supporting or opposing irradiation?
7. Is science education in the public and media at such a low level that consumers will never accept irradiation?

What Is HACCP?

HACCP is based on seven principles:

Principle 1: Do a hazard analysis to identify where problems may exist and actions that may be taken to remedy the problems.

Principle 2: Identify the critical control points (CCPs) (in 1 above) that are critical for the safety of the products.

Principle 3: Establish critical limits that define the difference between a safe and unsafe product.

Principle 4: Establish a system to monitor control of CCP.

Principle 5: Establish corrective actions to be taken when a CCP is not under control.

Principle 6: Establish procedures to verify that the HCCP system is working correctly.

Principle 7: Establish documentation for all processes and steps.

FOOD SAFETY IN THE HOME AND IN RESTAURANTS

Food safety at home or in a restaurant is ensured by hand washing, sanitation of the cooking surfaces and utensils, keeping uncooked meat separated from salads, a correct thawing technique, and cooking to an adequate temperature to destroy pathogens.

Hand Washing

Whether working in a restaurant or at home preparing a meal, we should wash our hands before we start food preparation, after preparing meats, and after using the bathroom. In addition, we should wash again before eating.

Many young children at the kindergarten level learn to wash their hands with a 20-second wash with both soap and water. However, surprisingly, only 68% of adults (74% women versus 61% men) in the United States wash their hands, even after using the restroom. The type of soap makes a difference in its effectiveness. E2-rated hand soaps, which is a USDA classification requiring equivalency to 50 ppm chlorine, are the most effective soaps at reducing the bacterial number on hands.

Sanitation of the Cooking Surfaces and Utensils

Wash cutting boards, utensils, and countertops with hot, soapy water containing bleach, and dry with paper towels or clean cloths. Then wash the cloths in the hot cycle often.

Keeping Uncooked Meat Separat]ed from Salads

Keep raw meat, poultry, and seafood separate from food that will be eaten raw such as salads, fruit, and vegetables when grocery shopping, in the refrigerator, and on cutting boards when preparing food. Vegetables to be eaten raw should be washed thoroughly.

Correct Thawing Technique

Food should be thawed in a refrigerator. If there is a need for quick thawing, the food can be submerged in cold water or thawed in a microwave.

Cooking to an Adequate Temperature to Destroy Pathogens

According to the USDA, food should be cooked to the following recommended temperatures to kill any contaminating pathogens (see fig. 24-7):

- Ground or cubed meat, including beef, pork, veal, and lamb: 71°C (160°F).
- Turkey and chicken: 74°C (165°F).
- Fresh beef, veal, and lamb, medium rare: 63°C (145°F).
- Fresh beef, veal, lamb, pork, and ham, medium: 71°C (160°F).
- Fresh beef, veal, lamb, and pork, well done: 77°C (170°F).
- Chicken and turkey, whole and other poultry: 82°C (180°F).
- Poultry breasts, roast: 77°C (170°F).
- Eggs should be cooked until the yolks and whites are firm.

FIGURE 24-7 Cooking meat to the recommended temperature is critically important to ensure that food-borne pathogens are destroyed. Reproduced by permission from Alexey Stiop. © 2010 by Shutterstock.com.

Traveler's Upset Stomach/Diarrhea

Travelers should avoid tap water, ice, and all raw food such as salads, uncooked vegetables, unpeeled fruit, and ice cream, together with raw or undercooked meat, fish, and shellfish.

Alternative Approaches to Improving Food Safety

There are alternate approaches being researched to eliminate or greatly reduce the load of pathogens on food and, therefore, improve food safety.

Bacteriophage

These are virus-like particles that require taking over the cellular systems of bacteria to propagate. These have an antibiotic effect.

Colicins

Some specific strains of *E. coli* and some related species when grown in culture produce proteins called *colicins*. These are toxic to susceptible bacteria. Colicins may have potential uses to reduce pathogen loads in food and, therefore, improve food safety.

REVIEW QUESTIONS

1. What are food-borne pathogens?

2. Why are food-borne pathogens a problem?

3. What is the magnitude of the problem in the United States?

4. In what populations in the United States are food-borne pathogens a greater problem?

5. Give three examples of bacterial food-borne pathogens.

6. Give an example of a viral food-borne pathogen.

7. How can food safety be improved before harvest?

8. How can food safety be improved after harvest?

9. What approaches can be used?

10. What is HACCP?

11. What is food irradiation? What does it do, and how does it work?

12. What foods have been approved by the FDA to be irradiated?

13. How can you improve the safety of food when either cooking at home or working at a restaurant?

REFERENCES AND FURTHER READING

FDA Center for Food Safety and Applied Nutrition. USDA Food Safety and Inspection Service. *Handwashing-related research findings*. Retrieved July 27, 2009, from http://www.foodsafety.gov/~dms/fsehandw.html

USDA. Economic Research Service. *Foodborne illness cost calculator*. Retrieved July 27, 2009, from http://www.ers.usda.gov/data/Foodborneillness/

USDA. Food Safety and Inspection Service. *Thermy*. Retrieved July 27, 2009, from www.fsis.usda.gov/thermy

Irradiation

Department of Health and Human Services. CDC and Prevention. *Food irradiation*. Retrieved July 27, 2009, from http://www.cdc.gov/ncidod/dbmd/diseaseinfo/foodirradiation.htm

Hecht, M. M. (1999). Scientific answers to irradiation bugaboos. *21st Century Science & Technology Magazine*. Retrieved July 27, 2009, from http://www.21stcenturysciencetech.com/hecht_irra.html

Organic Consumers Association. *Information on food irradiation*. Retrieved July 27, 2009, from http://www.organicconsumers.org/irradlink.cfm

U.S. Department of Health & Human Services. U.S. FDA. *Consumer information.* Retrieved July 27, 2009, from http://www.fda.gov/opacom/catalog/irradbro.html

Incidence of Food-Borne Diseases

Mead, P. S., Slutsker, L., Dietz, V., McCaig, L. F., Bresee, J. S., Shapiro, C., et al. (1999). CDC and Prevention. Food-related illness and death in the United States. *Emerging Infectious Diseases, 5*(5). Retrieved July 27, 2009, from http://www.cdc.gov/ncidod/eid/Vol5no5/mead.htm

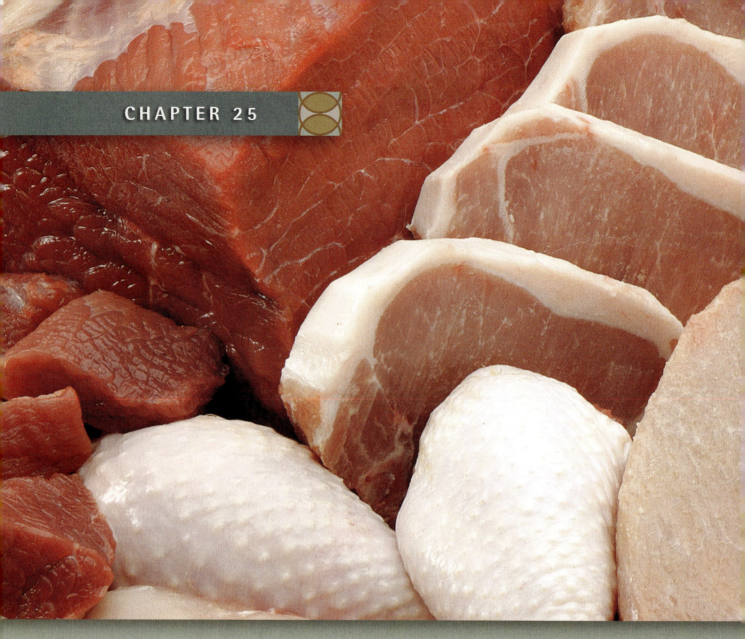

CHAPTER 25

Meats

OBJECTIVES

This chapter will consider the following:

- An overview of meat production and consumption in the United States

Meat production in the United States is a major industry, with $70 billion production in 2006. Meat is produced for the domestic market and for export. Table 25-1 summarizes the export and import of meat by the United States. It is readily apparent that the United States is a net exporter of meat, with pork and chicken exports more than making up for the imbalance in beef.

TABLE 25-1 Statistics on the export and import of meat by the United States in 2005

SPECIES	EXPORT	IMPORT	NET
Beef	0.35	1.80	-1.45
Pork	1.33	0.50	0.83
Lamb	0	0.09	-0.09
Chicken	2.65	0.015	2.64

Note: Data are in million tons of dressed carcasses.
Source: USDA's Economic Research Service.

Americans are the number-two country for consumption of meat in the world (see fig. 25-1 for an example of meat products). According to the U.S. Department of Agriculture's (USDA's) Economic Research Service, per capita, consumption of meats in 2005 was chicken 100.4 lb (60.4 lb boneless), beef 93.2 lb (62.4 lb boneless), pork 63.8 lb (46.5 lb boneless), turkey 16.6 lb, lamb 1.2 lb, and veal 0.6 lb. Note that these data are based on dressed carcasses, with saleable or boneless meat shown in parentheses where the data are available.

Meat is an important part of the human diet. In its "Food Pyramid," the USDA recommends the consumption of meat and meat products, or their equivalent. The recommendations are summarized in Table 25-2 (see also fig. 25-2).

MEAT QUALITY

The quality of meat is determined by a number of important factors: color (see fig. 25-3), amount of marbling or intramuscular fat giving a good eating experience

> **INTERESTING FACTOID**
>
> The United States is a major producer of red meat annually. According to the National Agricultural Statistics Service (http://usda.mannlib. cornell.edu/usda/nass/ LiveSlau//2000s/2007/ LiveSlau-01-26-2007.pdf), and based on its 2006 figures:
>
> - Beef, 26 billion lb (12 million metric tons)
> - Veal, 156 million lb (71,000 metric tons)
> - Pork, 21.0 billion lb (9.5 million metric tons)
> - Lamb and mutton, 185 million lb (84,000 metric tons)

FIGURE 25-1 The variety of deli meats, including sliced sausage, hams, roast beef, and turkey, is illustrated. These are essential to many sandwiches, e.g., using sliced bread or rolls, or submarines (hoagies). Courtesy of the Agricultural Research Service and USDA. Photo by Scott Bauer.

TABLE 25-2 Daily consumption of meat (or equivalent) recommended in USDA's food pyramid

	AGE (Y)	DAILY RECOMMENDATION AS COOKED WEIGHTS (OZ EQUIVALENTS)
Children	2–3	2
	4–8	3–4
Girls	9–18	5
Boys	9–13	5
	14–18	6
Women	19–30	5.5
	31 and older	5
Men	19–30	6.5
	30–50	6
	51 and older	5.5

Note: One ounce of cooked meat, including poultry or fish, is equivalent to one egg or ¼ cup cooked dry beans.

(see fig. 25-4), drip or exudate, physical characteristics such as shear, and palatability or eating properties.

Many would suggest that eating properties should be at the top of the list, but if the meat looks like it is the wrong color (too dark or brown), is too fat, or has "fluid" dripping off, it will not be purchased. Additional vitamin E in the feed the animal receives before slaughter can help the meat retain optimal color.

FIGURE 25-2 Food pyramid showing the USDA recommended diet.

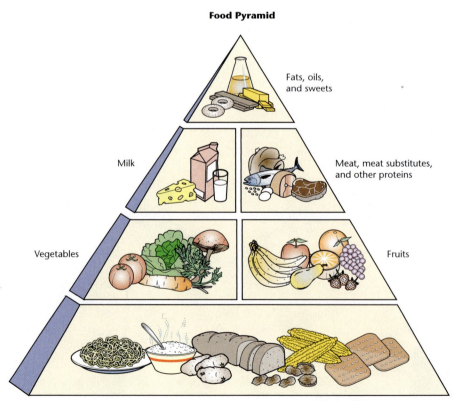

Food Pyramid

Fats, oils, and sweets

Milk

Meat, meat substitutes, and other proteins

Vegetables

Fruits

Breads, grains, and other starches

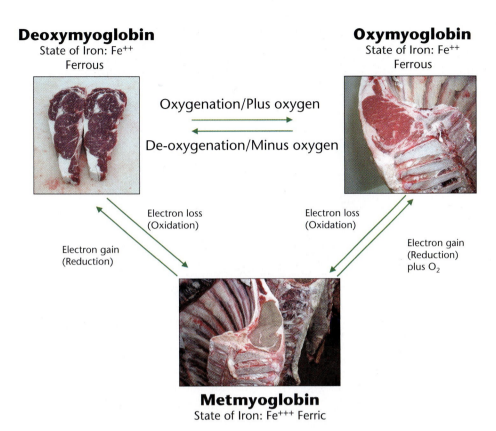

Deoxymyoglobin
State of Iron: Fe++
Ferrous

Oxygenation/Plus oxygen

De-oxygenation/Minus oxygen

Oxymyoglobin
State of Iron: Fe++
Ferrous

Electron loss
(Oxidation)

Electron loss
(Oxidation)

Electron gain
(Reduction)

Electron gain
(Reduction)
plus O_2

Metmyoglobin
State of Iron: Fe+++ Ferric

FIGURE 25-3 Consumers respond to the color of beef and pork. The color of meat is determined by the chemical form (the oxidative state) of the iron atom in the molecule myoglobin. Images show the changes in meat color with changes in the oxygenation (easily reversible binding of oxygen) or oxidation (changes in the oxidation state of the iron atom) in the myoglobin in meat. Images kindly provided by Steven Lonergan, Department of Animal Science, Iowa State University.

Porcine Stress Syndrome and Meat Quality (Pale, Soft, Exudative [PSE] Pork)

Porcine stress syndrome is where pigs exhibit increased susceptibility to stress, as can readily be established by their high sensitivity to the anesthetic gas halothane. The problem is that pigs with susceptibility to stress have very poor-quality meat, that is, PSE pork. The meat quality is poor based on the reduced color, high amount of drip or moisture loss, and rapid decrease in pH after death. This is a genetic condition, with the pigs homozygous for the trait (nn) showing a greater degree of PSE pork, and heterozygous pigs (Nn) being intermediate between the nn pigs and pigs (NN) not carrying the trait. Under some circumstances, however PSE pork can be found even in NN pigs.

The molecular basis of PSE pork is now known. It is genetic trait due to a mutation in a gene for the ryanodine receptor. Ryanodine receptors are a class of calcium-release channels in the muscle. There is a single nucleotide polymorphism in the ryanodine receptor 1 gene. This is the primary cause of classic porcine stress syndrome and the poor-quality meat.

Traditional Food Preservation

Traditionally, there have been a few techniques to preserve meat and fish, which easily spoil at room temperature. These techniques make the meat or fish inhospitable to the

FIGURE 25-4 The quality of the meat (beef) is determined by its eating quality (e.g., is it juicy?) when cooked. This in turn depends on the amount of intramuscular fat or marbling and the lack of fat that needs to be trimmed off. Note that there is a good deal of intramuscular fat. Images kindly provided by Steven Lonergan, Department of Animal Science, Iowa State University.

WORD ORIGINS

The origin of words used for processed animal products:

1. How they relate to the part of the carcass:
   - Bacon, which was originally from the back (*bakkon*) of the pig.
   - The canned spiced ground pork/ham, Spam, is a trademarked product of Hormel. The name may have been originally an acronym for the shoulder of pork and ham or from spiced h*am*.
2. How they relate to the method of food preservation:
   - Salami, which is from vulgar Latin *salamen,* from *salare* "to salt," and from the Latin *sal* (genus *salis*) "salt"
   - Sausage derives from vulgar Latin (through old French) *salsica* which means "seasoned with salt."

growth of pathogenic and other microorganisms by removing water or lowering the pH. The traditional approaches are curing by salting, drying, smoking, and pickling.

These techniques have not only preserved animal products but also developed a wonderful variety of different foods. Frequently, more than one approach is used with the same food. For instance, bacon is cured and then can be smoked, whereas ham is cured and then dried.

Curing

Curing is usually accomplished by salting, which consists of using dry salt or concentrated salt solution/brine or a brine containing salt "corns" to "pull" water out of the meat so that the meat has a high osmotic pressure, represents a hostile environment for microorganisms, and prevents autodegradation. Examples of curing include bacon, which is cured with brine; corn beef, which is cured in brine; ham, which is either dry salt cured or brine cured; and pastrami, which is beef that has been salt cured and cooked.

Drying

Examples of dried meat and fish include the following:

- Jerky, which is beef that is heat- or sun-dried to remove water
- Pemmican, which is dried pulverized meat (e.g., venison) with berries and fat

- Lutefisk, which is whitefish, such as cod, that is air-dried and treated with either soda lye or potash (a Scandinavian dish)

Smoking

Examples of smoked meat and fish include smoked bacon and ham, smoked salmon, smoked whitefish, and smoked herring or kipper.

Pickling

Pickling is used to preserve food by achieving an acid environment inhospitable to pathogenic bacteria. Examples of animal products include pickled herring, rollmops, and pickled eggs. The fish or eggs are immersed in vinegar (dilute acetic acid) to preserve them.

 ## PROCESSED MEATS

There are three major types of processed meat: ready to eat; shelf stable; and cooked, but not shelf stable. Common examples of processed meat are bacon (cured pork bellies in the United States), ham, and sausages.

Ham

There are many kinds of ham. Hams may be fresh, "cook before eating," cooked, picnic, and country types. According the USDA's Food Safety and Inspection Service, hams can be fresh, cured, or cured and smoked. Ham is the cured leg of pork and is usually pink. Types of processing to produce ham include the following:

- Dry cured, which is rubbed with salt together with sodium nitrite, and with some sugar. This is followed by drying and aging.
- Wet cured, which involves soaking in or injection of brine, traditionally followed by smoking.
- Prosciutto is ham produced by the technique of salting and drying to remove water, and originates from regions of central and northern Italy. Examples of prosciutto include Prosciutto di Parma (Palma ham) and Prosciutto di San Daniele.

Sausages

Sausages are traditionally ground meat with spices encased in the small intestine. They can be skinless or encased in artificial casing. In some sausages, the finely ground meat is in the form of an emulsion. Sausages may be eaten after cooking, or if cured, dried, or smoked, cooking may not be necessary.

There are many types of sausage (see fig. 25-5):

- Bologna, which is cooked, smoked, cured beef and pork that is salted and ready to eat.
- Bratwurst, which is fresh, and sometimes smoked and cooked.
- Braunschweiger, which is precooked, smoked liver, eggs, plus milk. This is ready to eat and is spreadable.
- Chorizo, which is smoked fatty pork.

---

3. How they relate to the geographic origin of different meats:
   - Bologna sausage or Bologna, which is sausage that originated in Bologna, Italy.
   - Frank or Frankfurter is sausage from Frankfurt, Germany.
   - Hamburger, which has no definite link to the town of Hamburg in Germany.
   - Vienna sausage, as the name implies, comes from the city of Vienna, as does weener, wiener, and wienerwurst.
   - Wurst, which is German for sausage.
4. Urban legend:
   - Hot dog is from the "urban legend" that hot dogs contained dog meat. If this was ever true, it has not been for more than 100 years.

NOMENCLATURE

Bratwurst is a German sausage originating from Thuringia. It is also popular in some states in the United States, such as Wisconsin and Iowa, which had high numbers of Germans who immigrated there in the nineteenth century. Bratwurst is made from finely chopped pork and/or beef. The term *brat* comes from the fine chopping.

FIGURE 25-5 An ever-popular form of meat is sausage. Sausages from pork in grocery store. Courtesy of the USDA. Photo by Ken Hammond.

- Hot dogs, which are cooked, smoked, pork or beef cured and ready to eat.
- Kielbasa, which is fresh, smoked beef and pork that is served cooked.
- Liverwurst (from the German leberwurst), which is made of ground pork and pig liver, and after curing and drying, it is ready to eat.
- Pepperoni, which is air-dried pork and beef that is usually ready to eat.
- Salami, which is hindquarter pork or beef, cured and dried, and can be ready to eat.
- Summer sausage (cervelat), which is pork dried, cured, and ready to eat.

REVIEW QUESTIONS

1. How important is meat exporting?
2. What is PSE pork?
3. How is meat evaluated for quality?
4. What are the traditional methods of preserving meat and other meat products?
5. What is the principle(s) underlying the curing, drying, and smoking of meats?
6. What are ham, jerky, corned beef, bacon, sausage, and Spam?
7. Give three examples of sausages. What are they made of?
8. Why is food pickled? Give examples of animal products that are pickled.

REFERENCES AND FURTHER READING

Food Consumption

USDA. Economic Research Service. *Food availability: Spreadsheets*. Retrieved July 27, 2009, from http://www.ers.usda.gov/data/foodconsumption/FoodAvailSpreadsheets.htm#dymfg

USDA's Food Pyramid

USDA. *Inside the pyramid*. Retrieved July 27, 2009, from http://www.mypyramid.gov/pyramid/milk.html

Types of Hams

USDA. Food Safety and Inspection Service. *Meat preparation*. Retrieved July 27, 2009, from http://www.fsis.usda.gov/factsheets/ham/index.asp

CHAPTER 26

Milk and Dairy Products

OBJECTIVES

This chapter will consider the following:

- An overview of milk and milk products
- Global milk production
- Events in milk history
- Liquid milk
  - Pasteurization
  - Refrigeration
  - Homogenization
  - Raw milk
- Butter and ice cream
- Cheeses
- Probiotics, including yogurt

AN OVERVIEW OF MILK AND MILK PRODUCTS

The United States is a major milk producer, and milk production is increasing. In 1997, 156 billion lb of milk were produced, with per-cow production of 16,800 lb of milk. In 2007, 186 billion lb of milk was produced, with per-cow production of 20,300 lb of milk.

Milk and milk products are a major part of the diet in the United States. The U.S. Department of Agriculture (USDA) includes milk together with milk products in the food pyramid. Because of the nutritive value of milk, the USDA recommends that adults and children 9 years and older need 3 cups of milk per day, and children younger than 9 years need 2 cups of milk per day. One cup of milk or yogurt equals 1½ oz of natural cheese equals 2 oz of processed cheese.

Milk and milk products can reduce the risk of low bone mass of osteoporosis throughout the human life cycle. The nutrients include calcium, potassium, iron, magnesium, vitamin D, and protein. Other health benefits of milk products are that they provide little fat to the diet when consumed in their low-fat or fat-free forms.

Categories for Milk

There are four usage categories for milk: class 1 for fluid milk, with 2% milk the largest market for fluid milk; class II for soft manufactured items such as yogurt or ice cream; class III for hard cheese; and class IV for butter and powdered milk.

The U.S. General Accounting Office reports that farmers receive 42% of the retail price of 2% milk. In April 2004, the USDA announced that it had increased the minimum price of milk paid to farmers to $1.69/gal (12 gal to the 100 wt).

 WORLD MILK PRODUCTION

Globally, there continue to be large increases in milk production, as can be seen from Table 26-1. The statistics produced by the United Nation's Food and Agriculture Organization combine milk produced from cattle and water buffalo, together with small contributions from sheep and goats.

TABLE 26-1 Leading milk-producing countries in the world, and comparison between 1990 and 2005

NO.	COUNTRY	1990	2005
1	India	53.7	91.4
2	United States	67.0	80.3
3	China	7.0	32.2
4	Russian Federation	a	31.4
5	Germany	31.3	28.5
6	France	26.8	26.1
7	Brazil	15.1	23.5
8	United Kingdom	15.2	14.6
9	New Zealand	7.5	14.5
10	Ukraine	a	13.7

Note: Milk production is given as million metric tons.
a: Not applicable because country did not exist in 1990.

INTERESTING FACTOID

What is *lactose intolerance*? Humans are said to be the only species in which the adults drink milk. The enzyme lactase is expressed in the small intestine in many but not all adults. Those adults who do not have lactase are lactose intolerant. The failure to digest lactose leads to unpleasant effects such as flatulence or gas, diarrhea with a foul-smelling stool, and bloating (abdominal distension). People with lactose intolerance should avoid dairy products that contain lactose.

There is a marked difference in the incidence of lactose intolerance in different races, with a low incidence in Caucasians (approximately 15%), higher incidence in African Americans (approximately 75%), and very high in Asians (>90%). It has been suggested that during human evolution, there was a high selective advantage to retain lactase expression into adulthood if cow's milk was available.

Today, 3% of the 70,000 dairy herds in the United States have more than 500 cows. These account for 36% of milk production. People are increasing cheese consumption. Liquid milk consumption is constant, although there has been a shift toward reduced fat milk. The 2020 projection is that large herds (>500 cows) will produce 85% of the milk supply.

LIQUID MILK: ITS CONSUMPTION AND TREATMENT

Of the U.S. production of about 20 billion gal of milk, 7 billion gal go to fluid milk products: whole milk, 2%, 1%, skimmed, and flavored milk (e.g., chocolate milk) (see Figure 26-1). The per capita consumption of milk for 2005 was 8 gal of whole milk and 15 gal of lower fat milk.

Pasteurization

Pasteurization is a technique of heating foods to reduce spoilage. It was developed initially by Louis Pasteur (see the box on pasteurization) originally for wine and then applied to milk. It greatly reduces the number of bacteria in milk. These can lead to spoilage, and some are human pathogens. Pasteurization destroys many of pathogens in milk, including *Campylobacter jejuni*, *Yersinia enterocolitica*, *Listeria monocytogenes*, *Salmonella* organisms, and *Escherichia coli* 0157:H7. Moreover, pasteurization greatly reduces the concentration of *Staphylococcus aureus* and, therefore, of the enterotoxin that it produces and that causes the very unpleasant symptoms of diarrhea and vomiting. *S. aureus* is a common cause of mastitis, which is an infection of the mammary gland, in dairy cattle and, therefore, can easily contaminate milk.

FIGURE 26-1 Milk in the dairy section of a grocery store. Courtesy of the USDA. Photo by Ken Hammond.

HISTORY OF MILK PASTEURIZATION IN THE UNITED STATES

Between 1850 and 1920, the incidence of early childhood death was high in the United States. One of the reasons was food-borne pathogens. However, pasteurization was opposed as an unnecessary expense. With pasteurization of milk and other public health measures, childhood mortality has declined to very low levels.

In 1863, Louis Pasteur in Paris, France, developed a method of treating wine by heating it to kill the microorganisms that caused the wine to become vinegar. The technique was soon applied to milk. In 1891, the first milk-processing plant in the United States with pasteurizing equipment was built in Bloomfield, New Jersey.

In 1908, Chicago was the first major U.S city to require pasteurization of milk. Between 1905 and 1914, Nathan Straus, a philanthropist in New York, provided pasteurized milk at a low cost for the poor and, particularly, poor children. He was a very strong advocate for the compulsory pasteurization of all milk.

In 1914, New York and Philadelphia required pasteurization of milk. In 1917, most major U.S. cities required that all milk be pasteurized.

Ultra-Pasteurization

Ultrahigh temperature or ultra-pasteurization involves heating milk to a high temperature (138°C or 250°F) for a fraction of a second. Ultrahigh-temperature milk can be stored at room temperatures for up to 90 days. After the sealed container is opened, it should be treated like any other milk and refrigerated.

Refrigeration

Milk should be stored at 3°C (38°F). This greatly reduces the multiplication of bacteria that lead to spoilage and of pathogenic bacteria.

Homogenization

Milk is an emulsion with the fat globules in an aqueous phase. When milk is left to stand, the fat rises to the top, forming a layer of cream. In homogenization, the diameter of fat globules is markedly decreased by passing the milk under pressure through a small hole. The fat globules have a greatly increased surface area to mass ratio, which results in a much more stable emulsion without the property of the cream separating from the skimmed milk.

Raw Milk (Nonpasteurized Milk)

Raw milk is an issue or "cause célèbre" for health food advocates. Some have gone as far as stating that pasteurized milk can result in allergies and even gone as far as saying that it causes heart disease and cancer. It is argued that there are beneficial bacteria in raw milk that allow people with lactose intolerance to drink it. Given the publicity and claims of problems with pasteurized milk, it is perhaps not surprising that there is increasing consumption of raw milk and products of raw milk in the United States. In many cases, the milk is certified as having below an established concentration of bacteria. However, the concentration of pathogens is not known. There have been documented cases of *Salmonella* from drinking raw milk, causing disease. The Food and Drug Administration has banned the interstate sales of raw milk, and in many states, selling raw milk is illegal.

 ## ICE CREAM AND BUTTER

Butter has a long history, having been produced for at least 3,500 years. In contrast, ice cream was only developed about 300 years ago. Frozen desserts have been around for over 2,000 years, made from mixing snow with fruit juice and/ or honey. One of the first events at which ice cream was served was the dinner for the inauguration of President James Madison in 1813. The first commercial ice cream production in the United States was in 1851.

Butter and ice cream consumption per capita in the United States was the following for 2005: butter 4.6 lb, ice cream 15.4 lb, low or reduced fat ice cream (ice milk) 5.9 lb, and frozen yogurt 1.3 lb.

Ice Cream

Frozen dairy desserts continue to be very popular in the United States (see Figure 26-2). However, there has been a marked decline in their consumption since World War II. It is argued that the decrease in frozen dessert consumption reflects a paradox: a concern for the health effects of animal fats and increased consumption of the premium or superpremium ice creams, such as Ben & Jerry's, Häagen-Dazs, and Cold Stone Creamery. These ice creams have higher fat contents than nonpremium brands.

FIGURE 26-2 Ice cream in the frozen food section of a grocery store. Courtesy of the USDA. Photo by Ken Hammond.

The fat content of frozen dairy desserts is as follows:

- Low- or reduced-fat ice cream: less than 10% butter fat
- Ice cream: 10% butter fat
- Premium and superpremium ice cream: 13–17% butter fat
- Frozen custard: 10% butter fat and 1.4% egg yolk

Butter

Since 2001, per capita butter consumption has been increasing. In contrast, per capita consumption of margarine continues to decline. In 2005, per capita margarine consumption decreased to below that of butter at 4.0 lb per year.

 ## CHEESES

Cheese consumption in the United States in 2005 was 31.4 lb per person per year. This is more than eight fold higher than the per capita consumption 100 years ago. The change in cheese consumption reflects the huge increases in the consumption of cheese-containing foods such as pizza (see fig. 26-3). The cheese consumed is of the hard varieties such as cheddar and Swiss and/or the cottage cheese; in 2005, consumption was 2.5 lb per person per year.

More than 1,000 different cheeses are produced around the world. Cheeses are made from the milk of cattle, sheep, goats, and buffalo. There is archaeological evidence from the Fertile Crescent that cheese has been produced for over 7,000 years. It is suggested that cheese was first produced by chance or "serendipity"

FIGURE 26–3 Consumption of pizza has been a major reason for the increases in cheese consumed in the United States. Courtesy of the USDA.

when milk was carried in the stomach of a young goat, sheep, or cow. The presence of rennet (an enzyme) from the stomach curdled the milk, contributing to solid chunks (curds) and liquid (whey).

Cheeses are made by essentially the same process of separating out much of the solid (curds) from liquid (whey) and bacterial fermentation together with the growth of molds. The different cheeses result from the different microorganisms used and the milk coming from different species. For instance, in Swiss cheese (or Emmental), a culture containing *Propionibacter* (bacteria producing propionic acid) is added. In Roquefort blue cheese, there are veins of the mold *Penicillium roqueforti*. Other blue cheeses, including Maytag blue produced in Iowa, also contain blue, blue-black, or blue-green molds.

Some cheeses are made from the whey. The dissolved proteins come out of solution when the whey is heated and, in some cases, also acidified by adding vinegar. An example is ricotta cheese, which is creamy but a high-protein/low-fat cheese, produced as a by-product (the whey) of producing mozzarella and other Italian cheeses.

The Production of Cheese

Essential to making cheese is separating the curds (solids) from whey (liquid with dissolved proteins). Critical to this is the addition of rennet. This is an enzyme preparation that has been used for thousands of years. The enzyme is a proteolytic enzyme called rennin or chymosin. Rennin acts on a milk protein, κ-casein, to generate a soluble small protein (in whey) and an insoluble larger protein that in the presence of calcium forms the curds.

FIGURE 26-4 Brie cheese. Courtesy of PD Photos.

FIGURE 26-5 Cheddar cheese. Courtesy of PD Photos.

FIGURE 26-6 Mozzarella cheese. Courtesy of PD Photos.

> **INTERESTING FACTOID**
>
> More than 50 years ago, children would eat "curds and whey" as a source of easily digestible protein. This is remembered in the nursery rhyme, "Little Miss Muffet sat on a tuffet, eating her curds and whey...."

Rennin or Rennet

There are several forms of rennet, including the following:

- Calf rennet extracted from the mucosa of the abomasum of calves.
- Plant-derived rennet (suitable for vegetarians).

- Recombinant-derived rennet, in which the bovine chymosin gene is placed into microorganisms. In 1989, the Food and Drug Administration affirmed that microbial chymosin is generally recognized as safe. Over 60% of U.S. hard cheese is made with genetically engineered chymosin.

PROBIOTICS AND MILK PRODUCTS

There is increasing evidence that certain bacteria improve the health of laboratory animal models, people, and livestock. Moreover, probiotic microorganisms can be helpful per se and reduce the lactose concentration in individuals with lactose intolerance. Examples of products containing probiotic organisms include "live" yogurt and sweet acidophilus milk.

Acidophilus Milk or Sweet Acidophilus Milk

Sweet acidophilus milk is milk that has been cultured with *Lactobacillus acidophilus*. It is believed that the milk is more easily digested and that gastrointestinal immunity is improved.

Yogurt

Yogurt is a semisolid fermented milk product. Specific microorganisms (*Lactobacillus bulgaricus* or occasionally *L. acidophilus* and *Streptococcus thermophilus*) ferment milk, converting the milk sugar lactose to lactic acid. The lactic acid in turn acts on the milk proteins, affecting their physical characteristics. Making yogurt originated several thousand years ago and was brought to Europe by nomadic people such as either the Bulgarians or Mongolians.

REVIEW QUESTIONS

1. According to the USDA's Food Pyramid, what is the recommended intake of milk—including milk products—by the following?

   - Children under 9 years old

   - Women and girls 9 years old and over

   - Men and boys 9 years old and over

2. What is 1 cup of milk or yogurt equivalent to nutritionally in natural cheese or processed cheese?

3. What is the basis of lactose intolerance?

4. What are the four categories of milk in the United States, and what are they used for?

5. How much milk is produced annually in the United States?

6. What is the per capita consumption of whole milk?

7. What is the per capita consumption of reduced-fat milk?

8. Name five events in the history of milk production and processing.

9. What is pasteurization?

10. What was the first product to be pasteurized?

11. When was pasteurization widely used in the United States?

12. What was the effect of this on public health?

13. How does pasteurization improve food safety related to milk and milk products?

14. What is ultrahigh temperature or ultra-pasteurization? How is milk processed by this different from pasteurized milk?

15. Why is milk homogenized in the United States?

16. Why is U.S. milk supplemented with vitamins D and A?

17. What is "raw" milk?

18. What is the per capita consumption of cheese?

19. What is the process of separating out much of the solid (curds) from liquid (whey)?

20. Why is this important in cheese making?

21. Why are bacteria and molds important in cheese making?

22. Name four cheeses made from cow's milk.

23. Where did they originate?

24. Name at least one cheese produced from sheep's milk and one from goat's milk.

25. Most cheeses are made from the curds; why is Ricotta cheese different?

26. What is rennet?

27. What is the role of rennet in cheese making?

28. How do people get rennet?

29. What is a probiotic?

30. What is acidophilus milk or sweet acidophilus milk?

31. What is the per capita consumption of butter?

32. Is butter consumption higher than that of margarine? If yes, why?

33. What is the per capita consumption of frozen dairy desserts?

34. What is the difference in the fat content of reduced-fat ice cream, ice cream, and premium ice cream?

REFERENCES AND FURTHER READING

Dairy Food Consumption

USDA. Economic Research Service. *Food availability: Spreadsheets*. Retrieved July 27, 2009, from http://www.ers.usda.gov/data/foodconsumption/FoodAvailSpreadsheets.htm#dymfg

Raw Nonpasteurized Milk

Bartlett, T. (2006, October 1). The raw deal. The FDA says it's dangerous. Selling it is illegal. So why does an avid band of devotees swear by the virtues of unpasteurized milk? *The Washington Post*. Retrieved July 27, 2009, from http://www.washingtonpost.com/wp-dyn/content/article/2006/09/27/AR2006092700108.html

Cole, W. (2007, March 13). Got raw milk? Be very quiet. *Time*. Retrieved July 27, 2009, from http://www.time.com/time/health/article/0,8599,1598525,00.html

Michigan State University Extension. *Milk pasteurization. Guarding against disease*. Retrieved July 27, 2009, from http://www.fcs.msue.msu.edu/ff/pdffiles/foodsafety2.pdf

Robinson, R. A. (2001, May 14). Fluid milk. Farm and retail prices and the factors that influence them. Retrieved July 27, 2009, from http://www.gao.gov/new.items/d01730t.pdf

USDA's Food Pyramid

USDA. *Inside the pyramid*. Retrieved July 27, 2009, from http://www.mypyramid.gov/pyramid/milk.html

CHAPTER 27

Alternative Animal Products

OBJECTIVES

This chapter will consider the following:

- Overview of alternative animal products
- Specialty animal products
- Other animal products

OVERVIEW OF ALTERNATIVE ANIMAL PRODUCTS

There is increasing consumption of alternative and specialty animal products. It is easy to understand with an increasing diversity of the population in the United States that there is increasing consumption and availability of halal and kosher foods for Muslim and Jewish populations, respectively. Moreover, there has been growing interest in organic and other specialty foods. By-products of animal production have long been important to the economics of animal production.

SPECIALTY ANIMAL PRODUCTS

There is increasing interest in specialty animal products. These products can be differentiated into products with different nutritional attributes, such as eggs with lower concentrations of cholesterol or higher concentrations of vitamins (e.g., vitamin E), or have differences based on the production methods used. In the case of the latter, this includes kosher products; halal products; organic meat, milk, and eggs; and specific production systems. These production systems might include cage-free or free-range eggs; Label d'Or or Label Rouge for chicken or Niman Ranch for beef, pork, and lamb; free-range poultry; and grass-fed cattle.

There is considerable price differential between those for specialty products and for conventional or commodity animal products. This may be because of the following issues:

- Perceived quality differences
- Real differences in taste or consistency
- Absence of antibiotics and other chemicals used in the production systems
- Religious rules

FIGURE 27–1 A label stating that a meat product is kosher for Passover. Reproduced by permission from Tony Matthews. 2010 by Shutterstock.com.

Kosher Products

Kosher foods have long been available in the United States. Kosher food is that allowed under Jewish law, which is based on Leviticus and other books in the Bible (see fig. 27-1). Among the foods that are permitted include the following:

- Beef, lamb, and chicken if slaughtered in an approved manner (i.e., with the carotid arteries cut to exsanguinate the animal) in the presence of a rabbi
- Fish with fins and scales
- Dairy and meat products that have been processed according to Jewish Talmudic law (Halacha)

What are not permitted are pork, shellfish, and fish without fins and scales such as shark and catfish. It is also not permitted to consume dairy products at the same time that meat is being consumed.

Halal Products

There is a rapidly growing market for halal foods with the increasing Muslim population in the United States and in Western Europe. Halal is food that is permitted by Islamic law. Halal foods are certified, for instance, by the Muslim Consumer Group. Among the foods that are permitted include beef, lamb, and chicken if slaughtered in an approved manner by a Muslim who speaks the name of Allah before cutting the animal's throat; and fish. Shrimp and some shellfish may be permitted depending on the branch and school of thought within Islam. What is not permitted in halal foods includes blood, pork, and alcohol.

Halal foods are labeled as such and represent a major value-added export opportunity for U.S. producers, with about 70% of Muslims worldwide keeping to the halal requirements.

Organic Production

There has been a tremendous growth in organic foods, with increases of 20% per year reported (see Table 27-1).

For instance, the Horizon organic line of dairy products includes ice cream, cheese, yogurt, cottage cheese, sour cream, smoothies, and cheese slices together with egg and infant formula. The number of certified organic chickens has increased four fold between 2000 and 2005. The premium on organic is between 150% and 300%.

The requirements for certification of organic animal products (see fig. 27-2) include the following:

- The livestock receive 100% feed from crops grown in an organic manner. Crops and forage can use improved plant varieties, but not genetically engineered plants. Neither synthetic fertilizers nor pesticides can be used in the production of the animal feed.
- Livestock must be provided access to pasture.
- Synthetic hormones and antibiotics are not permitted in production.
- Alternative health therapies, such as botanical remedies and manipulation, can be used.
- Vaccination, minerals, and vitamins are allowed.
- Synthetic parasiticides are not permitted.

Definition

Organic agriculture is defined by the National Organic Standards Board as "an ecological production management system that promotes and enhances biodiversity, biological cycles, and soil biological activity. It is based on minimal use of off-farm inputs and on management practices that restore, maintain, or enhance ecological harmony. The primary goal of organic agriculture is to optimize the health and productivity of interdependent communities of soil life, plants, animals and people."

TABLE 27-1 Organic agricultural production in the United States, and the changes between 1992 and 2005

USDA CERTIFIED ORGANIC PRODUCTION	1992	1997	2001	2005
Livestock animal (numbers in thousands)				
Beef cattle	6.80	4.43	15.2	36.1
Dairy cattle	2.26	12.8	48.7	87.1
Pigs	1.36	0.48	3.13	10.0
Sheep/lambs	1.22	0.70	4.20	4.47
Poultry animal (numbers in millions)				
Layer chickens	0.044	0.538	1.61	2.42
Meat (broiler) chickens	0.017	0.038	3.29	10.4
Turkeys (in thousands)	Approximately 0	0.75	98.6	144.1
Acreages				
Pasture and rangeland (million acres)	0.53	0.50	0.79	2.33
Cropland (million acres)	0.40	0.85	1.30	1.72

Source: Data are from the USDA Economics Research Service.

There are exceptions to the overall requirements. These include the following items that are permitted. Oxytocin is allowed for organic production to safeguard the health of cattle. The use of the parasiticide ivermectin is allowed for the treatment of dairy cattle and breeder stock, but not for animals going for slaughter. Antibiotics are permitted in organic livestock production systems in vaccines and semen as preservatives.

Quality of Organic Food

Because there is relatively little information supporting the contention that organic food is of a higher quality than conventionally produced food, the conventional scientific wisdom is that there is no difference. Recently, reputable studies such as the European Union's Quality Low Input Food project reported differences in the quality of organic food. For example, researchers reported that the concentrations of vitamin E and healthful fatty acids in milk are higher in organic milk and in cattle grazing on pasture than those in conventional systems. However, there were no differences between organic and conventional-grown vegetables (carrots, kale, peas, potatoes, and apples) grown in two consecutive years in a different study. The authors of a Danish study concluded that their "study does not support the belief that organically grown foodstuffs generally contain more major and trace elements than conventionally grown foodstuffs, nor does there appear to be an effect on the bioavailability of major and trace minerals."

Specific Production Systems

There are a multitude of animal products marketed as "natural meat," such as Niman Ranch or cage-free or free-range eggs, free-range poultry, and grass-fed cattle

FIGURE 27-2 Example of organic labeling approved by the U.S. Department of Agriculture. Courtesy of the U.S. Department of Agriculture.

ISSUES *for discussion*

ORGANIC AGRICULTURE AND ANIMAL WELFARE

1. Let us assume for a moment that a case can be made for not using antibiotics as growth promotants. If the farmer cannot use antibiotics therapeutically, how is a sick animal to be treated? What about a life-threatening disease? Would we not use antibiotics to treat infection in a sick person?

PUBLIC POLICY

1. Are organic standards based on sound science, or are they based on emotion?
2. Would we want to see that standard applied to food safety, human medicine, or veterinary care of our pets?
3. Could we feed the current population of the world using organic agriculture? Could we feed the projected increase in the world population? If the answers to either question are no, discuss whether or not it matters.
4. Organic production requires more land per unit produced. How do we reconcile this with issues of land availability?
5. Are the approaches of organic agriculture more environmentally sound? Is there less pollution such as contamination of groundwater with the nitrogen, phosphorus, and pathogens, for example, from manure, pesticides, hormones, and greenhouse gases?
6. Discuss whether organic meat or milk products may have advantages over commodity meat for nutrition or eating experience.
7. To what extent is nonorganic production paying sufficient attention to quality indices such as flavor and other aspects of the eating experience?

(see fig. 27-3). It is difficult to define what exactly the production systems require and when they either produce a superior product or provide an improvement to animal welfare.

Cage-Free Eggs

There is growing attention by U.S. retailers and restaurants to cage-free eggs, with almost 5% of eggs produced in the United States from cage-free hens. Ben & Jerry's and the Whole Foods supermarket chain are exclusively using cage-free eggs.

CO-PRODUCTS OR BY-PRODUCTS OF ANIMAL PRODUCTION

The economics of animal, and also plant, agriculture can be aided by the production of co-products or by-products. Leather has long been a valuable by-product of the production of meat. Other by-products include gelatin and animal glues. A recent example of a co-product is the synthesis of biodiesel from waste animal fat.

FIGURE 27-3 Examples of meat, milk, and eggs produced by specific production systems. *A,* Organic turkey. Reproduced by permission from Bochkarev Photography. © 2010 by Shutterstock.com. *B* and *C,* Cage-free or free-range chicken (reproduced by permission from LianeM. © 2010 by Shutterstock.com) and eggs (reproduced by permission from Stephen Aaron Rees. © 2010 by Shutterstock. com). *D,* Grass-fed cattle. Reproduced by permission from Matthew Jacques. © 2010 by Shutterstock.com.

A

B

**FIGURE 27-3
(CONTINUED)**

C

D

Leather

Global production of leather is 9 million tons of cattle rawhides. The United States is one of the biggest exporters of cattle hides, with $1 billion in sales per year. Major importers are where there is production of leather goods such as coats, jackets, and shoes in places like China, Southeast Asia, and Egypt. The leather is cleaned; soaked in lime to stretch; and tanned with vegetable tannin (traditional), chromium sulfate or aldehyde, or other chemicals. Tanning can be highly polluting.

EXAMPLES OF SPECIFIED PRODUCTION SYSTEMS AND SPECIALTY PRODUCTS

NIMAN RANCH

Niman Ranch uses what it describes as "traditional methods on family farms" to produce beef, pork, and lamb. These systems require that the animals are raised under specific conditions without antibiotics, hormones, or animal by-products and with vegetarian animal feeds.

Niman Ranch was started by Bill Niman in the early 1970s near San Francisco. Today, Niman Ranch has more than 500 independent growers. Menus in some very good restaurants include the statement that the meat is Niman Ranch because of its reputation for quality. Moreover, meat is sold by gourmet stores or on the Internet.

COLEMAN NATURAL FOODS

Coleman Natural Foods markets both natural and organic beef, pork, chicken, and sausages, with 35% of the products sold being organic and the rest natural. The company states that the natural products have the following attributes: natural, no antibiotics, no hormones used, no preservatives, and no animal by-products in the feed ("Always Vegetarian Fed").

Examples of its brands are Organic Rosie Petaluma's Organic Chicken; and Rocky the Range Chicken, which is a free-range chicken.

Gelatin

A total of 300,000 tons of gelatin is produced globally each year. This is used, for instance, by the food industry. Gelatin is produced by the hydrolysis of collagen, the protein in connective tissue. The raw material for gelatin production is cattle and pig hides and bones.

Animal Glues

There are two types of animal glues: hide glue and hoof glue. These are, not surprisingly, derived from the hides and hooves from cattle, pigs, and horses.

Biodiesel

Biodiesel is a renewable fuel. It can be used as the fuel in a diesel engine or blended with petroleum-based diesel. Biodiesel is produced from lipids (predominantly triglycerides) in algae, vegetable oils, animal fats, and recycled restaurant grease. Animal fat may come from rendering dead livestock or as what was thought of as a waste product, that is, animal fats removed during processing. Biodiesel is produced by a chemical process known as *transesterification* (see fig. 27-4). The glycerol (glycerine) is split from the fatty acids, and methyl groups are added to the fatty acids to form the fatty acid methyl esters that constitute biodiesel. Glycerol or glycerine is a valuable by-product. With the increases in biodiesel production, a surplus of glycerine is likely. This may become a significant energy source for pigs and poultry.

$$CH_2O. CO.(CH_2)n.CH_3$$
$$|$$
$$C\ H\ O\ . \rightarrow CO.(CH_2)n.CH_3 + 3\ CH_3OH\ 3\ CH_3O\ CO.(CH_2)n.CH_3 +$$
$$CHOH$$
$$|\qquad\qquad (3\ methanol)\ Fatty\ acid$$
$$CH_2O. CO.(CH_2)n.CH_3 \qquad methyl\ ester$$
Triglyceride

(shown with saturated fatty acids for simplicity)

$$CH_2OH$$
$$|$$
$$CH_2OH$$
Glycerol

FIGURE 27-4 Production of biodiesel from triglyceride fat.

Some large meat-processing companies are developing strategic partnerships with petroleum companies. For example, Tyson Renewable Energy has developed an alliance with ConocoPhillips to create the next generation of diesel fuels. A source of information on biodiesel is the National Biodiesel Board (http://www.biodiesel.org).

Other Animal Products

Premarin

Premarin is a pharmaceutical estrogen product isolated from the urine of pregnant horses (PREgnant MARes' urINe, or PMU). It is used for hormone replacement therapy. Over 9 million American postmenopausal women take Premarin to relieve the symptoms of menopause, including "hot flashes," night sweats, and vaginal dryness.

Premarin is manufactured by Wyeth-Ayerst, a division of American Home Products. Premarin is the third-most prescribed drug in the world. It is a complex of predominantly three estrogens: estrone sulfate (about 50%) and two unique horse estrogens, including equilin (3-hydroxyestra-1,3,5,7-tetraen-17-one) (25%) and equilenin (15%).

Essential to the production of Premarin is a reliable source of pregnant mare urine. There are about 700 farms with 80,000 horses producing the urine. The mares have to be maintained such that the urine can be readily collected.

ISSUES *for discussion*

PREMARIN

1. Does the treatment of mares justify the production of Premarin? Consider in your discussion the welfare of the horses, quality of life for the women, and economic effects on the farmers who are often in geographic regions where there is economic hardship (e.g., North Dakota).
2. Should synthetic versions of Premarin or alternative estrogens be used? Consider in your discussion whether these may be as effective.

REVIEW QUESTIONS

1. What are kosher foods?
2. How are animals killed in a manner consistent with kosher standards?
3. What are halal foods?
4. What is organic food, and how is it produced?
5. What aspects of organic livestock production are increasing?
6. What is natural food?
7. Give examples of natural food.
8. Why are cage-free eggs increasing?
9. What happens to the hides of animals?
10. Does the United States export animal hides?
11. What are some of the major leather-manufacturing countries?
12. Give examples of by-products and co-products of meat production.
13. How is biodiesel made from animal fat?
14. What is Premarin?
15. How is Premarin used?
16. What is the source of Premarin?

REFERENCES AND FURTHER READING

Kristensen, M., Østergaard, L. F., Halekoh, U., Jørgensen, H., Lauridsen, C., Brandt, K., et al. (2008). Effect of plant cultivation methods on content of major and trace elements in foodstuffs and retention in rats. *Journal of the Science of Food and Agriculture, 88*, 2161–2172.

Organic Farming

Delate, K., Iowa State University. (2000). *Fundamentals of organic agriculture*. Retrieved July 27, 2009, from http://www.extension.iastate.edu/Publications/PM1880.pdf

Karreman, H. J. (May 11, 2007). *When it comes to animal health and welfare, there are worse things than antibiotics.* Retrieved July 27, 2009, from http://www.newfarm.org/features/2007/0507/antibiotics/karreman.shtml

Organic Farming Research Foundation. Retrieved July 27, 2009, from http://ofrf.org/images/usmap_1280.jpg

U.S. Department of Agriculture (USDA). *Table 2: U.S. certified organic farmland acreage, livestock numbers, and farm operations, 1992-2005.* Retrieved July 27, 2009, from http://www.ers.usda.gov/Data/Organic/Data/Farmland%20livestock%20and%20farm%20ops%2092-05.xls

By-Products: Leather

World Leather Market. *The leather global value chain & the world leather footwear market.* Retrieved July 27, 2009, from http://www.factbook.net/leather_components.php

G

Gait, in horses, locomotion of, 66–67
Galactopoiesis, defined, 355
Galactose, structure of, 264f
Galloping, by horses, 67
Game birds, production of, 216–217
Gametogenesis, 320–321, 320f–322f
Gas(es), greenhouse, 429–432, 429t, 430f, 431t. See also Greenhouse gases
Gastrointestinal tract, of domestic animals, 240–244
  in digestion, 240–241
  functions of, 240–241
  regions of, functions of, 252t
  structure of, 241–244, 241f–244f, 244t
Geep(s), defined, 176
Geese, production of, 211–212, 213f, 214f
Gelatin, 482
Gelding(s), defined, 56, 187
Gender identification, in poultry reproduction, 201–202
Gene(s), 290, 292
  actions of. See Gene action
  candidate, defined, 289
  defined, 289, 291f
  sex-linked, defined, 290
  Sry, 316
Gene action, types of, 294–298, 295f, 295t, 297t
  animal color, 297–298, 297t
  breeding systems in, 296
  coat, 298
  hair, 298
  heritability, 295–296
  MAS, 296
  skin color, 298
  SNPs, 296
Gene expression, defined, 289
Gene ontology, 294
Genetic(s), 287–303
  animal, careers in, 40
  of animal color, 297–298, 297t
  cat, 80–81
  catfish, 230
  cattle, 111
  changes in, 293
  dog, 92–93
  fundamentals of, 289–294, 291f, 292t, 293f
  growth related to, 346
  of livestock
    applications of, 294
    chromosomes, 290, 291f, 292t
    cloning, 300, 301f
    comparative genomics, 292
    DNA, 290, 291f
    epigenetics, 293
    gene(s), 290, 292
    gene action, types of, 294–298, 295t, 297t. See also Gene action, types of
    genetic changes or mutations, 293
    genetic code, 290
    genome sequencing, 292, 293f
    impact on dairy production, 288
    microsatellites, 292

MSH, 298–300
  nuclear transfer, 300, 301f
  selection to improve poultry growth, 288–289, 289t
  terminology related to, 289–290, 291f
  transgenic animals, 300, 301f
  milk production effects of, 359
  molecular, in control of coat, hair, and skin color, 298
  mutations in, 293
  pig, 146, 154–157, 155f–158f
  poultry, 203, 203f
  rabbit, 193
  of sex determination, 317
Genetic code, 290
Genetic counseling, 299
Genome(s), defined, 290
Genome sequencing, 292, 293f
Genomics
  comparative, 292
  defined, 289
  fundamentals of, 289–294, 291f, 292t, 293f
  nutritional, 299
Genotype, defined, 289
Gerbil(s), 98f, 100
Germ cells, primordial, 316
  defined, 309
Gestation, 327–333, 327t, 328f–331f, 328t
GH. See Growth hormone (GH)
Gib(s), defined, 79
Gilt(s), defined, 145
Girl(s), defined, 187
Global warming, greenhouse gases and, 432
Glucagon, metabolism effects of, 257
Gluconeogenesis, defined, 257
Glucose, structure of, 264f
Glue(s), animal, 482
GnIH. See Gonadotropin inhibitory hormone (GnIH)
GnRH. See Gonadotropin-releasing hormone (GnRH)
Goat(s)
  billy, defined, 177
  breeds of, 177, 178f
  cheese from, 179
  classification of, 176, 177
  described, 176, 176f, 177f
  domestication of, 177
  external anatomy of, 176, 176f
  hair of, 179
  meat from, 179
  milk from, 177–178, 179t
  nanny, defined, 177
  pharmaceutical proteins from, 179
  pheromones in, 258
  production of, 169t, 176–179, 176f–178f, 179t
  reproduction by, 177
  terminology related to, 177
Golgi apparatus, 239, 239f
Gonad(s)
  defined, 309
  development of, 316
  function of, defined, 309
Gonadotropin inhibitory hormone (GnIH), 317